SHOE & LEATHER PETROLEUM CO

A CENTURY IN OIL

The "Shell" Transport and Trading Company
1897–1997

S S & Co.

A CENTURY IN

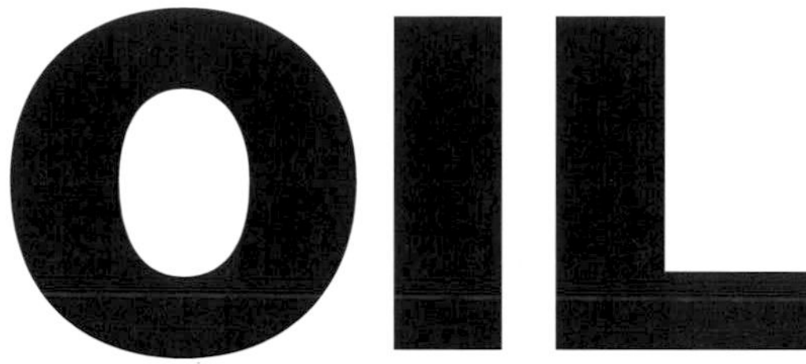

A Century in Oil

The "Shell" Transport and Trading Company
1897–1997

Stephen Howarth

Weidenfeld & Nicolson London

Shell companies have their own separate identities, but in this book the collective expression "Shell" and "Group" are sometimes used for convenience in contexts where the reference is to companies of the Royal Dutch/Shell Group in general, or where no useful purpose is served by identifing the particular Shell company or companies.

First Published in Great Britain, 1997
by George Weidenfeld & Nicolson Ltd
The Orion Publishing Group
5 Upper St Martin's Lane
London WC2H 9EA

A CIP catalogue record for this book is available from the British Library
ISBN 0 297 82247 0

Designed and Illustrated by: Terry Anthony
Typeset in: New Century Schoolbook

Acknowledgements
All pictures copyright Royal Dutch/Shell Group, except: BFI: 193; Bridgeman: 6, 11; Corbis: 27, 153, 233, 299, 335; E.T. Archive: 9; Calouste Gulbenkian Foundation: 147; Getty Image: 82, 113, 114, 115, 117, 159, 209, 215, 222, 234, 247, 272, 301, 309, 328, 333; *Illustrated London News*: 96, 111, 158; Mary Evans Picture Library: 10, 53; National Portrait Gallery: 59; Topham: 8, 81, 103, 170, 251; Tidmarsh & Sons: 176; RAC: 51; Victoria and Albert Museum: 238.
Every effort has been made to trace the copyright holders. Should any copyright holder not so acknowledged contact the publishers of this book, a correction will be made in the next possible edition.
'Murex' kindly reproduced by permission of the British Mercantile Marine Collection.
Lines from 'Four Quartets' by T. S. Eliot a reprinted by kind permission of Faber and Faber Ltd.
'Life on the Move' illustrations on page 240 by Hetty Barendregt-Backer.

CONTENTS

Ammonite

INTRODUCTION

This Vast Concern

No other single industry affected 20th-century civilization more rapidly or more profoundly than the oil industry. The century was nearly half over when the nuclear age began, and more than half over when the computer revolution swept the world. But throughout the 20th century – the most inventive hundred years that humanity has seen – oil was the great enabler, providing from one basic resource a rainbow range of products. The elementary ones are obvious: fuels for heating, lighting and engine-power, and lubricants for almost any moving mechanical object; the advanced ones – so advanced that their relationship with crude oil is often unrecognized by the public – range from textiles to the bases of perfumes.

This remarkable phenomenon, the oil industry, possesses another characteristic that is probably unique in commercial history: its beginning can be dated absolutely exactly – not just to a decade or a year but to a specific day, namely Sunday 28 August 1859.

The "Shell" Transport and Trading Company was formed only 38 years later, on 18 October 1897. Thus, by the time of its hundredth birthday in 1997, the company had been active for almost as long as the world-changing industry to which it belonged; and, as one of the parents of the Royal Dutch/Shell Group of Companies, it was responsible for businesses which in most of those 100 years had routinely produced 10–12% of the whole world's annual supply of crude oil.

In a phenomenal industry, the "Shell" could fairly claim to be a phenomenon in its own right.

Shell Transport continues to flourish into its second century. The history of its first century is very dramatic: sometimes serene and confident, sometimes disturbed and turbulent, and almost always exciting. And to trace that history, we need to go back not just to 1897, when the company was formed; nor to 1859, when the industry began; but to 1853, and a world without oil.

Opposite Page: Queen Victoria's Diamond Jubilee, 1897

It was a notable year – in some ways, even a momentous one. Events in 1853 not only helped to shape the rest of the 19th century, but also formed major parts of the cultural, political and technological foundations of the entire 20th century.

On 19 January, Verdi's opera *Il Trovatore* was performed for the first time. On 30 January, the French emperor Napoleon III wed Countess Eugenie de Montijo of Spain; a week later an uprising of Italians against their Austrian rulers was suppressed; a fortnight after that, Austria and Prussia signed a commercial treaty that was intended to last until 1865. Meanwhile, the Turkish Ottoman empire was rapidly crumbling: in February, the Tsar of Russia proposed that its remnants should be divided between his country and Great Britain. Britain disagreed.

In March, Ottoman forces withdrew from Montenegro in the Balkans, which they had occupied for a year. At the same time, there was civil war in China, and in the United States a lady named Amelia Bloomer spoke publicly in support of women's rights. Opposing a ruling that women must at all times keep their legs covered, she wore on the platform a short skirt with long Turkish-style pantaloons underneath, and unintentionally gave her name to them.

Amelia Bloomer

During the summer, French and British fleets assembled near the Dardanelles, preparing to protect Ottoman shipping and the holy places of Turkey from Tsarist claims; and in the Far East, Commodore Matthew Perry of the United States Navy steamed a squadron of four coal-powered warships into Tokyo Bay, requesting and requiring that the Japanese should end their centuries-long isolation and join the modern world.

In the autumn of the year, India's first railway was opened. Powered like Perry's ships by coal and steam, its engines puffed from Bombay to Thana - a distance of about 20 miles. East of Australia, France annexed the island of New Caledonia, and in West Africa, Britain's colony the Gold Coast (modern Ghana) was given its first legislative council. On America's west coast - where, four years on from its start, the Californian gold rush was in full swing - a Bavarian tailor named Levi Strauss was starting to sell tough, durable trousers with a French name, *genes*, to the miners. Simultaneously on the east coast, in New York, a German-born gentleman called Heinrich Steinweg (who would be remembered as Steinway) opened a piano-making factory. In Germany itself, Franz Lizst wrote his sonata in B minor; and in England, on 5 November 1853 in the East End of London, a baby boy was born to a Jewish family.

The next six years were years of great change, progress and turbulence for the world. In Japan, the immediate outcome of Commodore Perry's epochal visit was civil war, with the nation divided between those who hoped the status quo could be maintained and those who believed Japan must either modernize or become a colony of the West. In 1854, Britain fought

Florence Nightingale, 'the lady with the lamp', carrying the 'new light' of kerosene in her nightly rounds

Russia in the Crimea; at Balaclava, through a series of blunders, the Light Brigade was wiped out; and lamp in hand, Florence Nightingale brought a degree of humanity to the wards of wounded. This too was the year when the Pope declared the Virgin Mary to have been free of Original Sin, and when a canal was planned to link the Red Sea with the Mediterranean at Suez.

In 1855, on the Zambezi river, the explorer and missionary David Livingstone became the first non-African to see *Shongwenamutitma*, the mile-wide, 420-foot high 'boiling pot', which he promptly re-named as the Victoria Falls. In 1856, New Zealand was deemed capable of self-government; the Black Sea was made a neutral zone; and the French chemist Louis Pasteur declared that disease was generally spread by air-borne bacteria, rather than by poisonous vapours exhaled from the earth itself. As Britain re-opened the Opium War with China in 1857, the Indian mutiny against British rule began, and in New York, the first example of Elisha Otis's new invention, a passenger elevator, was installed in a department store. In 1858, in France, an Italian tried to assassinate Napoleon III and Empress Eugenie; the Bibliothèque Nationale was opened; and in the small town of Lourdes, the Virgin Mary appeared in a vision to a young girl. In London, Lionel de Rothschild became the first Jew to enter the House of Commons as a Member of Parliament; and in the Balkans, while the state of Montenegro was recognized by the major European powers, the prince of Serbia was deposed.

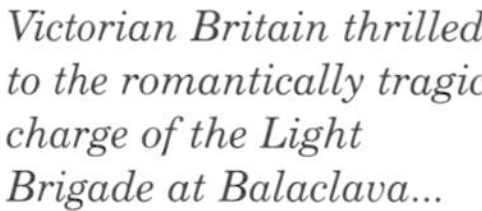

Victorian Britain thrilled to the romantically tragic charge of the Light Brigade at Balaclava...

The following year, 1859, brought publication in London both of Charles Darwin's book *The Origin of Species*, and the first part of *The Idylls of the King* by the Poet Laureate, Alfred Tennyson. Like his earlier grand, heroic and very English work *The Charge of the Light Brigade*, Tennyson's re-telling of the legends of King Arthur was immensely popular; in contrast, with its theory of evolution – understood as meaning that people were related to apes – Darwin's startling book produced horror and derision. This was also the year in which revolution broke out in Italy, as its various states sought to unite themselves into one independent nation; and in the United States of America, 83 years after their own Declaration of Independence, many landmark events occurred. Oregon became the 33rd state of the Union. The Massachusetts Institute of Technology was founded. For the first time, mail was transported by air, in a balloon flight taking only 19 hours and 40 minutes to travel the 812 miles from St Louis, Missouri, to Henderson in New York state. The campaign for the abolition of slavery was gathering strength, with many women (including the splendidly named Sojourner Truth, herself an escaped slave) prominent in its ranks; and on 28 August, near Titusville in the northern state of Pennsylvania, a self-styled colonel and former railway conductor named Edwin Drake drilled 69½ feet into the ground, and struck oil.

...and was deeply shocked by Darwin's theory of evolution

Out of all these mid-19th-century events, crowded into the brief space of six years, it would be hard to select the one which most affected the 20th century. Each did, although sometimes the 19th-century roots of 20th-century events are less obvious than the differences between the periods. For example, only a few 20th-century poets made fortunes from their writing, but (on what might loosely be called the cultural front), jeans became ubiquitous. In the chain of cause and effect, often the most visible links between the 19th and the 20th centuries are not so much cultural as martial. Through tortuous routes, the Balkan disturbances of the mid-19th century resulted in the outbreak of World War I – the Great War, the 'war to end all wars'. Following Perry's unwelcome visit, the civil strife in Japan resulted, logically, in the creation of the Imperial Japanese Navy. That in turn helped to bring about the 20-year-long Anglo-Japanese Alliance of 1902, the first full alliance Great Britain had made with any other nation for a whole century; and in 1941 – not as a simple act of revenge against America, but as the result of many years of exceedingly complex international politics – the Imperial Navy launched Japan's assault on Pearl Harbor, thereby dragging the United States unwillingly into World War II.

Clothes for the mining man: Levi Strauss's 'jeans'

In social matters, some might well cite the eventual abolition of slavery in the United States as the most important bequest of the 19th century to the 20th. Others might possibly add the progress in womens' rights, and the resulting greater equality (or reduced inequality) of individuals, regardless of race or gender. In the political spectrum, still others might point to

Below: Commodore Perry arrives in Japan, 8 July 1853

the fact that in the middle of the 19th century, Karl Marx – a refugee since the 1849 publication of his *Communist Manifesto* – was living in London, writing *Das Kapital*. Certainly, along with the downfall of Tsarist Russia, the establishment of international Communism must be very high on any list of global changes in the 20th century.

But it may well be that of all those mid-19th century events, the one which did most to set about shaping the 20th was Edwin Drake's discovery of oil on 28 August 1859. Similar experiments were taking place in Canada and Germany, but previously, it had been generally accepted that crude oil, or petroleum – a viscous liquid known to have uses (and therefore value) since ancient times – was something which oozed or dripped in small amounts from coal, like sap from wood. When Drake's rickety derrick bit into the soil of Pennsylvania, it was the first time anyone in America had shown that by drilling, instead of digging pits or skimming natural seepings from the surfaces of streams, oil could be produced in commercially viable quantities – which for Drake and his backers meant enough to fill 25 barrels, each of 42 US gallons, every day.

The fashionable 'shell box', foundation of the family business

All around the world, the standard unit for measuring large quantities of oil is still the 'barrel' of 42 US gallons. But nowadays most oil never sees the inside of an actual barrel, and global oil production has multiplied beyond anything Drake could have imagined. In 1865, within six years of his discovery, 7,000 barrels of oil were produced daily worldwide. Of that total, 6,800 barrels a day came from the United States, and an American reporter observed that 'To be deprived of it now would be setting us back a whole cycle of civilisation.' But no setting back was in line. By 1895, world production stood at 284,000 barrels a day. In 1945, it was just over 7.1 million; and in 1990, on any given day, about 60 million barrels were produced.

The range of products derived from crude oil is, if anything, yet more astonishing than those production figures, but even by themselves the figures indicate the depth of impact that oil had on the 20th century. There is little point in speculating what might have happened without oil, except to say that for better or worse, without it our world would have been different in almost every way.

Because of that, the birth on 5 November 1853 of the little East End Jewish boy takes on an unexpected significance in the history of the 20th century. At the time, the event meant nothing at all to anyone except those most immediately involved, but by 1890 the boy, Marcus

Samuel junior, had become a vigorous, imaginative and daring 36-year-old man. That was when he and his younger brother Sam decided to diversify the family business into oil shipping. And in doing so, they permanently transformed the economics of the oil industry.

Their father, Marcus Samuel senior, had been a merchant trading with the Middle East and the Orient, who had bought and sold everything from rice to exotic sea-shells. The shells were immensely popular in mid-Victorian Britain, especially in the form of 'shell boxes'. These small, ornately decorated trinket-boxes were very much the fashion, and by 1853, when Marcus junior was born, Mr Samuel's business was thriving – fortunately, because he already had a large dependent household: his wife Abigail, their five surviving children (four girls and a boy; two other girls had died young), and their maid Mary Ryan.

Mr Samuel died in 1870, having fathered his last child, Sam, in 1855. There was another older son (Joseph, known as Joe) but as will be seen, it was the youngest brothers, Marcus junior and Sam, who later accepted responsibility for the family business and went on to transform it completely: for in 1892, after two years' careful planning in absolute secrecy, they launched the first of a series of ten tankers of revolutionary design and began transporting illuminating oil – kerosene, or paraffin – to the Far East.

In honour of their late father's most popular merchandise, they invented a new brand-name for the oil, calling it 'Shell'. The venture prospered so much that just five years later, in 1897, something which had begun as a sideline was earning more than all their other businesses combined. To put it on a more organized footing, therefore, they founded another company, which they named The "Shell" Transport and Trading Company. At the age of 44, Marcus junior became its first chairman.

The Tennis Party

The company soon became commonly known as Shell Transport, or the "Shell". Somewhat whimsically, the latter name is always written with double inverted commas; but the company's development and history has been anything but whimsical. In 1907, a decade after its creation and still under Marcus's chairmanship, the "Shell" entered an historic alliance with the Royal Dutch Petroleum Company. While maintaining their separate identities, the two companies merged their interests in the proportions of 40% "Shell" and 60% Royal Dutch. Each became a parent company of the Royal Dutch/Shell Group of Companies, known as Shell – no inverted commas – and universally identified by its yellow pecten, or scallop shell, on a red ground. The name and logo have become two of the most famous trademarks in the world.

Before the alliance of 1907, the world's oil industry had been dominated by a single American company, Standard Oil. But only three years later, in 1910, the efforts of 'the Royal Dutch/Shell Group' had been so successful that Standard's dominion was irrevocably reduced. Everyone recognized that there were now two major oil corporations. Since then, others have come (and some have gone), but Shell has remained as one of the handful of 'oil majors'.

In 1942, referring only to the Group's fleet, a shipping digest described it as 'this vast concern' – adding, quite correctly, that it 'started in a very modest way'. It has not ceased to grow. By most measurements, Shell today is the largest oil organization in the world, and the largest enterprise of any kind in Europe. All statistics associated with it are enormous: it directly employs over 40,000 people in Europe, and over 100,000 worldwide; its annual net sterling income is counted in thousands of millions.

This book tells the story of the "Shell" in its first century, both in its ten initial years of independent existence and its subsequent long, constructive and predominantly happy relationship with Royal Dutch. It is a story of high risk and high reward; of careful calculation and occasionally astonishing oversight in business; of desperate competition and supreme self-confidence. There is exploration, in the harshest terrains and most hazardous seas; there is experiment and innovation, conspiracy and compliance, selfishness and selflessness, intrigue and honesty. It could almost be a novel, but it is all true.

The story is one which has affected, and does affect, the lives of us all; and that is why this book is not fundamentally a boardroom history, because the boardroom is not always the place where a company lives or dies: that place is the workface, where those who work for the company know that actually they are the company. To others, Shell can seem like some faceless monolith; but it is not. Today it is great, but its origins were modest; and like any other company, it was and is made up entirely of the strengths and weaknesses of its people. This is Shell's story.

Top left:
Marcus Samuel the elder, c.1860

Top right:
His wife Abigail, c.1860

Bottom left:
Their sons Marcus aged about 18...

Bottom right:
...and Sam aged about 17

DRAKE

CHAPTER ONE

Background: 1853–1891

In 1859 Edwin Drake's Pennsylvanian well introduced drilled oil to the Western world. In terms of both time and distance, however, it was very far from the place where oil was first discovered by drilling. Oil was first produced in China in around 200 BC. Though the Chinese used only bamboo poles with brass attachments, they managed to penetrate as deep as 3,500 feet into the earth – which makes Drake's 69½ feet seem a pretty slight effort, especially considering it occurred more than 2,000 years later. Moreover, even in 200 BC, knowledge of crude oil and some of its properties was already very old. As early as 3500 BC, in what is now Iraq, the Sumerians used asphalt – one of the component parts or fractions of crude oil – as a virtually indestructible adhesive for bricks and sealant for water-craft; by 3000 BC, in the Indus valley, the citizens of Harappa and Mohenjo-Daro used bitumen, another fraction, as a water-proofing agent for their baths; and from 2200 BC, following the Sumerians, the Babylonians built bridges, walls, tunnels, sewers, roads, the Hanging Gardens and the seven-storey Tower of Babel with asphalt as the bond.

Edwin Drake, originator in 1859 of the modern oil industry

In comparison to these ancient dates, biblical times seem almost recent. Noah caulked his ark with asphalt, and the infant Moses was set adrift in a bulrush boat waterproofed with it. Under the pharaohs, the Egyptians mummified their dead with, amongst other things, bitumen from the Dead Sea. In the 1st century BC, pre-empting Pasteur by two millennia, the Roman scholar Marcus Terentius Varro decided that some diseases were caused by creatures so tiny as to be invisible, and wrote of the disinfectant qualities of petroleum vapour. By the 2nd century AD, the Romans also burned crude oil as a fumigant (it was very good for killing infestations of caterpillars), and dressed damaged trees with it. By the early 10th century, Arab nations had worked out that it could be distilled. In Cairo, torches were lit with the product, and it is said that in the late 11th century, the equivalent of 1,400 barrels of petroleum distillate caught fire there, creating a monstrous blaze.

Opposite Page: Top-hatted Edwin Drake (styled a 'colonel' to impress the sceptical locals) stands in front of his historic well, 1861

Above:
Crude oil's versatility was recognized very early, especially in the Middle East. This limestone plaque, bordered with shells and bitumen, from Tell Aamar in Iraq, was made about 2600 BC…

Above right:
…The altar in the mausoleum of Bur-Sin (about 2220 BC) was covered with bitumen, proofing it against the offerings of liquid incense…

In the story of petroleum and the natural gases often associated with it, fire and conflict have often figured prominently, and sometimes simultaneously. In the 20th century, battles and even whole wars have been fought over the ownership of oil, and have been won or lost through the possession or lack of it. The same was true of the Middle Ages. Some time in the early part of the Roman Empire, it was discovered that if petroleum was blended judiciously with calcium oxide (lime) and then exposed to moisture, the result was spontaneous combustion. This had such obvious military value that it became a closely guarded secret, and after the fall of the western Roman Empire in 476 AD, it was partly because of this secret that the eastern or Byzantine Empire was able to maintain its dominion for nearly a full thousand years, until 1473. Put in fragile pottery containers similar in size to modern hand grenades, or pumped in jets like a modern flame-thrower, 'Greek fire' in the age of hand-to-hand fighting was a terrifying weapon, and was not superseded until the importation into Europe of gunpowder – like oil, another ancient Chinese discovery.

In more peaceful vein, mysterious eternal flames (caused by the ignition of gas seeping from the earth) have even prompted the foundation of fire- and light-worshipping cults and religions: Zoroastrianism is the main surviving one. But for centuries, for most people who have had access to it, by far the commonest reason for burning oil or gas has been to provide simple domestic lighting and heating. In 1272, passing through the town of Baku on the shores of the Caspian Sea, the great traveller Marco Polo noted that everyone used oil lamps, and that people came from miles around to buy oil. Half a millennium later, in 1795, Britain's first ambassador to Borneo wrote about the island's flourishing

Above left:
...The masonry of the ancient temple of Ur of the Chaldees was joined together with bitumen...

Above:
...and thousands of years later, in 1957, it was still possible to find coracles waterproofed with bitumen on the River Tigris

oil industry; reportedly, there were at least 500 wells (all dug by hand) supplying oil for lighting and heating to upwards of 7 million people.

Such comforts scarcely existed in Europe, even then, in the late 18th century; and Europe's earlier Dark Ages (the thousand years following the fall of Rome) were not only metaphorically but literally so. When the sun went down of an evening, that, for most people, was that. For the humblest, there was no light after nightfall, except from the moon, stars and possibly fires. Lamps made from rags or reeds dipped in animal fat were smoky, smelly and dim; candles made of lard or tallow were expensive; and both (especially in houses or huts built of wood and thatch) could be dangerous. So, generally, you went to bed at sunset and got up at sunrise, and around midwinter, joined your neighbours to rig up a fire festival. Whether or not this encouraged the sun to return, it was a sensible precaution (suppose, through its neglect, the spring did not come?) and it was fun: so much so that in some parts of the world the midwinter fire festival cheerfully survives.

However, except in a prolonged power-cut, it is hard to imagine how sharply the medieval European day was divided from its night. Yet when Europeans had little option but bed, their contemporaries in distant lands – particularly in the Middle East and Far East – could prolong the day at will. One might reasonably ask why it took the western world so long to cotton on.

In any social or economic sphere, the difference between having light after nightfall and having none is startling. It was not that Europeans in the Dark Ages wilfully ignored the possibilities of oil; those few who were educated certainly knew it existed, and were aware of some of its properties. The reason they did not establish an oil industry was simply

A Chinese drilling rig in the 3rd century AD. By then the Chinese already had 500 years' experience in drilling for oil

that there was no oil, or very little indeed, that was easily accessible. If there had been significant land-based oil deposits that could have been dug by hand near London, Paris, Rome or any other centre of western Europe's medieval civilization, they would have been exploited at the earliest possible date; and the entire world's history would have been different.

Today, of course, we know that Europe and the North Sea have many important oil and gas fields, and from early times petroleum seepage had been noted in several European locations, including the valley of the River Po in Italy. (Incidentally, it is interesting to note that the word 'oil' derives from the Latin *oleum* meaning olive oil – a commodity as central to the culture of ancient Rome as crude oil is to the modern world.) Yet although there were known western sources of crude, exploiting them was too expensive to be practical. As early as the Middle Ages, Europeans used rotary drilling to sink artesian wells, and in Grenoble in France in 1784, by a combination of rotary and percussion drilling, an exceptional hole was sunk 1,840 feet deep. However, these drillers were usually searching for water or for salt and other minerals, or latterly were extracting earth samples for civil engineering. Any discovery of oil was accidental, and the oil itself was an unwanted pollutant. Moreover, when Great Britain inaugurated the Industrial Revolution around 1750, machinery was designed to exploit the nation's known indigenous resources of wind, water and coal, so even then – in a world of steam technology – crude oil seemed useless.

It was not until 1828 that a Christian missionary named Imbert brought to Europe information about the ancient and much more efficient Chinese drilling techniques. By then, the search was on for cheaper and better lubricants, but, sensational as Imbert's information was, it did not actually help anyone to locate oil. Even today, and even by probing deep into the earth and below the sea-bed with three-dimensional seismic research, searching for oil is a hugely expensive task, with no guarantee of success. 'Wildcatting' (choosing a spot more or less at random and drilling into it) is generally a waste of time, effort and money; yet when the western world in the mid-19th century wondered where it might find a local source of oil, wildcatting was almost the only option. There were only two other possibilities: transporting oil (a hazardous occupation at the best of times) from a known source (and they were all far away), or drilling somewhere where oil was already seeping to the surface – such as at Drake's chosen site, Oil Creek in Pennsylvania.

In fact, Drake's venture came as near as possible to failure. The day he found oil, 28 August 1859, was a Sunday. What turned out to be the magic depth of 69½ feet was reached on Saturday 27 August, just before drilling stopped for the weekend. By then a letter from his backers was on its way, instructing him to cease operations altogether. It arrived in Titusville on Monday 29 August.

But on that Monday morning, surrounded by astonished local townsfolk, Drake and his small team were busy pumping and ladling their oil into any containers that could be found; before the strike, with money running out, they could not afford to stockpile barrels. When the letter was brought to Drake, he paused to read it, then put it away and carried on pumping. Quite rightly, he felt he could ignore the order to stop – for at last, the western world had found a commercially viable source of oil.

At the turn of the 19th and 20th centuries, the same principles applied; the only major difference was the introduction of heavy machinery like this Romanian percussion drill...

Near Titusville, the full-size replica of Drake's original rig is very humble in stature when compared to the oil and gas rigs of the North Sea, the Arabian Sea, the Gulf of Brunei and elsewhere. It is an ant to their elephants, a flea to their lions; but the effect it had on the 19th and 20th centuries was out of any proportion to its size.

A bewildering range and variety of products owe their origin to oil. Some are practically self-evident, in particular the fuels for cars, ships and aircraft: it is difficult to imagine a coal-powered aeroplane. But other oil-based products are much less obvious – chemicals, fertilizers, medicines, packaging materials, textiles – and even those represent only a small part of the colossal range of 20th-century essentials which are all either direct derivatives of crude oil or which owe their being to the oil industry.

...and this rotary drill at Baku in the Caucasus

Chemically, crude oil is a mix of hydrocarbons (molecules of hydrogen and carbon joined in numerous different patterns and quantities) which require physical or chemical processing to be turned into usable products. This processing separates molecules of different size and, with some of them, alters their composition and pattern, thus determining what kind of end product is created. Crude oil contains varying proportions, or 'fractions', of the different hydrocarbons, and separating them out into useful products is what refining is all about. The densest fractions are almost solid unless heated, and have such high specific gravities that, unlike most oils, they will actually sink in water. Asphalt and tar are two of these. Among the other lighter fractions, in descending order of specific gravity, are heating oils; diesel fuel; kerosene, or paraffin; naphtha; gasoline, or petrol; and at the lightest and most volatile end, petroleum gases, butane and propane. Finally, often found in association with crude oil, but not actually a product of it, there is the natural gas methane, which from its cleanness and efficiency has become a hugely important source of energy.

In the middle of the 19th century, however, the industrialized world's enormous vested interests in coal as an energy source meant there was scarcely any market for oil as a fuel; indeed, early oil producers generally threw away the heavier fractions of their crude and burned off the lighter gaseous ones, because (even though these together constituted 30–40% of the total) they were almost worthless. Light, heat and lubrication were all that were sought from oil; and all were found from the medium-density fractions. Lubricating oil was far better than animal fat for easing the works of high-friction machinery; but being relatively specialist and concealed by factory walls, lubricating oils did not impinge on the daily (or rather, nightly) lives of ordinary people nearly so much as kerosene, source of the 'new light'.

This colourless thin oil, less dense than water, can be extracted not only from petroleum, but also from coal, oil shale, and even wood. In the latter part of the 20th century, airlines became far and away its greatest consumers; permeating the local atmosphere, its distinctive and not entirely unpleasant odour could often be smelt some miles from an international airport. But in the latter part of the 19th century it was, world-wide, the principal source of artificial light.

By 1859, not only the principles of its extraction but also its actual production were very well established. In 1830, a German named Baron Karl von Reichenbach discovered the substance (at normal temperatures, a white waxy solid) and named it paraffin, from Latin words meaning 'too little related'. Chemically, it was one of the hydrocarbon series C_nH_{2n+2} – essentially the same thing as kerosene. Reichenbach chose its rather strange name because it appeared neutral, with very little affinity for other bodies. (Incidentally, it was the same Reichenbach who in 1833 discovered creosote, and later believed that somewhere between electricity, magnetism, heat and light, he had found the central life-force, which he named 'Od' – something largely forgotten except as a legitimate and occasionally useful word in Scrabble.)

A LAMPFUL OF OIL.

MAN'S ingenuity in the production of artificial light has spanned the gap between the primitive striking of flints and the brilliant electric glow of modern times. Though gas and electricity are the highest forms of this evolution, petroleum, soon after its introduction, as a cheap, portable, and brilliant illuminant, superseded all rivals as "the poor man's light." Whale and kindred oils had long occupied this position, but were about ready to resign it, as the pursuit of the whale had driven it to Northern latitudes, increasing the cost and scarcity of its products. The aid of chemistry was invoked to discover a substitute. This was found in the distillate of bituminous coals and shales, and its manufacture was largely increasing when the drill in Pennsylvania revealed vast quantities of a superior natural fluid. Refined petroleum literally "cast into the shade" all animal, vegetable, and other mineral oils, and its steady flame now not only burns in the frontiersman's cabin and the tenement-houses of the poor, but is the popular light in our villages and towns. Thirty-five years ago known only for its medicinal virtues, petroleum to-day is one of our great staple domestic products, and the fourth article in value of our exports.

Petroleum is a universal product, whose existence and burning properties have been known from the dawn of history. It is therefore very remarkable that its practical utilization should have

A gusher in 1886 – a spectacular and welcome sight, but wasteful and dangerous too

In its solid form, paraffin was (and is) very good for making candles; its liquid form was discovered about 17 years after Reichenbach's first effort. As often happens, different people in different parts of the world had been thinking on similar lines at about the same time. Of these, the two most important were James Young, an industrial chemist, and Dr Abraham Gesner, a geologist (and former medical student and horse exporter). In 1846, the Canadian-born Gesner succeeded in distilling an illuminating oil from bitumen. In 1847, courtesy of an English coal-mine owner who had found some petroleum in his mines, Young began experiments which soon resulted in the distillation of liquid paraffin. Because of the small amount of petroleum available to him, he then turned his attention to the more prolific oil-bearing shales of his native Scotland, again with success. In 1850 he was granted a British patent for the process, and in 1852 an American one. While Young experimented with shale, Gesner quite independently found and developed Albertite, a natural New Brunswick asphalt which produced what he called kerosene – an irregular word which he derived from the Greek *keros* (wax) and *elaion* (oil). At first, he was going to call the product keroselain, but kerosene sat more readily on the American ear, and under that name it was patented in America in 1853.

Certain difficulties ensued regarding the rival patents, but the important point for this narrative is that by 1859, the illuminating characteristics of kerosene or paraffin were well known, and the markets primed. If handled

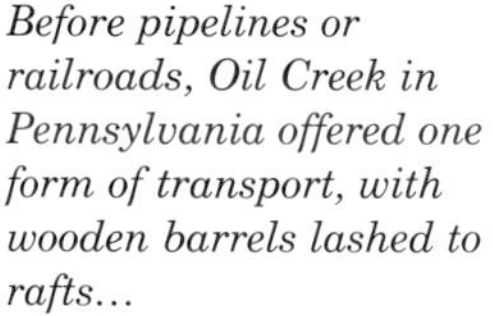

Before pipelines or railroads, Oil Creek in Pennsylvania offered one form of transport, with wooden barrels lashed to rafts...

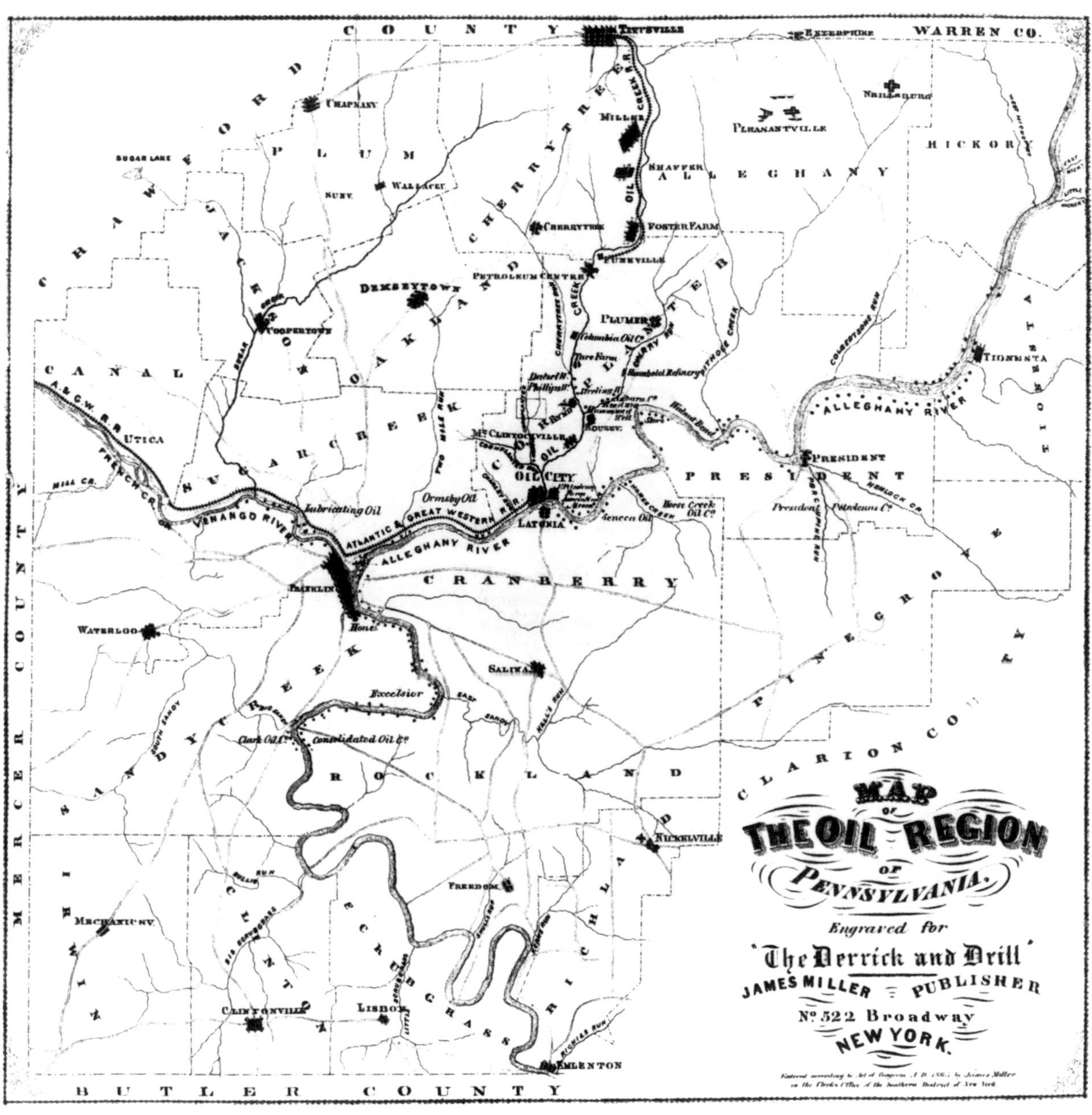

properly, the oil was generally safer than any other available form of lighting. Better, it was comparatively cheap; better still, it gave a wonderfully bright and steady illumination. People were therefore very willing to buy it; and as far as Edwin Drake's backers were concerned, the best fact of all was that it could be produced cheaply and easily from the petroleum which he had discovered in abundance. For this, there were ready markets, initially in the eastern United States, and later in Britain and continental Europe.

...but the Pennsylvanian 'Oil Regions', origin of the Western oil industry, were soon supplied with railroads, shown in this map from about 1865

Having found commercially viable oil, the next problem was to get it to the marketplaces. At first the storage unit, the barrel, was the overland transportation unit as well, with the clumsy awkward containers being lashed into carts, or onto rafts for river passages. But like oil, pipelines had been well known in the ancient world: it is said that as early as 1000 BC the Chinese used bamboo ones to bring natural gas into their homes, and the Greeks used ceramic pipes to transport water across long distances. (By using the siphon principle, they even managed to get water over mountains.) Pioneering American oil producers experimented, not very successfully, with wooden pipelines, but the Industrial Revolution's technology had already produced iron pipelines, and in 1865 the first wrought-iron oil pipeline was laid by one Samuel Van Syckel. It ran from Pithole City in Pennsylvania to a railway all of five miles away. With a diameter of only two inches and with relatively inefficient pumps, it could shift 800 barrels a day: a tiny little tube which is the forebear of all the industry's pipelines.

The first shipment of kerosene from the US to the UK took place in 1861, in a small sailing brig named *Elizabeth Watts*. When her crew learned the nature of their cargo, they all promptly deserted, and had to be replaced by trawling for drunks in the local bars. The crew's fear was quite reasonable: wooden barrels tended to leak, oil is inflammable, and fire at sea is one of the most terrible nightmares of any sailor. But the voyage passed safely, and America's petroleum exports – zero in 1860 – had soared to an annual value of $15 million by 1865, with most going to Great Britain.

From the moment the industry began, it was obvious great fortunes could be made. Immediately after Drake's discovery, thousands of strangers flocked to Titusville hoping to get rich quick, and the Pennsylvanian 'Oil Regions' were flooded with explorers, exploiters, speculators, venturers, conmen and cathouses in a way that could only be likened to the Californian gold rush. Sometimes the fortunate beneficiaries hardly had to lift a finger to become rich: in the course of just two years, investors in one outstanding well received $15,000 for every dollar they had put in, and in June 1865 an impoverished farmer near Pithole, a few miles from Titusville, was able to sell his land for $1.3 million. (One of the few comparable industries in recent years, mushrooming from nothing to a business and domestic essential, and making many millionaires along the way, is computing.) But if money could be made easily in the infant oil industry, it could also be lost. Pithole today is a ghost town; Drake himself shared the fate of many pioneers, and in 1880 died almost penniless, having subsisted for the last few years of his life on a pension provided (somewhat reluctantly) by the state of Pennsylvania. And, since the industry was brand new, it was virtually entirely unregulated, with its sole practical governance being that of the free market. At first it seemed obvious that the more oil you could extract, the more money you could make; but as lucky landowners counted fistfuls of dollars, and as fields and hills sprouted forests of ramshackle derricks, the market became rapidly saturated. Until the manufacture of oil lamps caught up, the oil's value plunged – from $20 a barrel when Drake began, to 10 cents only two years later.

This was the cause of many bankruptcies. It was, moreover, grotesquely inefficient; but very few people recognized that. One who did, and who became the first globally dominant force in the oil industry, was a strange man: hawk-nosed, eagle-eyed, cruelly methodical, upright in morals, ruthless in business, sternly God-fearing and, in his old age, generous to a fault – clichés, but true descriptions of John D. Rockefeller.

Genesis, chapter 6, verse 4: *There were giants in the earth in those days...mighty men which were of old, men of renown*. It is an apt quotation for the early days of the oil industry, and particularly so for this narrative: because from 1863, 'John D.', as Rockefeller was nick-named (one cannot say 'affectionately known'), set out to organize the chaos of America's oil – to structure it, unify it and rule over it as a personal fiefdom. Having started as a humble clerk and bookkeeper, his oil industry career began in 1863. His success in the industry was not a matter of luck, but of careful accounting, 'to unite', as he said, 'skill and capital in order to carry on a business of some magnitude and importance.' In every practical meaning, he succeeded.

John D. Rockefeller in 1905, aged 66. Creator of Standard Oil, he used all possible means – legal or not – to bring almost all the US oil industry under his personal control

In 1870, at the age of 31, he established the Standard Oil Company of Ohio. As a firm believer in a very Old Testament divinity, he understood the worth of compassion and generosity in the wider public sense: when he was very old and very rich, he gave away more than $500 million of his own money. But by the same token, neither compassion nor generosity figured for him in the narrow individual sense. The rules governing the Standard company (rules which reflected John D.'s own characteristics) were strict economy; beneficial mergers with competitors if possible; and if not, their ruthless elimination. John D. described this in a distasteful phrase: competitors would be given a 'good sweating' until they either went bust or sold out to his own company. If such merciless methods meant creating bitter enemies, prompting suicides and leaving families destitute, nevertheless it also meant that the industry was made efficient; and in Rockefeller's view, the end justified the means. Reaching out ever more widely and ever more strongly from its towering New York headquarters at 26 Broadway, Standard Oil became known as the Octopus; and through it, by the end of the 1870s, John D. had organized the American oil industry into his own image, controlling 90% of US oil refining and 80–85% of the US market, as well as dominating the entire transport system of the oil-producing regions – all without ever once forgetting the words of his favourite hymns. In worship as in life, Rockefeller always liked to lead the singing.

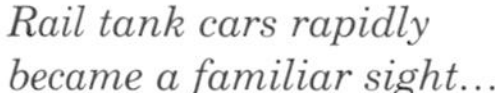

Rail tank cars rapidly became a familiar sight…

…and the departure of the first oil train from Bartlesville, Oklahoma, formed a suitably dynamic photograph

But those days were soon to pass, for he was not the only giant in the earth; nor was the north-eastern USA the only oil-producing region in the world. Very large known deposits existed in the Russian empire, particularly around Baku, on the western shores of the land-locked Caspian Sea, and were believed to exist in the East Indies, especially Sarawak and Borneo.

The Baku fields had been hand-dug since before Marco Polo's time. In 1872, in order that they could be more thoroughly exploited, they were released from their limiting status as a state monopoly. In 1874, the region's annual production was 600,000 barrels. Ten years later, that figure had multiplied 18 times over, to 10.8 million barrels, or nearly one-third of contemporary American production. The major part of the expansion took place under the leadership of Robert and Ludwig Nobel (brothers of Alfred, the inventor of dynamite and founder of the Nobel prizes). Their dominance in the Russian market was almost as great as Rockefeller's in the American: the Nobels' company refined half of all Russia's kerosene.

A vital factor in their success was that, like Rockefeller, they recognized the value of controlling an efficient, integrated transport system, from well-head to refinery to customer. An important difference, however, was that before they came along, there was no such system in place. Geography meant that initially, their transport expenses were very high indeed. Hemmed in by the 16,500-foot Caucasus mountains to the west and by the Caspian Sea in all other directions, Baku was extremely difficult to get

By 1890 Paraffin, fuel oil and lubricating oil were being carried by rail to Batum on the Black Sea and shipped to Europe and the Far East

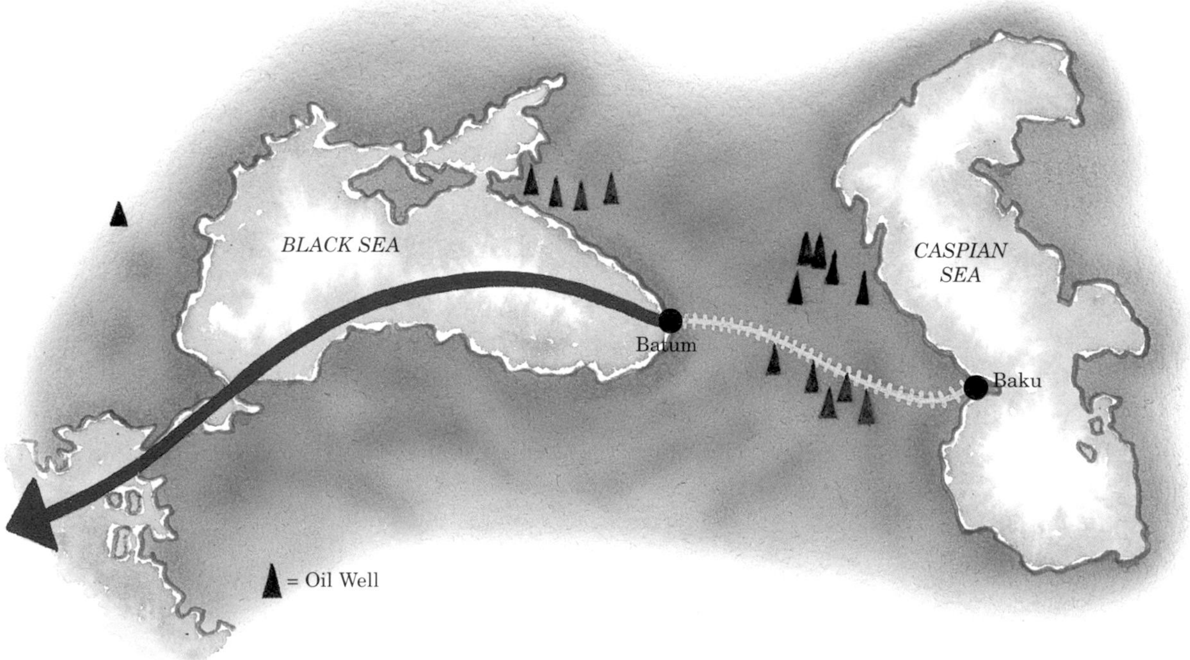

into, or out of. Although the Black Sea (which offered access to the Mediterranean) was less than 500 miles from Baku, the impassable Caucasus range meant that any oil which was to be exported had to cover a journey of more than 2,000 miles to the Baltic, the nearest accessible open sea, before it even left Russian territory. Besides that, a dearth of locally available timber at Baku meant that all out-going oil had to be carried in imported (and very costly) barrels – first by horse and cart seven miles to the refinery, then on a 600-mile inland sea voyage, then by river and lastly by rail.

As a final cost consideration, wooden barrels were very wasteful of space (because of their shape) and to a lesser extent of cargo, because they could leak. In the transatlantic trade, various people had improved upon this by packing their kerosene into cases, each containing two five-gallon tins, or by installing tanks into the holds of sailing ships. 'Case-oil' was quite easy to handle, and reached its end-user ready-packed in a suitable quantity; built-in tanks offered further savings. So, to reduce their own transport costs, the Nobels first built a pipeline from their wells to the refinery, which was far better than the horse-and-cart arrangement. They then had a small steam-ship built (aptly named *Zoroaster*) with tanks in her hold. However, even this still wasted valuable cargo space. How much better if the ship's hull was itself the tank!

Before long, *Zoroaster* had been converted in just such a manner, and became the forerunner of all modern bulk tankers. But she operated in the relatively restricted waters of an inland sea. Oil's inherent dangers (its inflammability and its fluidity, which in bad weather could destabilize and even upset a vessel) meant that, except experimentally in barges on rivers and short sea-voyages, no one yet dared to carry bulk oil in this way – with the skin of the tanker being the tank – on the open ocean.

With the Nobels controlling the route to the Baltic, the real break-through for independent producers at Baku occurred in 1883 when, funded by the Rothschilds of Paris, a railway was completed from Baku, through the Caucasus mountains to Batum on the Black Sea. With 1,500 miles lopped off the distance it had to travel, independently produced Russian case-oil could now compete in Europe with American oil.

Rockefeller's Standard responded in two ways: firstly, as it usually did, by cutting prices in the area of competition, and subsidizing the cuts by increases in its areas of monopoly; and secondly, with something much more unusual and imaginative. In 1885, its German subsidiary ordered a new ship, the 2,975-ton British-built SS *Glückauf* – the first authentic ocean-going steam tanker, capable of carrying bulk kerosene at about 25% less cost than case-oil. Launched in 1886, this ship was quickly followed by others: within six years there were as many as 80 like her plying the Atlantic. But they were extremely dangerous vessels.

The name *Glückauf* meant 'good luck', which was just as necessary as good seamanship; the ship's nickname, *Fliegauf* ('blow up'), said why.

Glückauf's forward-looking design promised much. In the year of her launch, however, the European oil markets were further complicated when the Paris Rothschilds decided to go into the Russian oil business themselves, using the railway they had financed to undercut both the Nobels' and the Russian independents' oil. Their new company, the Caspian and Black Sea Petroleum Company, soon became known by its Russian initials, 'Bnito', and was destined to play a crucial part in the foundation of the "Shell". Not to be outdone, the Nobels began laying their own trans-Caucasian pipeline, both speeding and shortening their oil's journey to the Black Sea by means of a 42-mile tunnel blasted through the mountains. It could help to have a brother who made dynamite, especially if you needed 400 tons of it.

The combined effect of these developments was soon apparent. In 1888 the United States supplied 78% of the world's exported kerosene trade, and Russia 22%. By 1891, the figures had altered to 71% and 29% respectively. Whether it was counted in roubles or dollars, a seven per cent shift meant a very large shift in the ownership of money; and at that very moment, another major shift was actually in the making, as a new and hitherto unsuspected player made ready to enter the field. Marcus Samuel junior, by this time 38 years old, knew that in daring to challenge the might of Rockefeller and the Octopus, he, like many others before him, could be destroyed. But instead, he became one of the giants of his age.

SS Glückauf *('Good luck'), nicknamed* Fliegauf *('Blow up'), was the first purpose-built ocean-going bulk oil tanker, providing part of the inspiration for the coup conceived by Marcus Samuel*

The Samuels' first tanker, the 5010-deadweight ton SS Murex

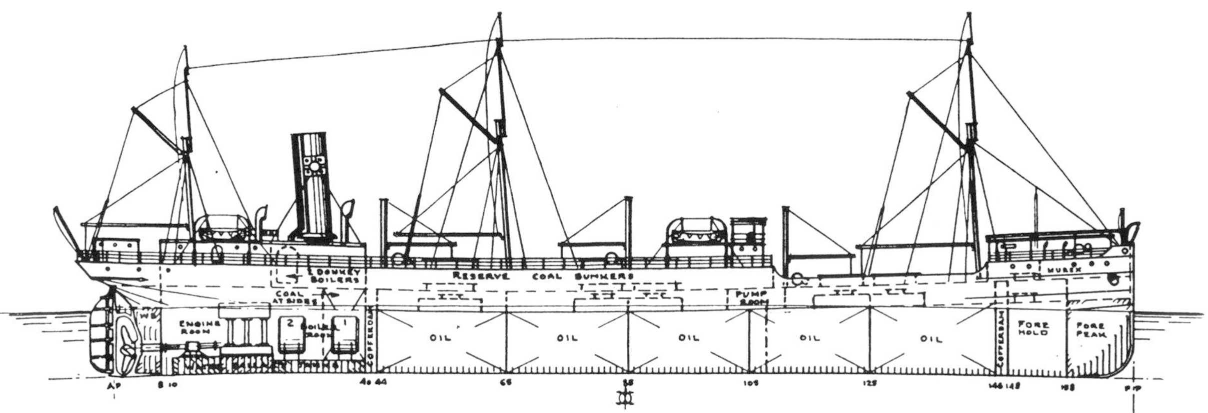

The key to the Samuels' coup was Flannery's design. This not only gained unprecedented approval for the carriage of bulk kerosene through the Suez Canal, but also incorporated such an efficient cleansing mechanism that on return voyages the tanks could be filled with foodstuffs such as rice or sugar, thus earning a profit both out and back

CHAPTER TWO

Origins: 1892–1898

In 1891, Marcus Samuel had recently returned to London from a trip to see his agents in the Far East. Visiting them – a normal business-like thing to do – provided a cover for the main purpose of his journey, which had been by way of Baku. What he had seen there and on the Caspian Sea had given him much food for thought as his travels progressed: so much so that for once, in a rare show of caution, he did not make his mind up at once, but sent his younger brother Sam to the Caucasus as well.

In the first half of the 1860s, their father's London-based business ('M. Samuel & Co. – Established 1834') had prospered very considerably, particularly in trading with Japan. That nation, after being opened to the western world by the US Navy, had begun its rapid rise to modernity mainly by trading with the United States; but when business was abruptly curtailed by the American Civil War, European merchants stepped willingly into the gap. When the war ended, so did the boom years; apart from anything else, the world-wide price of cotton crashed on the sudden release of accumulated American stock. However, Marcus the elder managed to ride the slump, and when he died was still moderately wealthy. He left his business to his eldest son, Joe, with instructions that the two younger sons, Marcus junior and Sam, should have a share in it later if they wanted. They did; what was more, Joe – never a very inspired businessman – did not really want it at all, so at some point in 1882–3 Marcus and Sam took it over altogether.

Sir Fortescue Flannery, Bart, MP

Their father also left the younger boys £2,500 each, given to them in 1878 when Marcus junior was 25 years old. This was a good amount of money, but what turned out to be the most valuable part of their father's bequest was something quite informal: his network of Far Eastern contacts and agents. Marcus junior soon met most of these personally, in two tours undertaken in 1873–4 and 1876–7. In 1878, 23-year-old Sam went to Japan and, as a counterpart to M. Samuel & Co., established Samuel Samuel & Co. in Yokohama, where he stayed for nearly a decade.

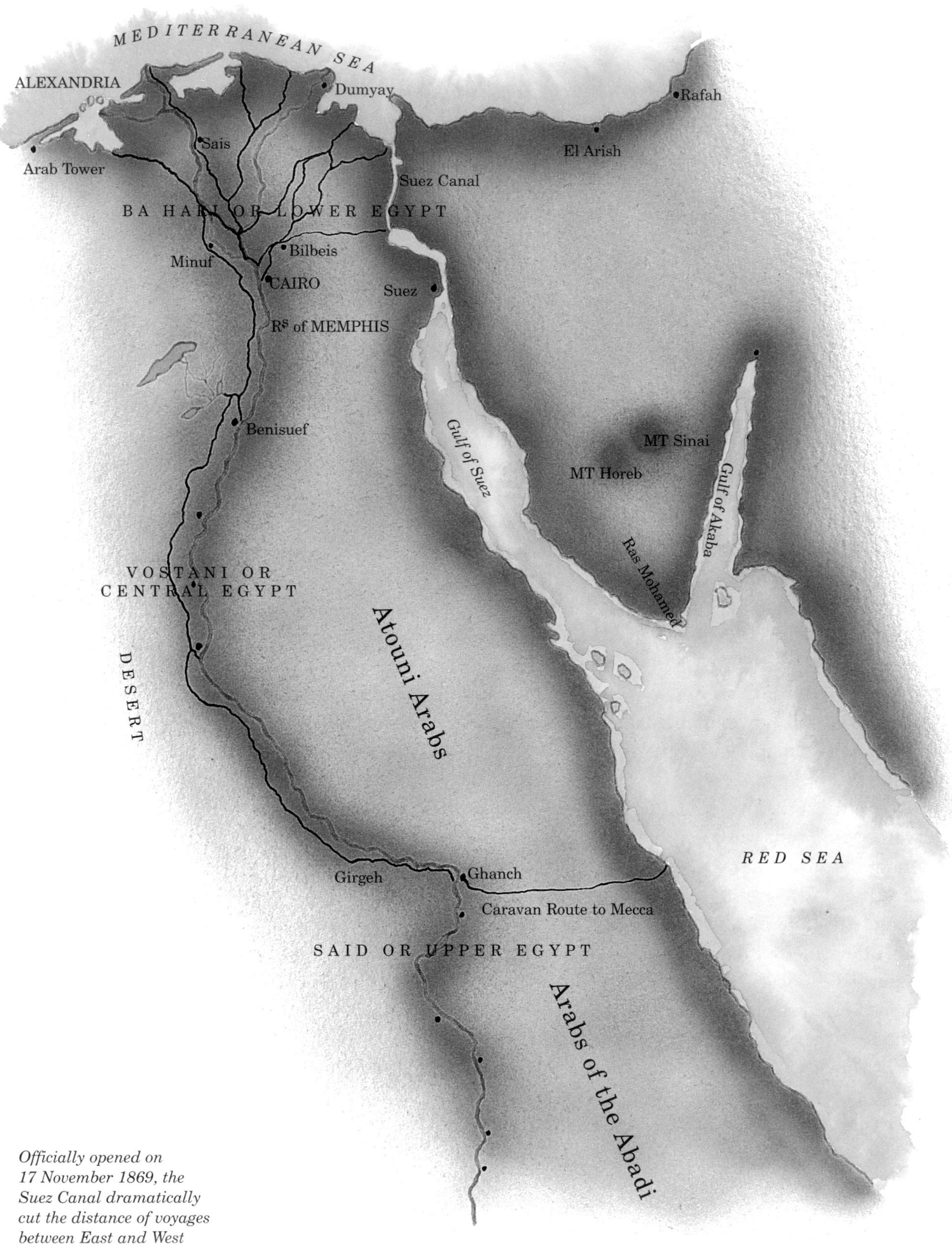

Officially opened on 17 November 1869, the Suez Canal dramatically cut the distance of voyages between East and West

Their father is credited with importing the first mechanical looms into Japan; under Sam's direction, the brothers continued this modernization, bringing British machinery, textiles and tools in, and taking rice, coal, silk, china, lacquerware and copper out, for sale in other parts of the Far and Middle East as well as in Britain and Europe. Simultaneously, under Marcus's direction, the London end of their operation traded in other commodities world-wide, particularly sugar, flour, wheat, tapioca and (of course) shells.

Jointly, the brothers made an important contribution to Japan's industrialization, and through that and their other interests, both were modestly rich before they were 30 years old; but of all their trading goods, none would compare to that which they began to buy and sell in 1886.

At first they purchased kerosene in small amounts either from Standard or from the great trading house of Jardine Matheson, for sale in Japan. Soon (there is some uncertainty about exactly when) they began to ship, or at least to consider shipping, case-oil. But certainly in 1888, Marcus asked Fortescue Flannery (later Sir Fortescue Flannery, baronet and Member of Parliament) to design a ship for him; and certainly in 1890, Marcus was recommended by Jardine Matheson to the Paris Rothschilds as the man best placed to implement a plan of theirs – to export their Caucasian kerosene to the Far East, for sale there in competition with Standard.

Mr Alderman Samuel, 1891

It was this idea which in 1890 took Marcus to Baku. As he travelled on from there to Japan, he slowly worked out the risks and ramifications that would be entailed. The risks were enormous: to succeed could bring greater wealth than anything he had imagined. But to fail would bring, in a word, ruin. That would be unbearable, not merely from a business point of view, but also because he was gradually achieving something he valued more than money: social recognition for himself and his family. He was hoping to be (and in 1891 was) elected an Alderman of the City of London, a rare distinction then for a Jew, and rarer still for one of his comparatively obscure and humble background. If all went well, in due course he could become Lord Mayor of London, the highest civic post in Britain. He was not prepared to forgo that possibility; but he was very attracted by the idea, especially because at Baku, he believed he had seen the key to making it work.

Having visited Baku for himself, Sam agreed. The key lay in making shipments not of case-oil, but of *bulk* oil, as the Nobels did on the Caspian. That, though, was only the key; it was not the lock. The lock held many levers, all of which would have to be turned simultaneously and as secretly as possible, or Standard would take pre-emptive action. First, the source of oil would have to be utterly reliable, and priced to reflect the refiners' savings on the cost of tinplate and cases. Second, not just one ship but a fleet would have to be built simultaneously, to ensure continuity of supply to the markets. Third, the ships would have to be much bigger than the Nobels' – bigger, indeed, than the *Glückauf*, and better, because the most economical route from the Black Sea to the Far East was via the Suez Canal, and there, the regulations governing the transport of oil in bulk were so stringent that no one (not even Standard, which had tried) had ever received permission to take such a cargo through. Fourth, it would be pointless to try the experiment in only one country, or here and there through the region; that would invite piece-meal destruction by Standard. Therefore, a complete network both of storage, distribution and sales would have to be in place before the first shipment of oil was sent.

Among other key personnel in the coup were Fred Lane, the ship-broker and consummate go-between, seen here in about 1890...

That, in summary, was the plan; and in even briefer summary, suffice it to say that outlandish as the entire scheme was, it worked in every particular. This was all the more astonishing because of the number of people involved – the Samuel brothers; Flannery, their ship-designer; the Rothschilds; Fred Lane, the ship-broker who originally put Flannery and the Rothschilds in touch with Marcus and Sam; their nephews Mark and Joe Abrahams, who were sent to find and develop tank-storage sites in the Far East; the personnel of several different British ship-yards; and the seven Eastern agents, from Bombay to China, whom Marcus drew into the plan.

With so many participants, secrecy could not be absolute, but it was very nearly: despite rumours in the press and questions in parliament, the first that the general public (including Standard Oil) knew for sure of the plan was on Saturday 28 May 1892, when the launch of 'the splendid steel screw-steamer *Murex*' was reported by the *Northern Daily Mail*. The launch in West Hartlepool had taken place at 4 a.m. that day (which shows something of the miraculous achievements of hot-metal printing presses), and as *The Times* added, there were other vessels soon to come: the 5,010-ton *Murex* was only 'the first of a fleet of steamers that is being built – for a syndicate of which Mr Alderman Samuel is the representative in this country – for the transportation of petroleum in bulk through the Suez Canal...'

The design of the ships was Fortescue Flannery's personal triumph. Every possible safety element was included, whether of equipment, lay-out or construction. Lloyd's of London rated the vessels 100.A.1, a first-class risk; the Canal authorities gave permission for them to pass through Suez; and in August 1892, laden with 4,000 tons of Caucasian kerosene, *Murex* did so – although even then, in natural caution, the Canal authorities insisted that her passage should be made under tow.

Nevertheless, the fact of her passage was a conquest, and short of criminal action, there was very little that any competitor could do about it;

the plan was too well laid and too well executed. By the end of 1895, just 40 months after that first transit, a total of 69 bulk-oil cargoes had passed through Suez, and all but four of them belonged to the Samuels.

This outstanding achievement is generally described as 'Marcus's coup', but it was really more than that. In the world of oil transportation it was a revolution, no less: the dramatic beginning of a permanent change. Yet for one brief period, the whole thing almost went completely wrong, for a small and rather ludicrous reason which no one – not even the Eastern agents – had thought of beforehand. The essence of the plan was to supply kerosene of at least as good quality as Standard's case-oil (which locally went under the name of Devoes) to customers bringing their own containers, and to pass on to them the savings made by bulk transport; yet with thousands of tons of good cheap kerosene available, no buyers could be found. Increasingly worried telegrams were sent to and fro, until at last someone divined the answer: in the eyes of the local populations, the blue Devoes tins (though rusty and battered after their long voyage from America) were themselves worth more than the difference in price, because when empty, they could be flattened out and made into all kinds of useful articles. A house could be roofed with them, and filled with bird-cages, bedpans, lamps, cooking pots and so forth, all made out of the old blue tins.

...and, also photographed about 1890, the brothers Mark and Joe Abrahams, nephews of Marcus and Sam Samuel

Once this simple domestic economy had been grasped, Marcus's response was immediate. If the buyers wanted tins, they could have them; and the tins would be assembled locally, near the Eastern ports, using Welsh tinplate exported for the purpose. It was a splendidly simple solution. Without compromising the principle of bulk transport, local employment was provided, creating goodwill towards the company; instead of being rusty, the tins were shiny and new; and even with their cost included, the bulk oil still undercut Devoes. Mark Abrahams, the Samuels' nephew, provided the imaginative finishing touch – the tins were painted an instantly identifiable bright red. It worked almost at once; within months, Oriental roofs, bird-cages and bedpans alike were changing from rusty blue to shiny red.

The lesson was clear. To outdo competition, the product must be at least as good, cheaper if possible, and somehow must also provide added value to the customer. Learned very

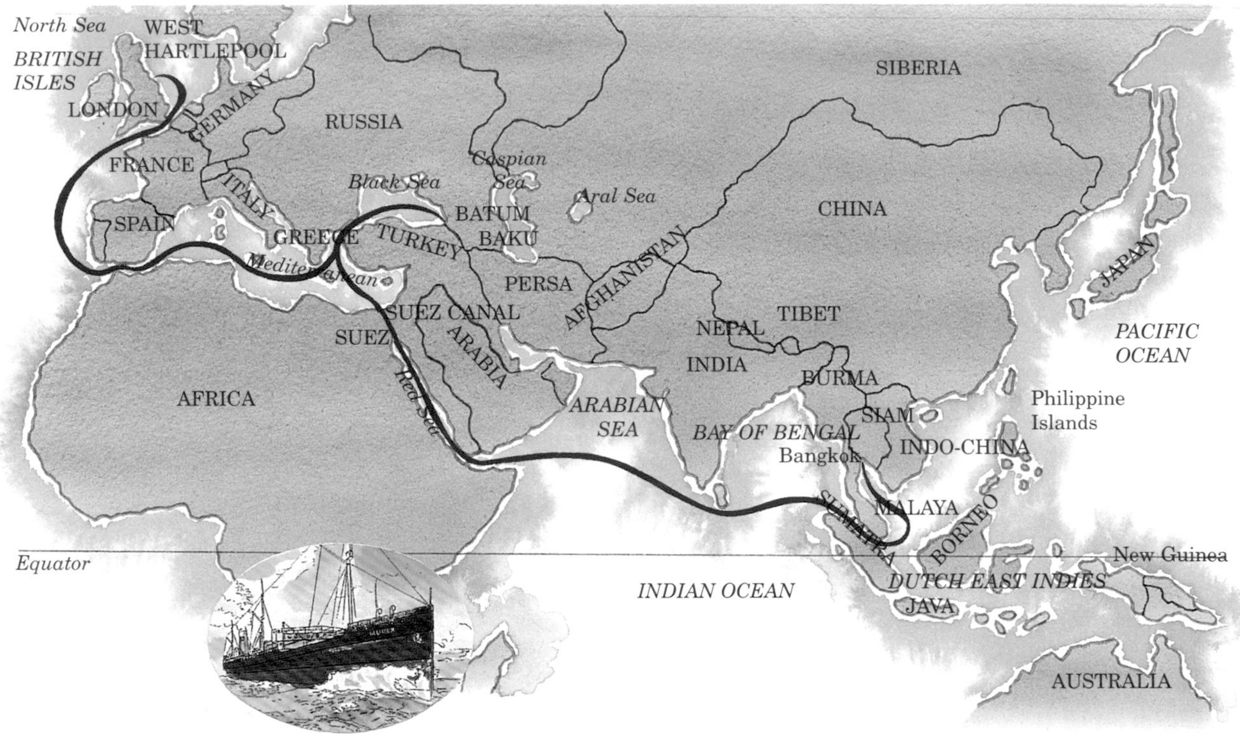

Marcus Samuel's coup: Murex's first voyage, 1892

early on, it was a lesson which, in the century since, Shell has almost always remembered, or forgotten at its peril.

Like the ships, which were all named after shells, the kerosene was called 'Shell Oil' from a very early date, while the traditional parts of the family businesses continued in parallel – in fact, often in the same ships. Most tankers at that time carried a paying cargo in only one direction, making their return voyage empty. However, Flannery's ingenious design for *Murex* and her sisters included a tank-cleaning system of such efficiency that even foodstuffs could be put into their tanks without being tainted; and since rice, flour, sugar and so on are commodities which can flow like liquids, the ships could be used to transport such goods back from the Far East to Europe. Thus, their return voyages were profitable too; and from them all Marcus and his colleagues soon amassed great wealth, each in proportion to his input.

This input did not necessarily have to be in cash. Indeed, one of the most intriguing aspects of the entire operation is the financial, for comparatively little cash was placed up front by anyone. William Gray's, the builders of *Murex*, agreed to such generous extended credit terms that Marcus placed eight of his first 14 ship-orders with them. Bnito, the Rothschilds' oil company, similarly agreed that the oil carried in the ships did not have to be paid for at the time of loading, but on bills of exchange 'at 3–4 months' sight' *after* it was sold in the Far East; and each of the seven Far Eastern agents involved agreed similar terms regarding the construction of settling tanks, storage tanks and both wholesale and retail points and methods of sale. For their own parts, Marcus and Sam essentially accepted final responsibility for the risk involved – no mean thing in itself. But of the whole amazing plan, perhaps the most adroit part is that its design meant the revolution did not have to be paid for until it was generating its own income and paying for itself.

It was certainly a pretty canny way of setting oneself up for several fortunes; but what must never be forgotten is that for all those involved, and especially the Samuel brothers, the element of personal risk was real, and very high. Similarly, the fact that such arrangements were possible underlines the degree of soundness attributed to the twin companies of M. Samuel and Co. and Samuel Samuel and Co. None of it could have come about without the earlier efforts of Marcus the elder in creating his network of agents, or without the continued efforts of his sons in maintaining what he always liked to call 'the good name of Marcus Samuel'. In the merchant world, the name of Marcus Samuel was good, and after the historic events of 1892 it would become better still – sometimes in unexpected ways. One of the consequences of the method of payment agreed with Bnito (by bills of exchange) was that the proceeds of sales were themselves often met by other bills of exchange from the Far East. In practical terms, this aspect of the business was a banking operation, and before long, Marcus junior found himself acting effectively as a merchant banker. Seeing this evolution, he and Sam decided to organize it properly: they set themselves up as co-owners of a private bank, later to become part of Hill Samuel.

In 1893, though, those developments still lay in the indeterminable future. For Marcus himself, the year held two main areas of interest. One was the obvious and immensely stimulating success of the new enterprise. The other was that when only just past his fortieth birthday, he was taken ill, diagnosed as suffering from cancer, and given less than a year to live.

Both diagnosis and prognosis were somewhat out; he did not have cancer, and he lived until 1927 – 34 years more, rather than one. But the experience shook him, not least because he realized that if anything did happen to him, his informal, personal business methods could leave his family vulnerable, just at a time when they were beginning to enjoy and grow accustomed to the fruits of those methods. He therefore decided to organize things somewhat more formally and methodically.

This decision was, for Marcus, a major mental turning-point: a recognition that he had successfully entered a much bigger game than any in which he had been previously involved, with potentially bigger rewards

Right:
Dotted around the world, workshops such as this one in France produced the tinplate for...

Below:
...Shell's distinctive red tins, soon to become world-famous themselves

Below:
From wells such as this one at Balik Papan in Borneo came the oil...

...which passed through refineries such as this at Pangkalan Brandan in northern Sumatra...

...to depots such as this one in Hiranumia, Japan...

Left:
...and brought great prosperity to all the members of the Samuels' Tank Syndicate. In 1895 The Mote at Bearsted in Kent became Marcus's new country home...

Above:
...and (from 1907 until his death 20 years later) 3 Hamilton Place became his town house

and certainly with bigger risks than any he had known hitherto. It seems plain that at this point he understood in his stomach that, for the sake of his family – having shifted the entire emphasis of his business from the fashion item of shells and the domestic staples of rice, wheat and flour, to the cut-throat and predatory world of oil – he could no longer continue in the sparky, somewhat disorganized way he had done in the past, relying on his personal charm, his assiduous nurturing of all his Eastern contacts, and agreements made on a handshake, a smile and a sense of mutual trust. Could the Russian suppliers – his sole suppliers – be relied upon? Not entirely: rail transport charges could and did vary from month to month, and already there were rumours that, having seen the profitability of the Rothschild-financed Baku-Batum railway, the Tsarist regime might nationalize it and might re-nationalize the Baku oil-fields as well. Could they not, if they chose, then make other arrangements to their benefit and his detriment? Yes; and that was too great a risk for himself, his business and his family. A more formal organization must come about.

Having passed that personal watershed, it was actually a quite simple matter to establish this organization. In the British press, frequent references to Marcus's 'syndicate' were often intended critically, to indicate something legal but perhaps not quite above board. But a syndicate was what it was – a group of individuals with shared interests, shared commitments and shared rewards, in addition to whatever extraneous and private projects they might care to pursue. And so in 1893, with minimum legal fuss and minimum upset to existing trading arrangements, a formally constituted syndicate was what it became: the Tank Syndicate.

Its members were the two Samuel brothers, Fred Lane the ship-broker, and the seven Far Eastern agents, sharing between them all profits and all losses too; for only in such a manner could they be sufficiently flexible to defend themselves in all areas against any threats, raids or incursions from the Octopus.

The Tank Syndicate was the immediate true forerunner of the "Shell". Almost at once it was provided with an unexpected fillip, in the shape of the Sino-Japanese War of 1894–5. Since Britain was neutral, the Syndicate was able to use its connections in the belligerent nations to trade with both sides, supplying grain, bags, railway machinery and equipment, weaponry, blankets and food. Japan's victory in 1895 brought overseas possessions to the nation: Formosa (Taiwan), and the adjacent Pescadores (Penghu) islands. Formosa then was an important producer of camphor (as used in mothballs, that mainstay of the Victorian wardrobe), and the industry was promptly made into a Japanese government monopoly – with its management given exclusively to what was now the leading British business in Japan, Samuel Samuel & Co.

This period, the middle 1890s, was one of the most productive and probably one of the happiest of Marcus's life. In his business life he seemed to have acquired the Midas touch, as his every interest turned to gold – even if Alfred Suart, one of his competitors, said that he conducted his businesses in the same way as he rode his horses: 'He always looked as if he might roll off, but never quite did.' And his home life had become quite idyllic. Most importantly, his family was complete, with two sons, two daughters, and a wife, Fanny, who was deeply devoted to him, as he was to her. Although prosperity alone is no guarantee of happiness, it certainly can help; and on the material side, if Marcus had never been really poor, he was now quite definitely rich. Take that point about horses: not everyone could afford to keep them simply for the pleasure of riding, but now he could, and he did, exercising them every weekday morning in London's Hyde Park. He could also afford ponies for his children, an Eton education for his sons, jewels for his wife, a large house in Portland Place (from which he would drive to the office at 31, Houndsditch, in his own carriage), and a great heap of a mansion (modestly called The Mote) in Kent, complete with a famous library, an equally famous stock of coarse fish in the lake, and 500 acres of parkland.

Not bad for a Jewish lad from the East End of London. If one aspired to be recognized as a gentleman, the trappings were essential, as well as enjoyable, and having acquired the necessary funds, Marcus had taken the sensible shortcut of acquiring all the trappings at one go: Lord Romney, former owner of The Mote, had sold him not only the house but everything in it – library, pictures, furniture and all. Perhaps it was all rather showy; but while Marcus enjoyed show, his pleasure in it was transparently

In 1898 the rescue of HMS Victorious, *stranded in the Suez Canal, earned Marcus Samuel his knighthood*

innocent, and like his brother Sam he was a generous man: and so very few people begrudged him his success. In 1894, from the position of Alderman, he was elected Sheriff of London, placing him definitely in line for the Lord Mayoralty, and in 1895 the ship-owners of London chose him as their representative on the Thames Conservancy Board.

The garden was, in short, extremely rosy; but at the heart of the rose lurks always the invisible worm. For Marcus in 1896, the worm was named Bnito. It was in his eyes an unreliable grouping. Transport fees from Baku to Batum were not so much a railroad as a roller-coaster, soaring and plunging unpredictably. The freedom of the entire Baku region was subject to Tsarist whim. If it remained his sole source of oil, then when the existing agreement ended in 1899, he and his Syndicate colleagues could find themselves almost literally over a barrel, bound to pay for it whatever price was dictated.

Half-way through the Bnito agreement, therefore, he began to consider alternative future sources, and it occurred to him that beyond the reach of roubles or the Standard's stranglehold, there was another well-known oil region in the world: the East Indies. His answer to the question of future supply was the motto of self-sufficiency: grow your own. The Syndicate should have its own oil, controlling the product at every stage, from pumping, through refining and transporting, to selling. Nephews Mark and Joe Abrahams were accordingly instructed to perfect the storage and distribution arrangements in India and the Far East; and when that was done, Mark was to go and discover oil in Borneo.

Borneo was chosen firstly because oil was known to exist there, and secondly because Marcus had met an elderly buccaneering Dutchman named Jacobus Menten, who had a drilling concession in the Bornese province of Kutei. The concession covered about 2,500 square miles of thick jungle, in which Menten, after several years' prospecting, had entirely failed to strike oil. What he had found, floating on streams, was a good deal of naturally seeping oil. He had sent samples to a Dutch geologist, who declared them quite unsuitable for making kerosene, describing them as 'a thick brown mess, thicker than treacle in Holland in winter'.

This was not the only regional option available to the Samuels' Tank Syndicate: in neighbouring Sumatra, where crude oil had been known for at least 400 years, there were other prospecting groups in need of support, including a small enterprise with a very big name – the 'Royal Dutch Company for the Working of Petroleum Wells in the Netherlands Indies'. However, Menten managed to convince Marcus, and at the end of 1895 returned in jubilation to continue his hunt in Borneo, backed by £1,200 of the Samuels' money, with the promise of a further £2,400 plus royalties if he actually found oil.

One of Royal Dutch's earliest installations: the distillation plant at Pangkalan Brandan, 1893

In October 1896, Mark Abrahams joined him. Mark (with his every move watched and reported by agents of Standard Oil) had by then completed his tank installations, and received his sole instruction in the techniques of oil production – a two-week stay in the Baku region. Menten for his part had recruited workers, cleared 18 acres of Kutei jungle, set up rudimentary accommodation, and decided to drill in a place called Sanga Sanga, 40 miles from a beautiful untouched bay called Balik Papan.

Although Mark Abrahams was only ten years younger than his Uncle Marcus, and although personally they got on very well, their professional relationship was never easy when Mark was away, largely because Marcus and Sam Samuel had no conception of the difficulties involved in searching for oil in a tropical jungle. At 31, Houndsditch, all their shipping operations – and their new attempts in oil discovery – were run from a single small room containing nothing more than a table, two chairs, a wall-mounted map of the world and a telephone, from which they would bombard Mark with futile and infuriating cable messages.

Nevertheless, on 5 February 1897, after drilling 150 feet into sandstone, Mark struck oil. He would actually have been very unlucky not to, because at the 'Black Spot', as the native Dyaks called it, oil was oozing to the surface of its own accord; but (as the Dutch geologist had warned) it was very heavy, and would be difficult to refine into kerosene. Still more galling was the news that at Moeara Enim – a concession not far away, in which Marcus could have bought an interest – a gusher of light crude, perfect for kerosene, had been struck. However, Marcus remained exceedingly optimistic, forecasting that in 1898 Kutei would produce 15–20 million cases. Partly because of this, in April 1897 talks ceased on what would have otherwise been an historic agreement: a union (not merely partnership, but complete amalgamation) with the third party in the East Indies, the seven-year-old Royal Dutch Petroleum Company.

This putative amalgamation could have had many mutual benefits. The Tank Syndicate had a distribution and marketing network, but only one well, producing oil of uncertain quantity and questionable quality. Royal Dutch, headed by Jean-Baptiste Auguste Kessler, had plenty of oil, but was only just setting up a proper sales system. From the day of his appointment in mid-1896, however, its young and very ambitious marketing director, 30-year-old Henri Deterding, vigorously opposed any transfer of company assets, and it was his opposition – just as much as Marcus's production forecast – which prevented an amalgamation.

Marcus did not mind very much, because just then another part of his growing empire produced most welcome fruit: the right to float in London, on behalf of the Japanese government, the first sterling loan to Japan. This went so well that other banking arrangements followed, providing funds which he poured into Kutei.

Riding high, he then moved the London office from Houndsditch to larger and more prestigious premises in Leadenhall Street, bought another seven ships, and, on 18 October 1897, created The "Shell" Transport and Trading Company.

This was almost as simple a process as the creation of the Tank Syndicate. Essentially, all that happened was that ownership of the fleet was transferred from the Samuel brothers to the company, while the company name was altered and its structure formalized, with an initial capital of £1.8 million in £100 shares. The original shareholders were the former members of the Syndicate, each receiving shares in proportion to his Syndicate input; and the Samuel brothers' personal proportion, representing the fleet's value plus some Kutei assets, was £1.2 million.

Mark Abrahams as British Vice-Consul in Borneo c.1898

The New Year, 1898, brought (for entirely unrelated reasons) calamity to Royal Dutch and increasing good fortune to Marcus. On New Year's Eve 1897 the Royal Dutch wells began to produce, instead of oil, a mixture of oil and water – a sure sign that their oil was running out. By the end of March 1898 most of their wells were dry. Shell's Sanga Sanga wells were then producing no more than 15 barrels a day, or a trifling 23,000 cases a year; but on 15 April, with his first well on a new site at Balik Papan, Mark Abrahams struck oil again, at 750 feet. Production multiplied more than eight-fold, to 130 barrels a day - equivalent to 200,000 cases annually.

Meanwhile, a notable event had taken place at the entrance to the Suez Canal. On 15 February, a British warship, HMS *Victorious*, had run immovably aground. When every available official means of towing her off (two naval ships, one P&O ship and sundry Canal vessels) had failed, the Admiralty began to think that despite her considerable value, she might have to be blown up in order to clear the Canal. However, a Shell tanker had waited by in case of need. Her services were now offered and accepted. The operation took 21 hours, but was successful, and highly profitable: for rescuing the stranded warship, the tanker's crew shared an Admiralty reward of £500, her master individually received the same, and ten times the amount was offered to her owners – which in practical terms meant to Marcus. He declined, saying truthfully that as a British patriot he was happy to have provided the service gratis; and so instead he was offered a knighthood. This he accepted with delight, and on 6 August the honour was conferred upon him by Queen Victoria in Osborne House, her home in the Isle of Wight.

At the same time, after the failure of its wells, Royal Dutch, like Shell, was having to buy Russian kerosene, and was selling it in competition with Shell. Kessler had feared Marcus would start a price war, which Royal Dutch could not have survived; but instead, much to his surprise, the two

No

REGISTERED OFFICES.
16, LEADENHALL ST E.C.

THE "SHELL" TRANSPORT & TRADING COMPANY, LIMITED.

INCORPORATED WITH LIMITED LIABILITY UNDER THE COMPANIES ACTS 1862 TO 1892.

CAPITAL £2,000,000

INCREASED FROM 1,800,000 JULY, 1900.

DIVIDED INTO 2,000,000 ORDINARY SHARES OF £1 EACH FULLY PAID.

DECEASED SEE BACK.

This is to Certify that [illegible] of [illegible] is the Registered Proprietor of one hundred [illegible] Fully paid Shares of £1 each in THE "SHELL" TRANSPORT & TRADING COMPANY, LIMITED, numbered as per margin subject to the Memorandum and Articles of Association of the said Company.

Given under the Common Seal of the Company the 30th day of August 1900.

THE COMMON SEAL OF THE SAID COMPANY WAS HERETO AFFIXED IN THE PRESENCE OF

DIRECTORS.

SECRETARY.

N.B. TRANSFERS MUST IN ALL CASES BE [illegible]OMPANIED BY [illegible]ARE CERTIFICATE.

On 18 October 1897, The "Shell" Transport and Trading Company came into existence

companies had started to talk again. Marcus, observing that the East was quite big enough for both companies, proposed that in mutual defence against Standard, they should agree not to undercut each other in the Far East. Naturally Kessler accepted. With that sales agreement achieved, and looking towards the expiry of his contract with Bnito late in 1900, Marcus then gained a supply agreement (on Christmas Eve 1898) with the Moeara Enim company. Under this agreement, Shell would handle all the company's exported oil; but because small amounts would be uneconomic, Marcus made sure he had an option to cancel the contract if, by 1 January 1901, the minimum annual supply had not reached 50,000 tons.

With the short-term future apparently fairly secure, he then turned his attention to another, potentially far more exciting, plan. Remembering the Nobels' Caspian fleet, he had thought of a better use for his heavy Kutei oil. *Zoroaster* had actually been fuelled by similar oil, the first ship in the world to be so powered. It dawned on him that, rather than wasting money trying to make kerosene, Shell's own ships could be made to run on Kutei oil – and so, come to that, could the ships of the Royal Navy, far and away the largest armed fleet in the world. Now *that* would be a market worth pursuing. Pursue it he did, and with such energy that over the next 15 years he became internationally acknowledged as the world's leading exponent of fuel oil. Marcus had often been called visionary, and certainly he was nothing if not forward-looking; but this particular enticing vision led, step by seemingly inexorable step, to his downfall.

Key Far Eastern sites in Shell's early years

CHAPTER THREE

Too Far, Too Fast: 1899–1906

On 27 February 1899, having arranged for a champagne supper for Government and company officials, Mark Abrahams set off on a 360-mile journey from Sanga Sanga across the straits to Macassar, the nearest telegraph station. He had important news for London – at 850 feet, well number 12 had struck a gusher of good light oil, ideal for kerosene. There was only one problem: it was on fire.

Earlier that same month, at Marcus's request, old Jacobus Menten, the original concession holder, had turned up to see how things were going. On arrival, he began writing a stream of letters to London, abusively criticizing every aspect of Mark's management. These caused such disquiet in Leadenhall Street that it was decided Sam Samuel should go personally to Borneo. It was a decision with far-reaching consequences.

Sam left England early in March and was away for nine whole months, leaving his brother Marcus in sole charge of the company. Of the two brothers, Marcus is always characterized as the eager, imaginative one, and Sam as the practical one, putting a brake on his partner's wilder ideas – a combination often found at the root of a successful business, and, in the Samuels' case, a pretty accurate impression. But for the combination to work, the two people must be together.

Sam in Borneo soon found that Menten's criticisms were no more than 'ravings', and that nephew Mark was doing as well as could possibly be expected under very adverse circumstances – circumstances which, as Sam now understood with embarrassment, included frequent and totally unhelpful interventions and instructions by cable from London. But during the same period, he had little close information about events at home; and when he returned to London and learned what Marcus had been doing in his absence, he was horrified.

The first "Shell" logo, 1900

On 12 December 1899 the brothers had a furious argument, Sam shouting that Marcus was a lunatic, an imbecile and must be out of his mind, and Marcus answering in an icy whisper that

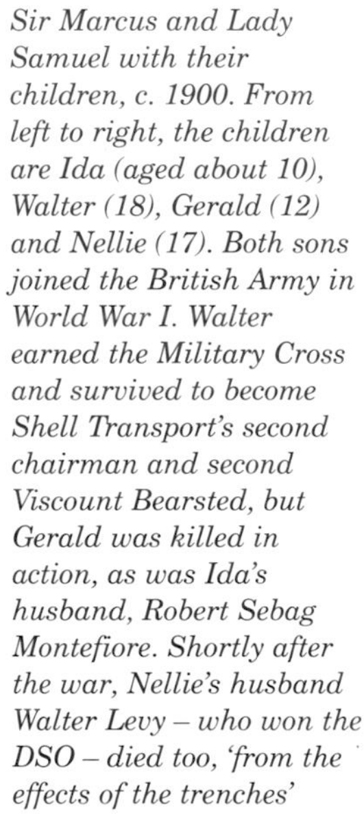

Sir Marcus and Lady Samuel with their children, c. 1900. From left to right, the children are Ida (aged about 10), Walter (18), Gerald (12) and Nellie (17). Both sons joined the British Army in World War I. Walter earned the Military Cross and survived to become Shell Transport's second chairman and second Viscount Bearsted, but Gerald was killed in action, as was Ida's husband, Robert Sebag Montefiore. Shortly after the war, Nellie's husband Walter Levy – who won the DSO – died too, 'from the effects of the trenches'

Sam was never satisfied. What he had done thus far was, he believed, completely logical; it simply was not yet finished.

Sam was not worried by the paper Marcus had delivered to the Royal Society of Arts in March 1899, extolling the virtues of liquid fuel. Nor was he worried by Marcus's immediately consequential friendship with the iconoclastic Vice Admiral Sir John Fisher. Sam was neither worried, nor particularly impressed, by the account of Marcus's visit on 17 June to the Automobile Club's first public road trials, where Marcus had decided that horseless carriages were the coming thing, and that especially in Britain, where he was a leading figure in the oil world, their fuel should be supplied by Shell Transport, rather than (as he himself had seen) by Standard. Nephew Mark Abrahams would be instructed to establish a petrol refinery at Balik Papan for this very purpose.

None of the above caused Sam direct concern. But while he had been away, Standard had approached both the Nobels and the Rothschilds with a proposal of alliance, and Marcus in response had decided Shell Transport must demonstrate strength, lest the potential monopoly should break him. Four new large tankers; further acquisitions in Borneo; the refining of petrol; increased oil storage capacity; and the purchase, on a rising market, of vast quantities of oil from any available source – such was his demonstration. When the Samuels' own fleet proved insufficient to carry the new stocks, he chartered; at one time in 1899 he had had 16 tankers in the Suez Canal simultaneously, either hastening out to the Far East, or hurrying back (empty, for there was no time to take in dry cargoes) to Russia. The outlay and risk alike were tremendous, and it was that which terrified Sam.

Interior of the Royal Automobile Club, founded just a few weeks after Shell Transport and Trading

Half a year later, at the AGM of 21 June 1900, it appeared that Sam's fears were groundless. The company's 252 shareholders learned from a no doubt rather smug Marcus that during his solitary leadership, profits had increased by 60% to more than £1,000 a day. The dividend for 1899 would be 8% of the shares' par value, and the interim dividend for the first half of 1900 5%. The trading value of the original £100 shares had tripled, and a Stock Market quotation was being sought. Meanwhile, to aid the buying and selling of Shell shares, Marcus proposed that their par value should be reduced to £1 each. To simplify share-dealing in the Far East, Marcus also proposed that the company should introduce a 'Warrants to Bearer' system, subject to a minimum of five shares per warrant. By this means, possession of a warrant would be proof of ownership of shares, and they could be traded anywhere. He further proposed that in addition to the company's existing £1.8 million capital, a further £200,000 should be raised by issuing new £1 shares at a premium of 30 shillings (£1.50) each; and not surprisingly, all his proposals were approved.

Opposite top:
Storage and buildings Tangku, Tientsin, China

Below:
In the anti-foreign Boxer rebellion of 1900, many of Shell's installations in China were looted and severely damaged

These figures are put into some perspective by one other: namely that in 1900, taking a spread of pay for unskilled and skilled workers, the average weekly wage for a British male manual worker was about one pound seven shillings, or in modern terminology about £1.35.

In the short term, Marcus's bullishness thus appeared entirely justified, and soon brought one positive outcome: much as he had planned, the impression of strength both helped to deter the creation of a Standard-Nobel-Rothschild monopoly, and was sufficient for him to be able to re-negotiate, in October 1900, the Bnito supply contract. Yet the AGM that year was barely over when everything began to fall apart. For reasons no one could have foreseen but against which Sam's innate caution would have guarded, the price of kerosene tumbled, and Shell Transport found itself in possession of enormous stocks which were only saleable at a loss. In the same year the Boxer rebellion in China severely damaged the company's large assets there, both financially and physically. Materials for a company installation at Tientsin were looted and appropriated by German military authorities, and because of the political troubles, large and potentially lucrative delivery contracts in Shanghai had to be sold at a loss, while the start of businesses at Canton and Hangkow were delayed by river piracy. Simultaneously in India, the Burmah Oil Company gained practical control of the kerosene market, negating Marcus's substantial investment there; and in the UK (although he did not realize it at once), Standard wrapped up most outlets for petrol, or 'motor spirit'.

Meanwhile in South Africa, the Boer War between the British government and Dutch settlers meant that a new marketing venture there could make little progress. Worse, the war caused such nationalism in Holland that the Dutch government decided only Dutch-flag ships could trade with their colonies in the East Indies, and simultaneously a general recession began.

As *The Economist* reported, 'From every part of the country and the world, we hear of shrinking trade and relaxation of industrial activity', and world-wide shipping freight rates collapsed. Shell Transport's fleet could not be employed, except at a loss; its enormous stocks of oil could not be sold, except at a loss. Sam's worst fears were coming true.

One thing more. On 14 December 1900, Royal Dutch's managing director Jean Kessler died. His successor was Henri Deterding, now 34, and as astute and ambitious as ever. When he was old, others called him 'the most powerful man in the world'. He preferred to call himself 'an international Oilman' (using a capital O, as he always did for the most important word in his life), but really he was much more a money man. Brought up in humble circumstances, one of five children raised from the age of six years by a widowed mother, he displayed as a teenager an

unusual gift for understanding figures, and became a bookkeeper for a bank in the Netherlands East Indies, thereby meeting J. B. A. Kessler, then managing director of Royal Dutch. The company was experiencing a serious cash-flow problem, and Deterding persuaded his superiors to accept its oil as collateral for a loan. This was 'a distinct innovation', he wrote later, 'for never before had the bank accepted Oil in this way', but it worked, and soon Kessler gave him a job. His 'lynx-eye for balance sheets' shortly made him indispensable, to the extent that Kessler named him as his successor, saying 'he knew no other man who could ensure the company's future prosperity in quite the same degree'. 'So', wrote Deterding, when Kessler died at sea in Naples, 'I became head of the entire Royal Dutch Company within ten years of my having been an obscure young bank clerk.'

The cordial relationship between Royal Dutch and Shell Transport had depended largely on the personal trust between Marcus and Kessler. Marcus's amiable view of the Dutch suffered a sharp reverse when the rule about Dutch-flag ships came into force, and both he and Fortescue Flannery made intemperate public speeches criticizing the Dutch government – criticisms which in turn put a distinct frostiness into communications from Royal Dutch. It remained to be seen how Marcus and Deterding would get on; and Deterding suspected, strongly and correctly, that far from being a vitally prosperous concern, Shell Transport was becoming a mere façade. His main interest now was how to exploit that suspected weakness for his company's benefit, because Royal Dutch was back in production.

After the failure of its Sumatran wells at the beginning of 1898, Royal Dutch had been driven almost to extinction: its share price on the

Left:
Royal Dutch's managing director J.B. August Kessler senior died in 1900...

Right:
...and was succeeded by the 34-year-old financial genius Henri Deterding

Amsterdam Bourse had tumbled from a high of 623 guilders that February to 250 in the October. Anyone with money and confidence could have snapped the company up at a bargain rate, but those with money had little confidence: Royal Dutch had drilled 110 wildcat wells which produced nothing, and there seemed no obvious reason why any further wells should be productive. 'You do not know what a sensation our setback has made in the country', Kessler had written then to Deterding. 'Everybody is talking about it, even old women and schoolchildren.... Truly, I am bowed down by cares.' However, hanging on grimly through 1899, he and his colleagues had adopted a more scientific approach, and for the first time in their company's history had decided to use geologists – a breed much distrusted by Marcus Samuel – in the search for oil. Led by Hugo Loudon (later a chairman of the company), the team concluded that the wildcat wells at Pangkalan Brandan were in the wrong place altogether, and recommended drilling about 100 miles north-east, at Perlak. Drilling there had begun on 22 December 1899, and as proof of the geologists' skill, oil was found just six days later. The discovery alone made Royal Dutch shares double in value in a single day. By mid-1900 the new wells were producing a respectable 5,000 barrels a day, with production continuing to rise steadily.

In contrast, Marcus's carrying contract with the Moeara Enim company in Borneo had proved almost entirely fruitless. After a brief period of apparently promising production, its wells ran dry, and the company informed Marcus that it could not fulfil its obligations to him. Rather than providing his ships with an annual minimum of 50,000 tons of cargo, it could not guarantee any cargoes at all for at least two more years.

Marcus knew that carrying Bnito cargoes alone could not support his fleet, so with a deepening sense of urgency, he hoped for the discovery of some major new source of oil. It was hardly a satisfactory way to end one century and begin another; but on 10 January 1901, a truly major new source *was* found – at a place in Texas called Spindletop.

The first strike there was unlike anything anyone had seen before in America: a roaring gusher, spouting from the earth with such force that it blasted the six-ton drill pipe straight out of the hole. From earlier shows of oil, the drillers had estimated they might find 50 barrels a day, but the thundering fountain they released gave *75,000* barrels daily. The first Marcus learned of this was from his newspaper. The fact that the find was on the other side of the world from his usual activities did not deter him; he had to get involved if his company was to continue. He had never heard of the field's main owner, Colonel James Guffey, but without waiting to consult geologists or even American lawyers, he immediately opened negotiations.

The first mesmerizing find at Spindletop in Texas, 10 January 1901. Marcus at once began to negotiate a stake in the business – one which would ultimately prove fatal to Shell Transport's independence

In addition to its prodigious flow, Spindletop had two considerable advantages for Marcus over the East Coast oil fields. One was that Texans did not like Standard in the least, and launched so many law-suits against the company and its president that eventually the Octopus gave up trying to get into the state. The other advantage was geographical: Texan oil could be pumped quickly and easily to the Gulf of Mexico for onward transport. In Leadenhall Street, the first half of 1901 passed in a state of almost unendurable tension; then at the AGM on 19 June, Marcus announced triumphantly that ten days later, Guffey and Shell Transport would sign.

The deal seemed excellent. For 21 years, Shell Transport would carry half of Guffey's production at a fixed rate of $1.75 a ton, plus 50% of the profits on sale, with an annual minimum delivery of 100,000 tons. This provided guaranteed freights, and oil which Marcus could legitimately sell in Europe. Next, therefore, he bought a £90,000 stake in a German company called Gehlig-Wachenheim (of which the Deutsche Bank was a major shareholder). Its assets included a refinery and storage facilities at Batum, whence kerosene had for many years been exported to Germany. From that base he established a new marketing company, known by its initials as PPAG. Operational in autumn 1901 and constituted in 1902, this marked the start of Shell's European marketing operations; and combined with the Guffey contract, it set in train a rapid sequence of further negotiations, bids and counter-bids.

Standard began to woo Marcus with flattering proposals and large offers. Understanding them to mean that Shell Transport could remain independent while working with the Octopus, he was not averse to the idea, and in October 1901 went to New York to discuss it.

Apart from anything else, it appeared that he would acquire worldwide control of his hobby-horse, liquid fuel. This was seductive talk indeed. Simultaneously, Fred Lane, the ship-broker (whom Deterding described years later as 'the cleverest man I have known in all my experience'), re-opened talks with Royal Dutch, and early in November presented Marcus with the draft of a possible agreement between Royal Dutch and the Shell. Marcus, believing a Standard-Shell union to be imminent, turned it down and instead proposed the creation with Royal Dutch of a new distributing company, the British-Dutch, chartering Shell tankers at the high rate of seven shillings a ton freight.

Now Deterding was on the rack: his was not a world-class concern, but Shell Transport was becoming one, and Standard already was one. Acceptance would mean steep expenses; refusal could invite total take-over by Standard-Shell. On 27 December, he accepted.

At about the same time, Standard clarified their terms to Marcus. They would indeed pay him handsomely (40 million dollars, of which about one-third would go to Marcus personally) and he would still have a high status within the company – but independence was impossible. Angered at the feeling that he had been misled, determined to keep Shell Transport British rather than be just part of an American company, and now strengthened both by Deterding's compliance and the Guffey contract, Marcus promptly terminated negotiations.

Entering 1902, he was once again a happy man, content in the choices he had made in business, and greatly buoyed at the end of January by a remarkable event: the signing of the Anglo-Japanese Alliance, Britain's first full-scale alliance with any foreign power for a full century. The very need for this was actually a significant milestone in the decline of Britain as a maritime power, but to Marcus, with his excellent Japanese connections, it represented a new and glittering door of business opportunity.

Henri Deterding, on the other hand, suffered a nervous breakdown in his efforts to disentangle Royal Dutch from the expensive chartering contract of 27 December. But that cost him only a month's recuperation in Monte Carlo. The events of the rest of 1902, which had started so auspiciously, cost Marcus much of his growing reputation and, soon, the independence of his company.

During a sandstorm over the Suez Canal in March 1902, a Shell tanker ran aground. Another, coming to help, took off too much kerosene, overflowed her own tanks, caught fire and was burned out.

Bad as this was, the consequence was worse, because it helped the Canal authorities to make up their minds about Marcus's current request to them. Having commenced refining petrol at Balik Papan, he wanted to bring it to the British market through the Canal in bulk cargoes, like kerosene, and had calculated the costs of this venture on the Suez route instead of the much more expensive route round the Cape of Good Hope. After the kerosene fire, however, the authorities refused to allow bulk petrol through the Canal, and the venture's basic sums were completely altered.

A lesson Marcus found almost impossible to learn was not to make an important commitment unless and until he was sure he could fulfil it. Of course, by his own lights, he had been; he had not imagined the possibility of refusal. Still, the venture went ahead via the Cape route, and the first bulk cargo of Shell petrol (the first such cargo from any company to cover such a distance) was brought to Britain, appropriately enough by *Murex*, the company's founding ship. Though it is said that the captain's hair turned white with worry during the voyage, the passage was safe, if expensive; and it was only when the cargo reached England that Marcus discovered that, by making exclusive contracts with most retail outlets, Standard had pre-empted him.

Even so, he remained confident the problem would be overcome; and in any case the time was approaching when a market of at least equal importance was to be tested in public: the Royal Navy and its possible conversion to liquid fuel.

To some naval officers the virtues of liquid fuel, as opposed to coal, were self-evident. Before being accosted by Admiral Fisher in the street in March 1899 after lecturing on the subject to the Royal Society of Arts, Marcus had taken a purely commercial view of the matter: liquid fuel was much cheaper than coal, providing 50% more heat per unit of cost and weight and (because it could be pumped instead of being shovelled) requiring far fewer men. Fisher pointed out the naval implications: without generating the tell-tale smoke that coal produced, an oil-fired ship with less weight and fewer men could be armed with larger, heavier guns. If in addition it was given relatively light defensive armour, it would also be correspondingly faster – and 'Speed *is* armour', he would say. 'Sea fighting is pure common sense. The first of all its necessities is SPEED, so as to be able to fight – *When* you like, *Where* you like, and *How* you like.'

Encouraged by meeting another character as vigorous, energetic and imaginative as himself, each man took warmly to the other, Fisher calling Marcus the 'Prime Mover', and Marcus calling Fisher 'the God-father of oil'. Together they used every means they could think of to ensure the navy's conversion to fuel oil. Marcus provided the Lords of the Admiralty with all the information he possessed on the subject, offered them the use of his ships for trials, and in March 1902 even offered them seats on

Shell Transport's board.

But the Admiralty, a traditionally conservative body, refused everything, for reasons of varying validity. First, they said, wholesale conversion of the navy would require a world-wide chain of supply (true, said Marcus, but his ships could do it). Second, Shell Transport was a foreign company, a subsidiary of Standard, dependent on foreign sources

Admiral Sir John Fisher, who as First Sea Lord shook the Royal Navy out of its Victorian complacency

of supply (not true, but with enough perceived truth to be difficult to dislodge). Third, fuel oil was insufficiently tested (not true at all, as witness Marcus's own ships, which had been using Kutei oil since 1898). Eventually, though, probably through irritation at his tenacity, the Admiralty agreed to a public trial of fuel oil in one of their own ships, HMS *Hannibal*. The date was set for 26 June 1902.

Before then, Marcus and Henri Deterding had met – for the first time on 11 January 1902 and again on 6 April, both meetings being engineered by that consummate go-between Fred Lane.

Marcus was a knight, Lord Mayor-elect and 13 years older than Deterding. However, Deterding was not in the least over-awed. At the first meeting, using his considerable personal charm – a quality to which Marcus was often susceptible – he persuaded the Englishman that their agreement of 27 December (potentially so costly for Royal Dutch) was not yet binding, since it had not been ratified by the board of Royal Dutch.

A 'Four-Wheeled Petroleum Gig'

The second meeting went still further in shaping a new agreement, which would create a joint marketing company in the Far East. A man of Marcus's standing could hardly be expected to take a back seat, but in Deterding's own words, 'Mine is a personality which does not readily submerge itself.' So to make the most of each man's different abilities, the pair quickly decided that Marcus would be chairman and Deterding managing director, responsible for the company's day-to-day efficiency.

Karl Benz

On 24 January 1902, in between those first two encounters of two men who would subsequently affect each other's life so deeply, Shell Transport and Trading held an Extraordinary General Meeting. Since the change in par valuation of the company's shares, its number of registered shareholders had more than tripled to 785, with 48,770 shares being held on Warrants to Bearer.

Marcus informed shareholders of a few new statistics concerning their company. Storage capacity in Borneo was 65,000 tons, with a further 12,000 tons' worth being built; world-wide storage capacity was 220,000 tons (of which 130,000 were on freehold land and the balance on leased Government or public body land), with a further 260,000 tons' capacity being built; 31 installations were completed and owned in Egypt, India, China, Japan and Australia, with 11 more being built; upwards of 320 subsidiary depots were owned in the Eastern Hemisphere towns with railway wagons, bullock carts and so forth for onward transport – in short, £3.5 million in tangible assets. He also took the opportunity to propose that the company should raise another £1 million capital by issuing 100,000 Preference Shares at £10, on 5% a year; and he spelled out the company's current policy. It was two-fold: first, to get production as near to the consumer as possible; second, to create a business in Europe, South America, the West Indies and South Africa similar in scale to that in the East. All was approved.

Marcus also mentioned that, having substantial liquid fuel stocks stored in the Far East, the company was in a position to offer ship-owners comprehensive facilities throughout the world (which was a constructive, if somewhat misleading, presentation of the problem created by the stockpiles he had earlier accumulated), and he outlined the situation regarding Standard and Royal Dutch. Of course, ethically he could not tell shareholders what Standard's rejected offer had been, since its terms were confidential, so he simply announced that negotiations had ended, relations remained friendly, and that discussions were continuing with Royal Dutch. But no one can say how the shareholders would have reacted if he had told them the terms he had refused.

On 23 May he signed a new draft contract with Royal Dutch. Almost at once, Deterding set about changing it again, for in Europe another opportunity had arisen – one which he found irresistible and believed might be unrepeatable. Its causes were complex, stemming from Marcus's presence in Europe through the PPAG, but essentially what Deterding now had in mind was to draw the Paris Rothschilds into the proposed British-Dutch alliance. Visiting the Rothschilds in early June, he and Fred Lane soon persuaded the Frenchmen, and Deterding returned to London in high excitement. Putting the idea to Marcus, he explained that with all three groups as equal partners under Marcus's chairmanship and his

(Deterding's) direction, Texan oil from Shell Transport and Russian oil from the PPAG could dominate Europe, while the Far East could be dominated by East Indian oil from both Royal Dutch and Shell Transport, and by Russian oil from Bnito.

There was, he confided, something else Sir Marcus should know: he, Deterding, personally had acquired concessions in the new oil fields of Romania. Steaua Romana, the major company there, was controlled by the Deutsche Bank; through Gehlig-Wachenheim, Sir Marcus was already an associate of the Bank; and Romanian oil could be sold through central Europe more cheaply than Texan or even Russian oil – an elegant tying-in of mutual interests to mutual benefit. Try that on Standard!

To Deterding's astonishment, Marcus did not instantly agree. Instead, behaving more like his brother Sam, he put up many detailed small criticisms. In business, he was always a man of instinct rather than analysis, and it appears these criticisms reflected his instinct, which was to refuse: he had never much liked dealing with the Rothschilds, and so far from being tempting, the Romanian prospect put him off. Religious faith for many people is more a nominal than a practical matter, but being Jewish was to Marcus a matter of great importance. He was scheduled to become Lord Mayor on 29 September, and had already planned that his inaugural parade in November would go through the Jewish community in the East End of his birth, where no official Lord Mayoral visit had ever been paid. He was proud to live in a country where Jews could live free of fear, and rise in society; and one of the few things he knew about Romania was that, at that very time, his co-religionists were being generally persecuted, with many being imprisoned and killed. So although in this instance its rule was expressed in cerebral terms, his heart ruled his head as usual, and he hesitated.

Deterding was incredulous, and when back in Holland had to be admonished by Fred Lane. Writing to him on 20 June from London, where he was busy working on Marcus and drafting yet another contract, Lane said, 'I think everything is going satisfactorily: but you must have patience: and on no account must you worry yourself, because you will unfit yourself for business again.'

On 25 June 1902, Shell Transport shareholders assembled for their AGM – a somewhat downbeat affair, because profits had slipped from £376,000 in 1900 to £254,000 in 1901; the price of oil was still falling; competition with Royal Dutch and Standard had been damaging; £100,000 had been spent in Australia, but would bring no result for some time; and though several cargoes had been brought from Spindletop, their financial benefit would not be visible until the following year. The only immediate bright spots were the likelihood of some sort of alliance with Royal Dutch, and – slated for the very next day – the Royal Navy's public trial of fuel oil.

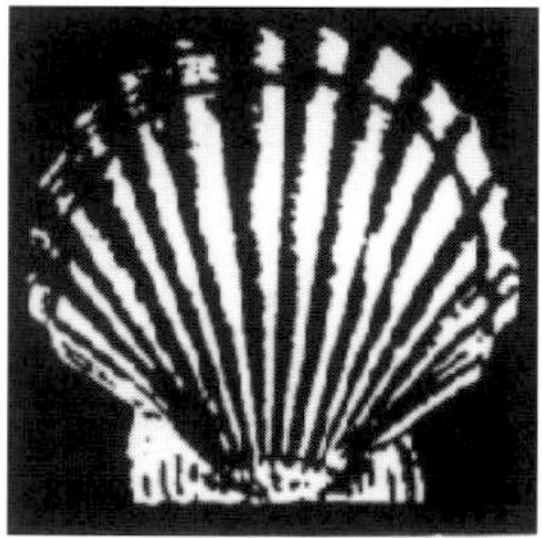

In 1904 the scallop shell or pecten replaced Shell Transport's first logo. With variations and refinements it has remained in use ever since, becoming one of the best-known corporate symbols in the world

Liquid fuel versus coal: the catastrophic trial with HMS Hannibal, *26 June 1902. It was later alleged that this disastrous performance set back the Royal Navy's acceptance of liquid fuel by a decade and contributed to Britain's low level of readiness at the outbreak of World War I*

If it went well, the world's largest single fuel oil market would soon, probably, be supplied by Shell Transport.

The trial on which so much depended was a humiliating catastrophe. Admiral Fisher watched with Marcus in Portsmouth, as did hundreds of other people: it was well publicized. What neither man knew was that (unlike Shell ships, which used modern and virtually smoke-free atomizing burners in their engines), HMS *Hannibal* was fitted with out-dated and notoriously smoky vaporizing burners.

It is not very likely that this was a deliberate ploy on the part of the conservative Admiralty, but any experienced marine engineer could have said at once what would happen. Leaving harbour, the warship trailed light white smoke from her furnaces fired with Welsh coal. At a signal, the switch to liquid fuel was made; and within minutes, the funnel discharged large volumes of black soot, followed by thick clouds of black smoke, which kept on coming. The ship completely vanished from view.

If the Admiralty did not actually say 'Told you so', they certainly felt it; this single demonstration delayed the introduction of fuel oil into the fleet for nearly 10 years. When the 'trial' finished, Admiral Fisher went to Germany to take the cure at Marienbad spa, and Marcus tottered back to London. There, the following day (27 June), he picked up the Deterding-Rothschild contract which lay waiting on his desk, and signed it.

The company established by this contract was named the Asiatic Petroleum Company, and would be a Far Eastern marketing arm for the three contracting parties, all holding equal stakes. The contract stipulated terms of charter for Shell Transport tankers, and because of this, Marcus viewed it as a guaranteed source of employment for part of his fleet – an eastern lifeline, just as Guffey was in the west. Indeed, in Shell Transport's own archives for the period, it is noted that with this contract, 'competition from Royal Dutch came to an end'.

But Marcus was wrong. Shell Transport and Royal Dutch were now linked, but *only* in Far Eastern marketing; in every other way, they remained competitors. Deterding understood this, and as far as the Asiatic was concerned, what mattered to him was the most economical and profitable running of the new company. He was not *obliged* to use Marcus's

Sir Marcus Samuel as Lord Mayor of London, the highest civic post in the land, 1902–3

H.E. TIDMARSH.
ST LAWRENCE JEWRY.
AND THE GUILDHALL.

ships, and as soon as the Asiatic agreement was signed, he began chartering cheaper vessels whenever possible.

Marcus protested, but to no avail; then in August 1902 came disastrous news from America. Over-production at Spindletop had drained the driving pressure of natural gas, and Guffey's wells were running dry. This effectively sundered Shell Transport's western lifeline. At the same time, its eastern lifeline – the anticipated charters by the Asiatic – was proving fragile at best. Half the fleet had to be laid up for lack of use, and the four big tankers intended for Guffey's trade were converted to carry cattle instead.

The deepening trials of business life became all rather too much for Marcus, especially when placed in contrast to the achievements of his civic career. As Lord Mayor, his inaugural parade was a success (although Deterding, present as his guest, in private derisively likened it to a circus). Thereafter, the demands of civic office were so great that in carrying them out conscientiously, Marcus gave less and less time to his companies. For 9½ weeks after he took office as Lord Mayor, he was not seen in Leadenhall Street at all. It was not until 4 December that (in his own words) he was able to 'put in an hour at the office'. Over the next few weeks he went in again from time to time, learning bad news on almost all occasions.

Opposite page: The Guildhall, hall of the corporation of the City of London, with the church of St Lawrence Jewry Within (1676) on the left. The Guildhall's original foundation date is unknown. Rebuilt in 1411 above a more ancient crypt, it was restored in 1789 and 1870, and is the scene of the corporation's state banquets. Painting by Henry Edward Tidmarsh (1885–1939)

On one day, Sam informed him that earnings were so poor, a dividend of only 2½% could be allowed. On another, Sam advised him that estimated figures for 1902 indicated the company had been making a daily loss of almost £500 – a far cry from the £1,000 daily profit reported at the AGM of 1900; and on 27 December 1902, Marcus found a letter from Fred Lane awaiting him in the office. It was long, systematic and highly emotional – as well it might be, for after nearly 20 years' friendship and collaboration, Lane was resigning from Shell Transport's board.

In doing so he laid at Marcus's door many accusations. The word 'reckless' occurred repeatedly; the people in charge of the business (which meant, primarily, Marcus) were 'either too busy to devote their mind and time to it, or too incompetent'; hundreds of thousands of pounds had been squandered; the company was run with a 'happy-go-lucky frame of mind' unique in Lane's business experience, and the 'most absolute ignorance' of bulk-oil distribution. If treated on its merits, said Lane, the company 'would undoubtedly have to be written off as a failure'. His summary was a harsh indictment: 'A great splash has been made, and the situation capitalized; but it cannot last, and the bubble will burst'.

There was much more in the same vein, and most of it was painfully true. Yet soon after, Marcus took another decision which in the short term at least was a further business calamity: he absolved Guffey's oil company of its contractual obligations. An American lawyer looked at the original contract, called it 'incredibly neglectful', and declared that a lawsuit in

America would have little chance of success. He added that if a suit were served in England, it would be a different matter; but when Guffey's chairman, the banker Andrew Mellon, personally came to The Mote (and thereby risked being sued), Marcus – for no other reason than sympathy and good nature – let him off the hook.

During the 18 months to June 1903, Shell Transport dividends averaged less than 1% of par value, and after Marcus's year as Lord Mayor was over, the press launched into him.

'However the directors may try to excuse themselves,' said *The Financier* in one of the less virulent attacks, 'they cannot escape the reproach of exceedingly poor management.' Accounts from the Asiatic were consistently overdue and incomplete, weakening and delaying the presentation of Shell accounts ('as usual belated... their shareholders are sick of excuses'). At the end of 1904, Royal Dutch shareholders received a 50% dividend. From Shell Transport, 'Needless to say, there is no dividend'. That particular newspaper analysis concluded firmly: 'We have no faith in the future of the Shell'. In September 1905, the *New York Herald* stated that the company had been 'well-nigh driven to the wall'. Marcus sued and won, but the allegation was very close to the truth. It was not that he had been inactive: on the contrary, if anything he was probably far too busy. But very few of his exertions bore fruit.

THE RIGHT HON^BLE SIR MARCVS SAMVEL K^NT LORD MAYOR
GEORGE WYATT TRVSCOTT ESQ^RE ALDERMAN
THOMAS HENRY BROOKE=HITCHING ESQ^RE
SHERIFFS
GREETING TO
WHO IS INVITED
TO THE BANQVET AT THE GVILDHALL ON MONDAY THE 10TH
DAY OF NOVEMBER 1902
Lewis N. Myers
CHAIRMAN

At a crucial time in Shell Transport's life the demands of office kept Marcus Samuel from giving proper attention to his business affairs

Shell Transport had achieved two notable firsts in 1903, when it introduced the importation both of fuel oil and motor spirit to Australia. In the same year, analysis of its heavy Kutei oil showed the oil to be exceptionally rich in toluol, the main constituent of the explosive trinitrotoluene. Once again, Marcus approached the Admiralty with an offer of supply; once again, they refused, saying they could get enough from British coal.

In 1904 there were problems too with the government of India, which excluded Shell Transport from a concession in Burma (then ruled by India), and levied unusual duties on its kerosene in the sub-continent, making it a penny a gallon more expensive than competing kerosene.

It was also in 1904 that a venture in which Shell Transport had a majority interest was established at Xanthi in Greece: the London Oil Development Company, under the leadership of Mark Abrahams, back from Borneo. This soon failed, having drilled only two wells; the first was abandoned at 678 feet on 11 August that year, and the second at 1,240 feet in mid-summer 1905, when London Oil renounced its concessions and went into voluntary liquidation.

In the midst of these difficulties, a small but important event took place which would literally mark the "Shell" for evermore. On 10 October 1900 the company had registered its first logo. Naturally, it was a shell, but it was an uninspired little device, like a squashed mussel. Few people would recognize it today. However, in 1904 the scallop shell, or pecten, was introduced. Its first design was intricate and detailed, even fussy; yet there it was, and it has stayed. Like Shell's products, it has been developed, refined and improved over the decades, but has remained true to the original concept and is immediately recognizable.

For Shell Transport in 1904, the one bright commercial light was its creation, on 10 May, of the General Petroleum Company. Sam Samuel was chairman and Henry Benjamin, Marcus's brother-in-law, was one of its directors. General Petroleum was to run the storage and distribution of Shell products in Europe, and to that end it took ownership of Shell tank farms in the region. On 18 January 1905, this company established an advertising budget for the coming six months of up to £250, and by 4 October could report some success: approval for a motor spirit canning factory in Manchester and mixing tanks at Purfleet, the sale of 6,000 tons of Romanian gas-oil to the Gas, Light and Coke Company at 2¼d a gallon, and the contract to supply 500,000 gallons of motor spirit to the London and District Motor Bus Company. These agreements were soon followed by the sale of 1,000 tons of Texan oil to Oldham Council and 4,000 tons of Romanian oil to Manchester Corporation, at 2¾d a gallon.

But General Petroleum's very success simply emphasized the weakness of Shell Transport. The press began to ponder what was actually left of it. Part of its assets had gone to the Asiatic; now another part had gone to General Petroleum. 'Therefore,' said the *Petroleum Review* with an obvious but sharp pun, 'all that now remains is the 'shell', and the one question that will naturally appeal to our readers is, "Is the 'shell' sound?" ' The answer was No. By 1906 Marcus found himself in an inescapable financial dilemma that brought him and Shell Transport face to face with ruin. In a last-ditch attempt to stave off disaster, he sold his six best tankers and his PPAG shares at a bargain rate to the Deutsche Bank, but even that was not enough. Only one option remained: Shell Transport *must* achieve a union, in one form or another, with Royal Dutch.

Over the winter and spring of 1905–6, Marcus negotiated directly with Deterding, reporting developments back to his board. Initially he tried to achieve 50:50 terms. Deterding immediately rejected this, pointing out that Shell Transport shares had slumped from a trading peak of three pounds to twenty-three shillings (£1.15), with a recent 'high' dividend of 5% of their par value. Royal Dutch's dividend for the same period was a luxurious 73% of par.

It was not, Deterding explained, that he was unwilling to unite the companies somehow; but he would only do so on his terms – 60:40 in

JANUARY 1, 1907.

SHELL TRANSPORT AND TRADING COMPANY.

EXTRAORDINARY GENERAL MEETING.

An extraordinary general meeting of the shareholders of the Shell Transport and Trading Company (Limited) was held at Winchester House, Old Broad-street, E.C., yesterday, Sir Marcus Samuel presiding. There was a moderate attendance of shareholders.

Public admission of private trauma: the notice summoning shareholders to an extraordinary general meeting, 1 January 1907, to vote upon the proposed alliance with Royal Dutch

favour of Royal Dutch. Marcus protested that this would mean Shell Transport's property and interests would be managed by foreigners, which his shareholders would not accept; but subsequent minutes of the meetings of the Shell board of management show, rather pathetically, how he and his colleagues came slowly to recognize the impossibility of doing anything other than whatever Deterding said.

Royal Dutch had no capital liabilities. On 27 February 1906, knowing how deeply in debt their own company was, Marcus's board decided that if a 60:40 division were to be accepted, they should try to get an equal division of the first £600,000 profit from 'the new combine', with 60:40 coming into effect thereafter. On 13 March, Marcus reported that the furthest Deterding would go was an equal division of the first £300,000. The board again said Marcus must continue to press for the best terms possible. But Deterding was becoming harder: by 27 March he offered either everything at 60:40 from the start, or the first £300,000 divided equally, followed by proportions of 65:35.

With the ground slipping from under his feet, Marcus tried for an equal division of the first £100,000, and then 60:40. Deterding refused, but did agree that the Shell should receive the first £50,000 a year of its share of combined profits before anything went to Royal Dutch. On 3 April, Marcus wrote a long letter to Deterding summarizing their agreement so far. Deterding's reply confirmed this, but also warned Marcus that the Royal Dutch board had not yet been consulted and might not ratify the 'extreme conditions'.

They would not. On 24 April Marcus reported failure again: Deterding was now insisting that before anything else could happen, Shell Transport should 'liquidate their outstanding liabilities amounting to about £585,000.' Marcus broke off discussions. Though this can only be seen as a show of bravado, his board loyally 'approved the Chairman's decision not to concede these points'; but inevitably he had to return.

By 1 May, he had done so; yet he was still worried, for he remembered the experience of the Asiatic, when Deterding had exploited their mutual marketing agreement to the benefit of Royal Dutch and the detriment of Shell Transport. He therefore said that Royal Dutch must provide some absolute guarantee that its management would not be prejudicial to Shell Transport. It was a reasonable condition which Deterding met squarely: to demonstrate his good faith, Royal Dutch would take a large stake in Shell Transport, buying 25% of its shares at a good premium, and would not sell them without Shell Transport's permission.

This was more than fair; but because Deterding still insisted that the subsequent division of profits between the companies must be 60:40 in favour of Royal Dutch, some of Marcus's colleagues still hesitated. On 22 May he told them he was going back to The Hague and 'hoped to be able to give satisfactory news at the next Board meeting. He, however, stated that he would probably find it necessary to give way to the Royal Dutch Company...'

In The Hague he found nothing had changed. Though the offer still stood, it would not be improved upon, and he had to make up his mind at once. Deterding said briskly to him, 'I am at present in a generous mood. I have made you this offer, but if you leave this room without accepting it, the offer is off.'

There was nothing more Marcus could do. He accepted. Yet he did so with pain and humility. When he was young, slim, good-looking and exuberant with all his career before him, his namesake and father, the shell merchant, had enjoined the Samuel boys to be 'united, loving and considerate and keep the good name of Marcus Samuel from reproach'. But the shell business had gone years back, sold off to a relative – how quaint it seemed, and long ago! – and now...

Well, now his reputation was in tatters; he was middle-aged, stout and had lost his looks; and the tankers and the very company itself, all named after those shells, were passing into other hands.

CHAPTER FOUR

Two into One: 1907–1914

Thus, after long negotiation, the terms were settled. It was an unusual agreement. Although at the time everyone involved referred to it as an amalgamation, the term has fallen out of use, for the two companies were not intended to become one, and never have done. Another common contemporary phrase – 'the new combine' – was somewhat more accurate. The companies would remain separate, but would unite their interests, sharing risk and rewards according to a cut of 60% to Royal Dutch and 40% to Shell Transport.

Someone once observed it was rather like a dynastic marriage, and in recent decades the word 'alliance' has become accepted as the best description of the relationship. But whatever the terminology, the agreement had to be ratified by shareholders in Shell Transport and Trading. Marcus rather doubted they would be willing to do so or that enough of them would wish to sell their shares, even at a premium, for Royal Dutch to buy its agreed 25% 'good faith' stake in Shell Transport. He was in for a salutary surprise.

Shell Transport shares were trading at 23 shillings; Deterding offered 30 shillings apiece, ex dividend; and shareholders came flocking. Within weeks, more than 70% of Shell Transport's shares had been offered for sale to the Dutch. So much for not wanting foreign management. It seemed the shareholders did not mind who managed the company, so long as it was not Marcus.

He had already had to face private censure. At a board meeting of the Asiatic Petroleum Company on 1 June 1906, Deterding moved a resolution 'condemning recent correspondence conducted without the approval of the Executive Committee' between Marcus and certain company agents as being 'detrimental to the interests of the company'. Fred Lane, as a director of Asiatic, seconded the motion. Three board-members voted against it: Marcus himself, his brother-in-law Henry Benjamin and a newly appointed director, 29-year-old Robert Waley Cohen (whose name would figure in Shell's history for many

Opposite top:
Shell tanker Vulcanus, built 1910

Opposite bottom:
1st crude distillation battery, Suez 1913

decades), but five members voted in support, and the motion was passed.

That embarrassment had at least been behind closed doors. Inevitably the alliance of Royal Dutch and Shell Transport was completely public, and much more difficult for Marcus to bear. Faced with its enthusiastic endorsement by his shareholders, Marcus described himself to the press as 'a disappointed man', but it was a considerable understatement; privately he regarded himself as a failure and a spent force.

The press shared the latter view, with regret: *The Petroleum Review* wrote a laudatory article that sounded almost like an obituary, so firmly did it consign him to the past. As a consolation prize he bought himself a 650-ton motor yacht, *Lady Torfrida*, and went away for a four-month cruise; but, well within another year (to the surprise of many people, not least himself), he had returned to the forefront of the oil industry, and was evidently much happier than he had been for a long time.

The fact was that by its alliance with Royal Dutch, Shell Transport and Trading had not merely survived to trade another day; it had entered a completely new era. Few people anticipated this (Marcus certainly did not), but the change was swift, dramatic and enduring, and the powerhouse behind it was Henri Deterding. The reasons lie partly in the way he and Marcus worked together, and partly in the nature – the structure and characteristic business practices – of the new Royal Dutch/Shell Group of Companies, or 'the Group', as it soon became widely known. Without understanding these it would be impossible to understand Shell's subsequent history, so it is worth looking at them more closely.

First, the working relationship between Marcus and Deterding. Marcus remained chairman of Shell Transport, and Deterding remained managing director of Royal Dutch; but under their agreement Deterding also became a director of Shell Transport. When Marcus departed in *Lady Torfrida*, it was with the impression that his star had been permanently eclipsed by Deterding, the dynamic victor in their negotiations. Yet though he did not realize it then, Marcus had attributes which Deterding needed for the good of the company: his name, his status as a former Lord Mayor, his connections in the oil world and London's civic life, and his Britishness. Deterding was too astute to ignore or dissipate these advantages and instead cultivated them. Step by step he nurtured the restoration of Marcus's bruised self-respect. In public, whenever Shell Transport was being discussed, he always referred respectfully to Marcus as 'our Chairman'; in private he was scrupulous in consulting Marcus on every major decision. These were considerations which Marcus appreciated and to which he responded – not because he was weak enough to be bought off by flattery, but because he realized Deterding now genuinely had Shell Transport's interests at heart and that he himself still had a positive contribution to make. If Marcus was no longer absolute ruler, he was instead in a much more durable position: one very akin to that of a constitutional monarch, with Deterding as his prime minister in charge of the executive. To an extent this meant Marcus became only a figurehead, albeit a valuable one, and it was rumoured that if he said 'Yes' and Deterding said 'No', the 'No' would usually prevail; but that was not the point. The point was that despite their long battle, the two men now wished to establish a constructive *modus vivendi*, and both were willing to make the necessary effort.

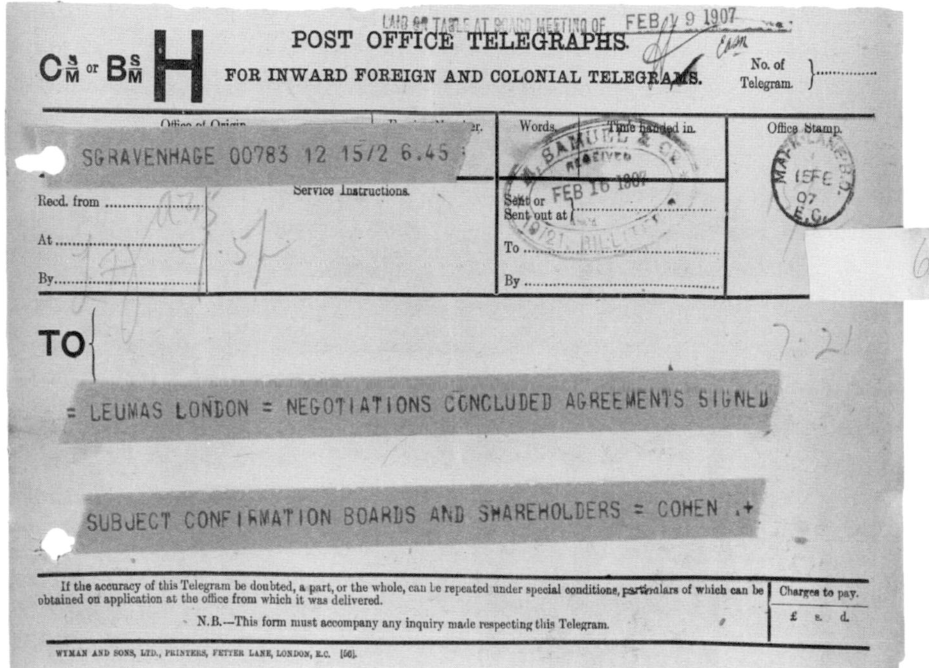

LAID ON TABLE AT BOARD MEETING OF FEB 19 1907

C S/M or B S/M H

POST OFFICE TELEGRAPHS.

FOR INWARD FOREIGN AND COLONIAL TELEGRAMS.

No. of Telegram.

SGRAVENHAGE 00783 12 15/2 6.45

Service Instructions.

Recd. from ...
At ...
By ...

Words. Time handed in.

Sent or Sent out at ...
To ...
By ...

Office Stamp.

TO = LEUMAS LONDON = NEGOTIATIONS CONCLUDED AGREEMENTS SIGNED

SUBJECT CONFIRMATION BOARDS AND SHAREHOLDERS = COHEN .+

If the accuracy of this Telegram be doubted, a part, or the whole, can be repeated under special conditions, particulars of which can be obtained on application at the office from which it was delivered.

N.B.—This form must accompany any inquiry made respecting this Telegram.

Charges to pay. £ s. d.

WYMAN AND SONS, LTD., PRINTERS, FETTER LANE, LONDON, E.C. [56].

15 February 1907: Waley Cohen telegrams news of the historic agreements, 'subject confirmation Boards and shareholders', from which the Royal Dutch / Shell Group sprang.

In short, following the alliance, the watchword between Marcus and Deterding was co-operation rather than competition. Ever since, the same has remained true for Shell Transport and Royal Dutch, to an outstanding degree. In 1951, having worked for Shell for 50 years, Sir Robert Waley Cohen remarked: 'As far as I can remember, no vote has ever had to be taken at a combined Board meeting. All our decisions have been unanimous and our interests have been identical in every way.' Recalling how he defended Marcus against Deterding's motion of censure in 1906, that is a fairly remarkable statement. It is explained by the nature of the allied companies – their structure and way of doing business.

Perhaps the most unusual aspect of this structure is that technically the Royal Dutch/Shell Group does not exist: there is no such legal entity. When the alliance came into effect on 1 January 1907, neither Royal Dutch nor Shell Transport ceased to exist. Instead, when they merged their interests while retaining their separate identities, each became a holding company rather than an operating company. In other words, rather than carrying out operations themselves, both began to co-own and derive their income from a variety of operating companies. The first two of these, created for the purpose, were the Anglo-Saxon Petroleum Company, based in London, and the Bataafsche Petroleum Maatschappij (the Batavian Petroleum Company, always known as the Bataafsche), based in The Hague. Together the new operating companies took over virtually the entire assets of their holding companies, Royal Dutch and Shell Transport. Anglo-Saxon owned and ran the transport and storage facilities; the Bataafsche owned and ran the oil lands and refineries; and both new companies were wholly owned by the holding companies in the established 60:40 ratio.

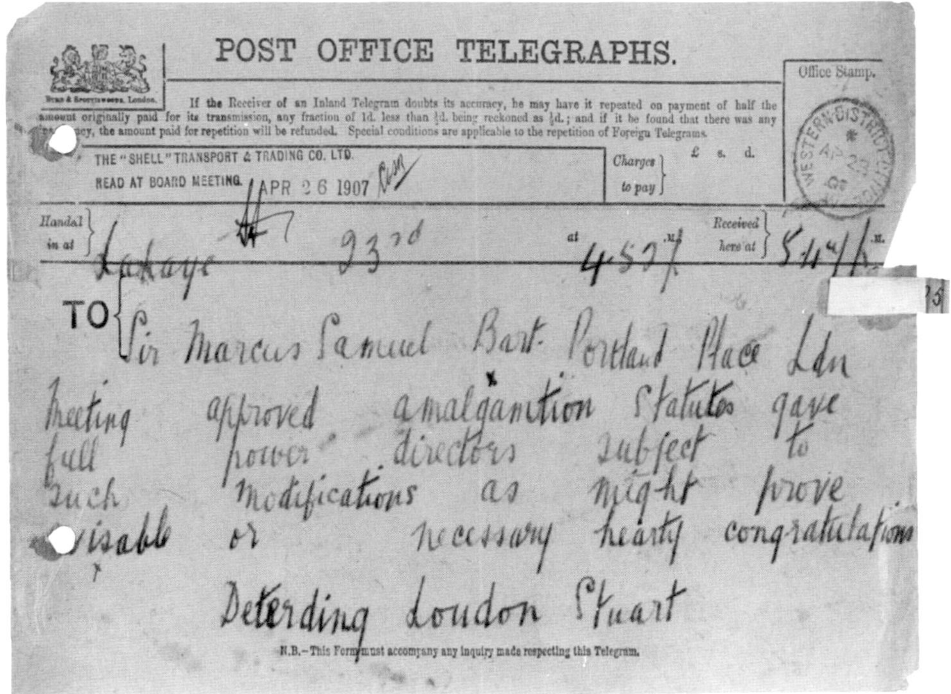

POST OFFICE TELEGRAPHS.

Office Stamp.

If the Receiver of an Inland Telegram doubts its accuracy, he may have it repeated on payment of half the amount originally paid for its transmission, any fraction of 1d. less than ½d. being reckoned as ½d.; and if it be found that there was any inaccuracy, the amount paid for repetition will be refunded. Special conditions are applicable to the repetition of Foreign Telegrams.

THE "SHELL" TRANSPORT & TRADING CO. LTD.
READ AT BOARD MEETING APR 26 1907

Charges to pay £ s. d.

Handed in at Lahaye 23rd at 4.57 Received here at 5.41

TO Sir Marcus Samuel Bart Portland Place Ldn

meeting approved amalgamation statutes gave full power directors subject to such modifications as might prove advisable or necessary hearty congratulations

Deterding London Stuart

N.B.—This Form must accompany any inquiry made respecting this Telegram.

23 April 1907: Henri Deterding telegrams his congratulations to Sir Marcus Samuel on the 'amalgamation' of their companies – a bitter message to read at first, but one which proved to be just the beginning of far greater success.

Since then, the two-tier formation of holding companies and operating companies has altered again. Shell Transport and Royal Dutch are now termed the parent companies. Beneath them are three holding companies: in the UK, the Shell Petroleum Company Ltd; in the Netherlands, Shell Petroleum NV; and in the United States, Shell Petroleum Inc. Then come the service companies, and (whether by creation or acquisition, in whole or in part) operating companies – now many more than two – working in over a hundred different countries around the world. Each of these myriad companies is a separate legal entity, and together they make up the Group, even though that does not exist in law.

Similarly, because Shell Transport and Royal Dutch have always kept their separate identities, there has never been a single Group head office: instead both sets of headquarters were retained, in London and The Hague, with shared financial and commercial affairs being focused in London and technical matters in The Hague. Likewise, there has never been a Group board of management: instead, working from their separate offices and meeting in either city, there is the Committee of Managing Directors, known as the CMD.

This uncommon but highly effective arrangement stems from and perpetuates the original concept of doing business in a co-operative rather than a competitive manner. Marcus showed this attitude long before the alliance. In 1898, when Royal Dutch feared a take-over by him, he put their minds at rest with the remark, 'The East is quite big enough for both the Royal Dutch and the "Shell" line.' In 1899, writing to another oil company in the Netherlands East Indies, he confirmed and amplified this: 'I renew the assurance I gave you that it is not part of our policy to inflict either loss or annoyance upon friendly competitors, but that our aim would certainly be to act in accord with them.' The truth of his words and constancy of his policy was strikingly demonstrated when, on 1 September 1903, he accepted a new agreement with Andrew Mellon, chairman of the Guffey Oil Company, in the wake of Spindletop's failure to supply. Had Marcus been a vindictive or exploitative man, he could have ruined Mellon; but that was not his way. Nor was it Deterding's way; in his memoirs, the Dutchman wrote:

> Unlike some of our past rivals, it could never be said of us that we grew to mammoth size by gobbling up competitors after beating them into helplessness....Always our Royal Dutch-Shell policy has been to create goodwill by giving all concerns amalgamating with us not only a definite share in the future profits of the business but a voice in the management. Always, we have taken them in as our partners, and the wisdom of this policy is shown in that never has any one of our many trading agreements, fixed upon these lines, been repudiated or even disputed afterwards. To crush a rival is to make an enemy.

Deterding took this further in a magazine interview. Referring to 'the Americans' (and there was no doubt about which company he meant), he observed:

> Their system is to crush competition in order to be able to reduce prices....Because of this, many producers are at a standstill, many refineries are ruined, while subsidiary industry cannot make

> progress. Everywhere we come we bring our experience, our work and our capital, and we are happy when we are received as sincere and faithful allies, who succeed in finding a satisfactory profit for ourselves, as well as assuring prosperity and progress for our neighbours, thanks to the natural riches of the country, the work of the population side by side with us, and a community of interests and reciprocal good feeling.

Even Marcus's biographer (not a great admirer of the Dutchman) commented that Deterding's words 'were carefully measured and they faithfully expressed the policy of the Group. It has remained their policy to this day.'

The inclination towards mutually beneficial co-operation may sound Utopian, but it is real, with a sound commercial basis and purpose. Without it, Shell's rapid expansion after the alliance would have been impossibly expensive. A move into a new part of the world or a new product area was done by finding allies rather than making enemies. Sometimes a new wholly-owned operating company would be formed for the purpose; sometimes an interest, majority or part, would be bought in an existing one. Sometimes nothing was bought except products; sometimes just the management; sometimes there was an outright purchase. There were as many variations as there were deals to be done, and no fixed pattern except avoiding (in Deterding's words) 'the boa-constrictor's method of strangling, followed by swallowing'.

In close parallel to the philosophy of co-operation runs another characteristic inherited from Shell's first leaders: candour. Of course Shell has and keeps business secrets, but as Deterding said as early as 1934,

> Our Royal Dutch-Shell operations would never have succeeded as they did if we had tried to keep any part of our general working policy a secret....there is no better way of winning the confidence of your shareholders than to make them understand as you go along every possible detail of just how any business, into which they have put their money, has been run.

The habit of a relatively high degree of openness has brought business benefits – unquantifiable perhaps, but real nonetheless. Rather than meeting its occasional critics with a wall of silence, Shell is customarily open, accepting blame if it is justified, and, when it is not, attempting to inform and educate. The latter does not always work, but the attempt is made, and is recognized by most of the public as an honest effort. More simply and no less importantly, prospective partners in any venture are naturally more willing to trust and deal with an organization which has earned a reputation for integrity.

All together, the value of this method is so well established that it continues to be part of the published Group philosophy, and is sure to remain so. In Shell's *Statement of General Business Principles*, the Group is described as

> typified by decentralized, diversified and widespread operations, within which operating companies have wide freedom of action. However, the upholding of the Shell reputation is a common bond which can be maintained only by honesty and integrity in all activities. This reputation is a vital resource, the protection of which is of fundamental importance....These principles have served Shell companies well for many years and will continue to do so in the future.

This provides a solid basis for self-respect and a clear conscience, and also makes good business sense. In an atmosphere of mutual benefit, trading is more enjoyable – and is much more likely to succeed.

If proof of this were needed, it was provided by the dramatic financial turn-around effected by Deterding in his first year of management. In 1906 Shell Transport's assets had totalled just over £507,000 and its liabilities more than £1 million. Whichever way the sums were done the answer was the same: the company owed its creditors twice what it was worth. But in 1907, under Deterding's guidance, *all* liabilities were discharged. With evident delight, Marcus told shareholders: 'Your

By 1907, motorized sight-seeing trips around London were available four times a week. It is interesting to note how this expensive service was targeted at tourists: at a time when the average British working man might earn 30 shillings a week, the round-trip fare (given on the bottom information board) was either four shillings or one dollar. The vehicle's maximum permitted speed was 12 miles an hour; ninety years later, the average speed of motor traffic in the metropolis was exactly the same

company for the first time in its history has no debts of any description outstanding – a matter upon which I heartily congratulate you.' Still more impressively, earnings for dividend amounted to £1.5 million a year, 'and', he said, 'are not equalled by the Bank of England.' Shell Transport had become 'a great company. I suppose no one will dispute that...'

From these experiences, Marcus learned something in the summer of 1907 which he had not expected: that two into one did go, and that 'the new combine' was greater than the sum of its parts. Together he and Henri Deterding had resurrected a dying business which now could face the world openly, while making considerable amounts of money. At Shell Transport's 1908 AGM, after declaring that 'no chairman ever had a better board to work with', Marcus described Deterding as 'a gentleman who is nothing less than a genius'. And the change in fortune was permanent. Until World War II, Shell Transport expressed its dividends to shareholders as a percentage of the shares' par value. At their lowest ebb, shareholders had received a wretched 1% – no wonder so many of them rushed to sell to Deterding, who provided his shareholders with 70% or more of par! But when Royal Dutch had bought enough Shell Transport shares and wanted no more, those who were disappointed soon found they had actually been the lucky ones. A glance at the dividend figures shows why. 1907: 15%. 1908: 20%. 1909: 22½%. 1910: 22½% again. 1911: 20%. 1912: 30%. 1913: 35%. 1914: 35% again...

It was enough to make anyone a little light-headed – even the 20% of 1908 was described by one delighted shareholder as 'prodigious'. The shares' trading value increased almost as prodigiously: with par at £1 (20 shillings), they rose from the doldrum level of 23 shillings in 1906 to the point when, in 1910, the directors agreed that a new issue should be made 'at a price of not less than 95 shillings per share'.

In four years, Royal Dutch's investment in Shell Transport had more than tripled in value. If Shell shareholders who had sold at 30 shillings now kicked themselves, those who had failed to sell breathed a sigh of relief, while those who had resisted the temptation congratulated themselves; and at every AGM the mere mention of Deterding's name brought cheers.

His ability apparently to magic money out of the air did not cease with dividends. On 12 August 1908, Mr James Kennedy, Shell Transport's company secretary, recorded an unprecedented question in his Minutes book: what should Sir Marcus and his fellow directors do with 'the Company's surplus cash'?

Marcus might well have felt he needed a translation of the phrase; but the surplus, and indeed the entire financial turn-around, had been created simply by Deterding's application of sensible business practices – for example, the legitimate avoidance of double taxation, a subject which arose in 1908 during discussions on paying the 1907 dividend. If dividends received by a British company from a foreign company were used in turn to pay dividends to the British company's shareholders, then 'provided it did not receive such dividends in the United Kingdom', the British company did not have to pay Income Tax on them. Marcus alone might not have bothered with such a technical but elementary detail, yet it meant a lot of money. On behalf of Shell Transport, the Bataafsche paid dividends which nominally belonged to Shell Transport (as one of its holding companies) direct to Shell Transport shareholders. Shareholders were happy; the Inland Revenue had no quarrel with the arrangement; and Shell Transport saved nearly £19,000.

The first dozen years following Shell Transport's alliance with Royal Dutch provided many opportunities to demonstrate the quality of their products, including record-breaking races, flights and journeys of exploration. In 1907, rally-driving from Peking to Paris was not always easy – but the winner, Prince Borghese, made it on Shell motor spirit! At Brooklands racetrack in Britain, the same fuel brought winners aplenty

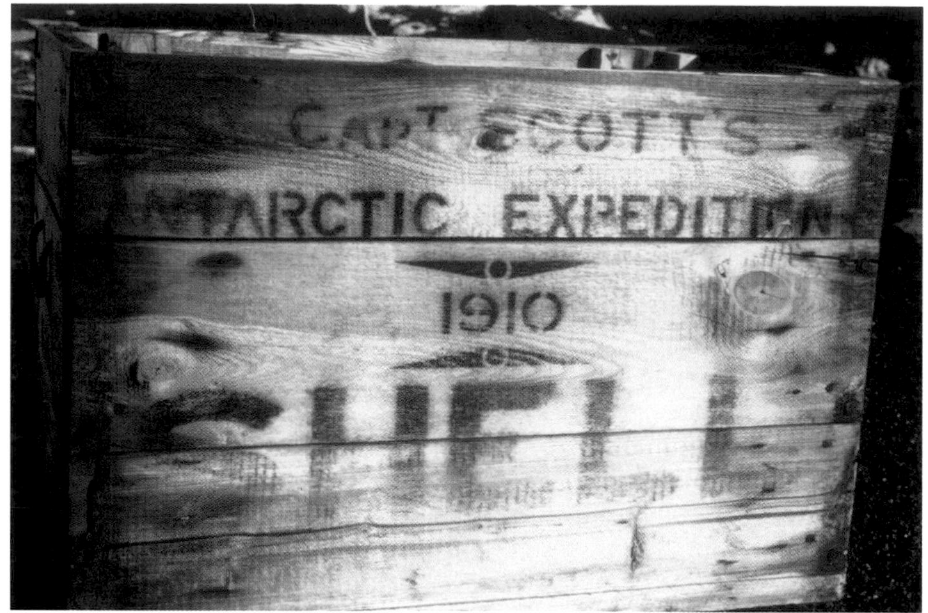

...and even in the icy wastes of Antarctica, Shell spirit was chosen as the best possible to meet the task by both the Shackleton and (a little later) the Scott expeditions

Louis Blériot's inaugural cross-Channel flight was made on Shell. This series of pictures shows a photograph and drawing of his aircraft, his triumphant return to London, and his disagreeable experience when being measured for his own waxwork

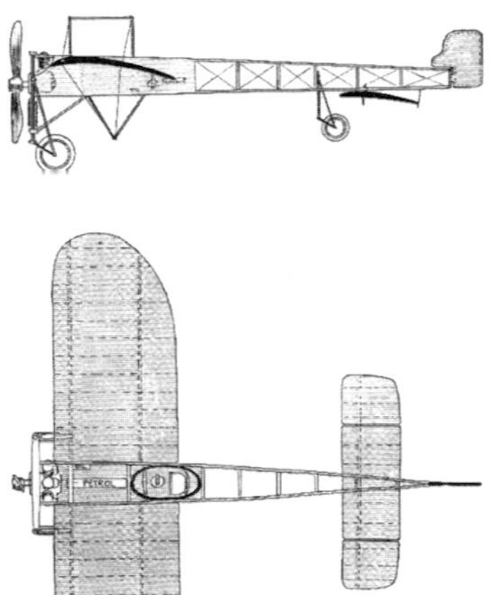

To a man like Deterding, unravelling the problems besetting Shell Transport was second nature. The £19,000 saved on Shell Transport's income tax in 1907–8 was just part of what Deterding meant when he wrote to Marcus, during their negotiations, of the 'enormous savings' that would be possible after alliance. By routinely acting on such opportunities, he helped to rebuild Shell Transport's fortunes rapidly: from the crippling half-million-pound debt of 1906, in 1914 it had just over £2 million clear profit, with a reserve fund that had already topped £3 million, an insurance fund of £833,000, and a separate War Risks insurance fund of £250,000.

These sums were not achieved merely by juggling numbers in different account books. The oil industry's focus was shifting rapidly away from its

original main product, kerosene, to the one which would dominate it for many decades: gasoline, petrol, or (in the phrase of the time) motor spirit.

After their alliance both Royal Dutch and Shell Transport profited greatly from this major new development. Before then, however, the internal combustion engine had affected them very differently. It is worth stepping back in time to see how.

The global reign of kerosene, the 'new light' and the oil industry's foundation, was brief: just 30 years or so. Hard on the heels of Drake's discovery in 1859 came the invention, in 1879, of the incandescent electric light bulb. Priced to compete with kerosene and gaslight, yet safer and easier to use, this even newer light spread rapidly through the capitals and cities of the industrialized world, posing a severe threat for oil producers structured to the kerosene trade. Virtually simultaneously, however, motor vehicles made their appearance: in 1885 Gottlieb Daimler built the first successful motorcycle and Karl Benz the first successful motorcar. From those flirtations sprang the 20th century's most environmentally and commercially significant 'love affair', coming to first flower (not surprisingly) in that nation of great distances and independent minds, the United States. There, in 1895, there were 300 motorcars; in 1905, 78,000; in 1910, 459,000; and in 1914, over 1.7 million.

Opposite page: Next came Henry Farman's record-breaking flight of 112 miles in 1906; a full-face image of Claude Graham-White's subsequent 113-mile flight, together with the fuelling of Graham-White's air-craft; and lastly that most historic couple Alcock and Brown, pilots of the first trans-Atlantic flight, 1919

As the market for kerosene declined, so the market for motorcar fuel expanded. In the years prior to the alliance, this suited Royal Dutch very well and Shell Transport not at all. Crude oils are not all the same, and by a quirk of geological fate, one of Royal Dutch's Sumatran wells produced oil which, even untreated, was perfectly good gasoline. For several years, when Standard was unable to keep up with America's gasoline demands, Royal Dutch was able to sell its competitor between five and eight million gallons of the fuel annually, sent across the Pacific to the US west coast. With production and transport being cheap and easy, and the market accessible and profitable, this was a practically ideal position.

Shell Transport's position was almost the exact opposite. Its Borneo wells produced oils suitable for ships' liquid fuel and (as the chemist H. O. Jones demonstrated) for a very good motor spirit – but only when refined. Construction of Shell Transport's first refinery began in Balik Papan in 1898. Experience would show that refineries were better placed close to the market rather than close to the source of production, but even that first small, primitive, palm-thatched affair was a costly investment which could only offer a long-term return. Added to that, Marcus had assumed the Suez Canal authorities would allow him to bring bulk motor spirit through the Canal, just as they permitted bulk kerosene. When they would not, he had to accept the huge expense of carrying the refined product to the UK via the Cape of Good Hope. If the need for refining (an

In 1910, Cadillac in the USA introduced the 'closed automobile'...

outlay which Royal Dutch did not have to make) was bad luck, the unexpectedly high cost of carriage was culpably bad forecasting; and added to *that* was what can only be called a woefully inept lack of market research – Marcus's ignorance of the fact that Standard had tied up most UK retail outlets in exclusive contracts. Having pumped the oil, refined it and transported it to Britain, he could only store it, not sell it.

Although it was as much a matter of luck as judgement, this contrast in fortunes – the ability or otherwise to supply a profitable market – was central to the decline of Shell Transport, the simultaneous rise of Royal Dutch, and the subsequent alliance of them both. By the time of the alliance, although too late to prevent it, Shell Transport had reached an agreement whereby 'Shell Spirit' was being sold through British Petroleum to UK motorists; and on 5 July 1907 (by ironic coincidence, just a few months after the alliance) the Suez Canal was finally opened to the bulk transport of motor spirit. If this had happened five years earlier, there is no saying what might have happened, but it would certainly have changed Shell Transport's history.

Yet there was no use bemoaning lost opportunities. Between them, the wells and refineries of the allied companies provided every product the market desired. Taking bulk motor spirit through Suez made its transport from the Far East so much quicker and cheaper that it effectively doubled the economic efficiency of Shell Transport's heart – the tanker fleet – and greatly enhanced the marketing of Shell Spirit in Britain and continental Europe. The product was exceptionally good, and sales surged: by 1909 Marcus felt able to describe Shell Spirit as 'a household word for petrol'. Indeed, the brand was so successful that in a hugely significant move the

...but even in 1912, delivering the fuel was still a horse-and-cart matter!

Group decided to abandon Royal Dutch's 'Crown Oil' trade name and use the name 'Shell Spirit' for all its petrol. It was a plain commercial decision governed by the product's popularity, but it was probably the single most satisfying event for Marcus after the alliance; after all, his own 'good name' and that of his father were inextricably entwined with the word 'shell'.

In that exciting era of new mobility, independent confirmation of quality was, as it still is, an invaluable advertisement. One spectacular example was in 1907, when Prince Borghese won the historic Peking to Paris motor rally on Shell Spirit. Another, from a name which already counted for much in the motoring world, was the 'spontaneous testimony to its purity and excellence' by the Honourable Charles Rolls. Yet another, again in 1907, was an early adventure into motor racing, when a Napier car made a record-breaking run at Brooklands race-track on Shell Spirit. There may not have been many records to break then, but as soon as there were records, Shell products began to break them. Shell Spirit worked equally well in sub-zero temperatures; the explorer Sir Robert Shackleton used it in his 1907–9 Antarctic expedition, as did Captain Scott in 1910, with Marcus pointing out to motorists that the explorers' fuel was exactly the same as 'the ordinary Shell, such as is in everyday use'. And the public was incomparably thrilled by the achievements of the internal combustion engine in aviation. On 27 July 1909 Louis Blériot made the first cross-Channel flight in a heavier-than-air machine; in the same year Henry Farman flew a record-breaking 112 miles; in 1910, Claude Graham-White flew from London to Manchester. Graham-White also flew from London to Paris, and Rolls across the Channel, and the report of the AGM quoted both Marcus's speech and his shareholders' reaction: 'It is gratifying to know that in all the great flights accomplished, Shell Spirit has been chosen by every aviator in the United Kingdom. (Cheers.)' Alcock and Brown's pioneering transatlantic flight was yet to come, in 1919, but that too was done on Shell Spirit.

In 1912, Marcus made a prescient observation about petrol: 'To my mind...its greatest importance is the inevitable use that must be made of it for Army transport, for aeroplanes and hydroplanes.' Moreover, he knew better than most people that the phenomenon of the internal combustion engine was not confined to land and air. In 1910 Shell launched *Vulcanus*, the world's first ocean-going diesel-driven tanker, and at the end of November 1911 Marcus wrote to an old friend, 'How right you were then and how right you are now!' 'Then' was a decade earlier, the year 1901; the old friend was Admiral Fisher. 'The development of the internal Combustion engine is the greatest the world has ever seen,' Marcus continued, 'for so surely as I write these lines it will supersede steam and that with almost tragic rapidity...'

Winston Churchill

From 1904 to 1910 Fisher served as First Sea Lord, the Royal Navy's professional head. From that exalted position he pushed the navy's modernization in every way possible, including its conversion at last to liquid fuel instead of coal. With Germany readying its sea arm to fight on liquid fuel, the British Admiralty was persuaded, but Winston Churchill as First Lord of the Admiralty (the Navy's political master) wanted to be sure supplies could be found in British territories. From that, in the years leading to World War I, sprang a large part of Shell's new exploration strategy: wherever possible, sources of crude oil should be found within the British Empire.

The countries chosen were Sarawak, New Zealand, Trinidad and Egypt, all then under British authority. Yet they were no tale of unbroken success. Exploration in New Zealand's North Island, begun in 1910, was abandoned in 1915 and not renewed until the middle 1950s. Trinidad (which Shell entered in July 1913 with a 25% stake in United British Oilfields) had long been known as a source of petroleum: it held natural lakes of asphalt. In the event, however, it was slow to develop. But Sarawak was productive, and Egypt prolific. The rajah of Sarawak (an Englishman named Sir Charles Brooke) granted Shell a free and exclusive right to explore for oil throughout his 48,000-square-mile kingdom, stipulating that if any were found, a plentiful reserve should be kept for Admiralty use. On 10 August 1910 oil was struck near the mouth of the Miri river, fortuitously containing a full 70% suitable for liquid fuel. By the end of the year the field was producing 90 barrels a day, a figure which rose steadily to 1,260 barrels in the summer of 1914.

Egypt became the scene of Shell activity in July 1911, when Anglo-Saxon and a company called Red Sea Oilfields agreed to form Anglo-Egyptian Oilfields Ltd., managed by Anglo-Saxon with Mark

NEW BORNEO OILFIELD.

FIND IN BRITISH TERRITORY.

From a Correspondent.

SINGAPORE, Monday.

After passing through two good oil pays in Sarawak, oil has been struck in large quantities at a depth of 860ft by the bore-masters operating on account of the Anglo-Saxon Petroleum Company (Ltd.), of London. It is reported here that the development of the territory will be proceeded with immediately.

The Anglo-Saxon Petroleum Company, of which Sir Marcus Samuel is chairman, is one of the group of companies associated with the Shell Transport and Trading Company (Limited). The experience gained by the managers of that company is a guarantee that no time will be lost in developing this new British oilfield, which is of great extent. The importance of the discovery cannot be over-rated, in view of the part which petroleum will play in the future owing to the rapid development of the internal combustion engine.

It is understood that a pipe line of no less than 110 miles will be constructed in order to convey the oil from the field to a suitable shipping port; whilst a large refinery and great tank storage will also be provided. An industry will thus be created which will add materially to the prosperity already attained in Sarawak under the rule of his Highness, Rajah Brooke.

The Daily Telegraph *reports on oil being struck at Miri, Sarawak*

Sir Charles Brooke (1829–1917), the White Rajah of Sarawak

Abrahams in charge. After a year or so, Marcus thought this a costly mistake: money was being spent in exploration without anything coming back. Deterding recorded somewhat cattily that 'As expenses grew and grew on so poor an outward showing, my shrewd colleague...advised discontinuance'. But in 1913 oil was struck at Hurghada on the shores of the Red Sea, a refinery was built at Suez, and, through 'relying purely on my instinct', the company became one of Shell's 'most uniformly prosperous'.

With these successes inside the Empire, it seemed Marcus's long-standing desire to supply the Royal Navy with liquid fuel must now come true; but despite all the effort, the Admiralty turned him down yet again. Because oil had strategic and therefore political implications, both Churchill and the government balked at contracting for supplies from a concern which was now viewed as being 60% Dutch, however much and however well it might explore and produce within the British Empire, where it was subject to British law. Instead, the government bought a controlling 51% share in a new company, Anglo-Persian Oilfields, operating solely within British influence. After this rebuff, Marcus, not surprisingly, gave up on the Admiralty for the time being, and instead turned some of his attention to establishing a new enterprise (the Flower Motor Ship Company) independent of Shell, to experiment in and improve the technology of liquid fuel for ships. Simultaneously he kept a benevolent and interested eye on Henri Deterding's considerable plans for Shell's expansion outside the British Empire. These were considerable.

'Profits in their true sense', said Deterding, 'are simply the reward of foresight and courage – the foresight to see where opportunities exist to meet mankind's needs more adequately and more cheaply than before; and the courage to risk one's energies and one's savings in exploiting those opportunities.'

This was the period when Shell began truly to circle the world in every direction: the trade mark was registered and marketing companies were formed through almost the whole of Europe, and in places as far apart as Jamaica and French Indo-China, Siam and Scandinavia, the Philippines and Peru. Elsewhere, too, an enormous amount of new exploration and production took place – in Russia, Romania, Venezuela, Mexico and even in Standard's home territory, the United States. Risky ventures can by definition fail, and the sorry stories of Russia and Romania are best left to the next chapter: in wartime, disaster overtook both. But the western hemisphere proved very fruitful.

1909

Shell's entry into the United States was created, paradoxically, by the giant Standard Oil. In 1910, having built what appeared to be sufficient gasoline refineries to cater for American motorists, Standard cancelled its existing arrangements for 5–8 million gallons' annual supply from Shell's Sumatran wells. This was not a sum to be ignored, so Deterding acted on thoughts he had already had: Standard's domestic market should be invaded, because it was by raising prices there, in its area of functional monopoly, that Standard could afford to wage price wars elsewhere around the world.

As Shell prepared this impertinent intrusion, an event took place which made news all around the world. In 1911, after years of defending itself against myriad accusations of illegal monopolistic methods of trading, Standard was forced by the United States' Supreme Court to dissolve itself and separate its constituent parts into more than 30 competing companies. Theoretically, this dramatic legal judgement made Shell Transport and its Dutch ally the largest oil company in the world, at a single stroke; and it seemed to many oil people, especially in smaller companies, that the sudden elimination as a cohesive unit of their harshest competitor must markedly improve their own chances of survival and growth. Certainly the dissolution of the Octopus did provide a fillip to every other oil company, including Shell, but that was about all.

Henri Deterding, one of the few to foretell accurately how limited the effect of the dissolution would be, observed that Standard's separated companies retained their old names and, whether they were in California, New Jersey or wherever, continued to sell oil under the same brand name. Overall, therefore, he considered they were unlikely to enter into genuine competition with each other; and for at least a decade this proved true.

Important as it was, Standard's sentence thus made much less immediate difference than might have been supposed – and none at all as far as Shell's strategy was concerned. If marketing were to begin in the US, it could logically begin in Canada too, so in 1911 sites were chosen for five North American ocean terminals: firstly at Montreal in eastern Canada and at Vancouver in the west, and then, moving down the US west coast, at locations near Seattle, Portland and San Francisco. (There was one slightly disconcerting discovery: in Los Angeles just a few months earlier, a 'Shell Petroleum Company' had been set up by someone else altogether, one William Nelson Shell, so for a while the new Royal Dutch/Shell firm had to be called the American Gasoline Company. In 1914 it was renamed the Shell Company of California, Inc.)

The first Shell cargo of Sumatran gasoline (over a million gallons) reached the Richmond Beach terminal, near Seattle, on 16 September 1912. Shell's retail outlets on the Pacific coast grew swiftly in number, from 25 that year to 208 in 1914. These varied greatly in size, from the relatively large complex of Richmond Beach down to general stores stocking a few of Shell's red cans of petrol – a situation mirrored in Britain, where sales were governed by the Petroleum (Hawkers) Act.

With the United States being (in Marcus's words) 'the land of petrol', it was obvious that Shell should invest in production too, and in 1912 Mark Abrahams was sent from Egypt to Oklahoma, where he bought ten small oil-producing companies. Put together, these became the Roxana Petroleum Company (named, by Mark's wife, after the wife of Alexander the Great), which in its first three years of operation alone produced over 1.6 million barrels of oil. Added to that, a major event in 1913 was the purchase at Coalinga (halfway between San Francisco and Los Angeles) of California Oilfields Ltd, which pushed Shell's total North American oil production up from 723,000 barrels in 1913 to 4.7 million barrels in 1914.

All this oil needed to be transported and distributed: hence the 28 railroad tank cars and 43 tank trucks which figured in the company's inventory in 1914, and the creation that same year of the Valley Pipeline Company, established to build a 170-mile pipeline from Coalinga to Martinez on San Francisco Bay – the site of that last essential, a refinery.

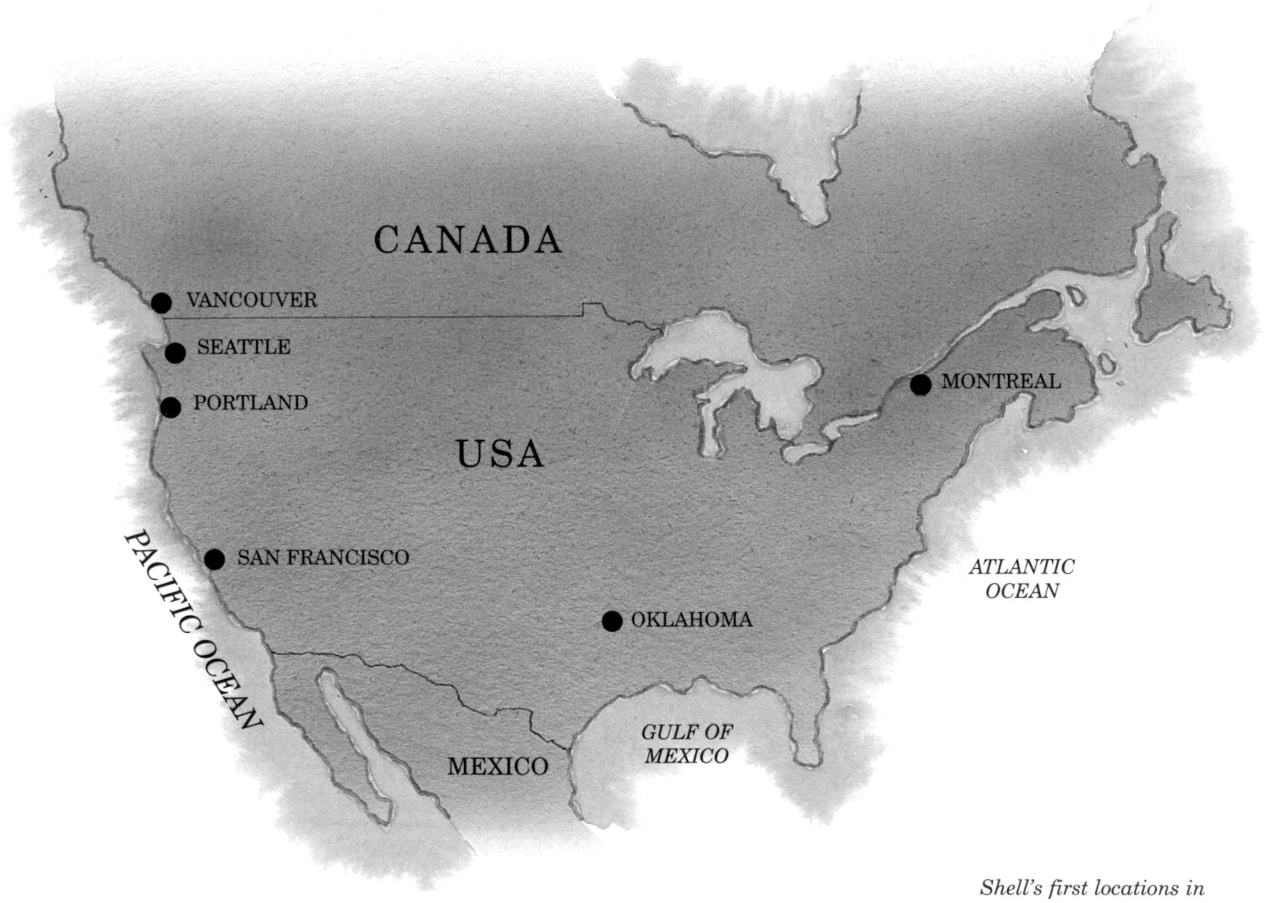

Shell's first locations in North America

In 1911, despite – or because of – its successes, such a 'Jewish family business' still evoked semi-humourous responses.

Thus, in only four years, Shell installed in the US most of the fundamentals of a completely integrated oil company: production, transport, refining, distribution and marketing. The only aspect absent for the time being was exploration; but plenty of that was happening elsewhere.

Deterding understood the value of geologists in hunting for oil; without them, the Sumatran wells which had rescued Royal Dutch in 1898 would probably not have been found. Marcus's scepticism about the profession had been overcome, and he was not too proud to recognize their worth; now at AGMs he would refer to them as 'our experts'. Sometimes, though, even geologists could not say where to drill for oil, only where not to. Such was the case in Mexico, which had been identified as a worthwhile source of oil at the turn of the century, and where in 1912 Shell acquired leases on 1.45 million acres. Geological reports ruled out some areas, which was a help, but could not pronounce definitely in favour of any particular place. When Deterding read the reports, he remarked: 'It more or less seems that in Mexico it is always a question of wildcatting.' In the course of 1913 eleven wells were drilled, and to begin with, it seemed they had struck lucky: the first, started on 20 March, produced 100 barrels a day. That was not considered a bad quantity; it was certainly commercially viable, if followed by others.

But all the other ten were dry. Worse: just a few weeks earlier, in February 1913, the Mexican revolution had begun. President Francisco Madero was deposed and shot by General Victoriano Huerta, who, in October, declared himself dictator. The United States did not like this. Early in 1914 came news Shell had long hoped for: on 11 January its eleventh wildcat Mexican well gushed a simply astonishing 100,000 barrels a day. Then in April 1914, US troops invaded Mexico. In July, Huerta was expelled from office; the country collapsed into anarchy and civil war; and under those conditions the Mexican oil camps had to be abandoned. Considering the capital outlay that Mexico represented, this was (as Marcus said) 'thoroughly unsatisfactory'. It was perhaps as well he could not see the future, for in just a few weeks every peacetime plan would be altered, and in the years to come many peacetime prospects were changed or lost for ever.

CHAPTER FIVE

The Great War: 1914–1918

In 1914 the balance of political power in Europe depended on a complex, fragile web of international alliances and guarantees. On 28 June, the assassination of Austria's Archduke Franz Ferdinand by a Serbian nationalist in Sarajevo sent a shock through every filament of that web, and one by one the nations declared war against each other. On Tuesday 4 August German forces invaded Belgium on their way to France. The action brought an ultimatum from London: if the invaders did not withdraw by midnight Berlin time, Britain would declare war against Germany. But everyone knew there was little hope the ultimatum would be accepted.

Shell Transport and Trading had recently moved its London head offices into new premises at St Helen's Court, and there Sir Marcus Samuel and his co-directors met to discuss the news. The company secretary noted down the directors' first consideration:

Opposite:
Sarajevo 1914: the murder of Archduke Franz Ferdinand triggers the Great War

'*Employees*. Having regard to the state of war in Europe, it was decided:-
1. That subject to the heads of departments retaining sufficient men to carry on with the work, employees be allowed to leave to join their colours, whether British or foreign, either compulsorily or voluntarily, that their posts be kept open for them, and that their salaries be paid for at least three months, at the end of which time the matter is to come up again for consideration.'

Walter Samuel, Marcus Samuel's first child and elder son (born 1882), as a Second Lieutenant in the British Army, 1915. Walter succeeded his father both as chairman of Shell Transport and Trading (1921) and as Viscount Bearsted (1927)

It was not an unusual attitude for the directors to adopt; Shell already had a reputation for giving first thought to its employees and their welfare. This was manifest in many ways, from good salaries and security of employment down to the free cups of afternoon tea provided for office staff – a pleasant custom instituted by Marcus personally in May 1906. Now facing imminent war, it seemed to him and his colleagues entirely natural to guarantee their employees' jobs and pay, and only fair to allow them to fight for their country, whichever it was. The same applied to holidays: if any employee lost out, then 'everything that was fair should be done'.

To modern minds there is something poignant in that minuted decision: it would stand 'for at least three months'. Distant but audible is a small sad echo of the general belief at the time, that the war would be over by Christmas. Instead it lasted more than four years. Throughout, those employees who could not join up did all in their power for their fighting colleagues' welfare, sending them monthly parcels of food, clothing and little luxuries, and keeping in touch via an in-house magazine, *St Helen's Court Bulletin*, swiftly created for the purpose (its first issue was on 21 November 1914). Such reminders of normality, small as they were, were highly valued by their recipients in the Forces: 'I find it difficult', wrote one, 'adequately to express my appreciation of the kind interest you take in us fellows.' 'It is particularly pleasant', another agreed, 'when with all our new pals, to have so many remembrances of old friends and times as the *Bulletin* always contains. Very kindest regards to you and all friends at St Helen's Court...'

When the subject of paying employees in the Services duly came up for review, it was decided that all those whose company salary was up to £2 a week should be given, as an addition to their Service pay, half-salary until further notice. Employees with higher salaries would be considered individually. By the end of the war 1,050 Shell employees world-wide (including Marcus's sons Walter and Gerald) had joined up from the Allied nations, and rightly felt themselves to be well looked after by their employers and colleagues.

One hundred and nine of them died on active service; Gerald Samuel was one, killed in action at Verdun on 7 June 1916. Inscribed on the Group's Roll of Honour, those 109 names show the wide variety of services they joined: the Royal Navy and its Reserves; the nascent Royal Naval Air

Service; the Royal Flying Corps and its successor the Royal Air Force; the American, the New Zealand and the Australian Expeditionary Forces; and no fewer than 69 different regiments or battalions of the British Army.

For all employees and for the dependants of those who were killed, another benefit was provided: money from the Provident Fund, established on 1 January 1913. 'If I am to be remembered for any achievement', said Deterding later, 'I would rather that it were for my pioneer-work in establishing this than for anything else.' (Actually it was not a new idea, nor unique to Shell; for example, the shipping company P&O had set up a similar fund some fifty years earlier.) The Provident

St Helen's Court, above: Shell's London base from 1913 until 1962 – exterior of the building...

...and interior (left) of the entrance. Both these photographs were taken in 1949. In the 36 preceding years two world wars had been fought; on the right-hand side of the interior picture Shell's Roll of Honour is partially visible

Fund was built up by placing in it 10% of each employee's salary, plus the equivalent from the company, plus further sums from the company as available each financial year. From this the company was able to give each employee a lump sum payment on retirement, resignation or dismissal; and if the employee died while still in company service, the money was given to his dependants. 'So,' said Deterding, 'no employee leaves our service without being assured of a nest-egg.' Waxing lyrical, he added: 'Only think of it – twenty per cent of his earnings during the whole time he has been with us, and substantially increased by accumulated interest through judicious investment!' However, this was not a pension, only a one-off payment. Deterding believed it to be better than a pension for the dependants of employees, because the support of a pension would end with the employee's death; and as always he had a hard business reason too – 'I dislike pensions...because I hold that no company is justified in incurring liabilities, not definitely terminable, even for its employees.' In that light the Provident Fund does not sound so generous, but no one seems to have

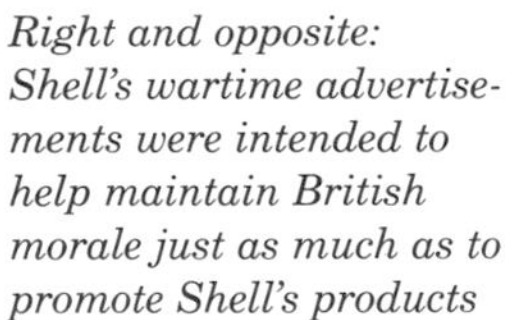

Right and opposite: Shell's wartime advertisements were intended to help maintain British morale just as much as to promote Shell's products

an official or semi-official character connected with the war. It is felt by many that the generosity of the Press in this matter has been in no small measure abused.

Why, it is asked, should newspaper proprietors be required to supply for

WHAT PETROL IS DOING

TWO BIG GUNS

a few resent the prevalent assumption that it is their duty to supply it at a loss.

It was quite refreshing to find that the authorities were prepared to spend a little money on advertising the war-loan prospectus. In most cases they got remarkably good value for their expenditure, but no one grudged them that.

Among advertisers who have been

minded receiving a large lump sum instead of a regular smaller payment; and the company's minute books show that in fact, if widows of former employees needed regular financial help, it was usually forthcoming.

The Great War introduced Shell to a new and difficult experience, resulting from its already multinational character. The lesser part was that many shareholders were foreign, and some were German. What would happen to their dividends? To pass them on would constitute trading with the enemy. The greater part was that the company was a private merchant concern, not a government one, with first responsibility to its shareholders to sell its products where they would fetch the best price. But what if that was against the national interest? Interweaving with both these problems was a third: namely, that in wartime a parent company of a multinational group might find that one or more of its offspring, whether wholly or partly owned, was part of the enemy's economy. If so, then the offspring would not only serve the enemy, but might provide in dividends part of the parent's income – a nice moral maze. Lastly (but not at all least, for it encapsulated the entire quandary) there came on the personal level the sad and sobering thought that an employee of one nationality might have to fight – might even kill – an employee of another.

In 1914–18 most of these dilemmas were solved with relative ease. German shareholders' dividends would be banked for the duration and paid when peace returned. Equally simple, for both Marcus and Deterding, was the question of how a private company should conduct its affairs in wartime. As far as Royal Dutch was concerned, this was largely answered by its national government. The Netherlands remained officially neutral throughout the conflict, and banned all oil exports from the country; but unofficially its attitude was biased towards the Allies. Henri Deterding, who had based himself in London since before the Royal Dutch/Shell alliance, was totally and openly biased towards the Allies; indeed, he identified with their cause so much that he became a naturalized Briton and like Marcus's sons, his son joined the British Army. So, for Deterding and

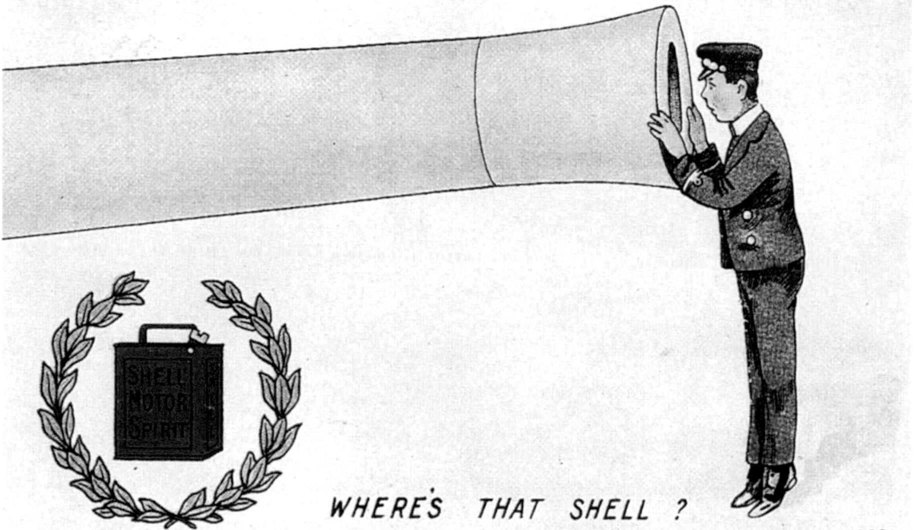

Left:
An example of an early Shell advertisement c. 1912

Marcus, the question of Shell Transport's stance in wartime was really answered before it was raised. With an attitude which governed their entire wartime activity, even when it lost money for the company, they regarded Shell as 'a Government agency' in all but name.

The question of doing business with the enemy was rather trickier, at least in public eyes. One aspect of this was dealt with fairly readily. Before the war, the Admiralty had not been interested in Shell's Borneo oil, even though it was proved to contain an unusually high percentage of toluol, the prime constituent of the high explosive TNT. However, the German Army had ordered large quantities. As part of normal peacetime business, the orders were accepted, and because the product was not wanted in Britain, a special refinery was built in Rotterdam, to reduce transport costs. Shell had already ceased selling oil to the German government before the outbreak of war, and with the Dutch ban on oil exports, the refinery was shut down. But Standard, as a neutral American company, continued to sell oil to Germany and Britain alike, and for some weeks unfounded rumours circulated in Britain that Shell was doing the same and was becoming 'a serious National menace'. The company felt obliged to issue a statement of denial, which was not universally believed at once. An anonymous journalist asked three questions: what contracts were in force between Shell and Germany or Austria at the outbreak, what deliveries were made within a month preceding, and what deliveries had been made since? Sam Samuel responded firmly, '(1) None, (2) None, (3) None', and challenged the writer – whom he suspected was in a competitor's pay – to provide contrary evidence. There was no reply.

If that was disposed of as a scurrilous fantasy, another business link with Germany – a real one – was not. Since 1907 the distribution of Shell petrol in Great Britain had been handled by British Petroleum – at the time, a marketing company which (strange as it may sound now) was German-controlled. The distribution contract ran until 31 December 1916, but at the beginning of the war, when victory was expected in a few months, the situation did not seem too bad. Two years later it was a different matter. The reputation of Shell's product did not suffer; at the AGM in 1916, Marcus told shareholders:

> Were I to disclose to you the communications as to the misery suffered by unfortunate motorists, who could not obtain Shell Spirit and have been driven to use other sorts, you would realize how certain it is that on a return of normal conditions the vast majority of motorists will use Shell Spirit wherever they can obtain it.

'Hear, hear', said the shareholders. However, the connection with the German-controlled British Petroleum was well known in Britain, and damaged public good will towards Shell Transport, if not towards its fuel. Marcus recognized this as 'the natural prejudice existing against a petrol sold through a German-owned organization', yet despite company efforts and despite the state of war, there was no legal way to revoke the distribution contract. All that could be done was to wait until it expired and in the meantime prepare a new marketing organization. This was in hand: 'we see daylight now, and, after January 1 [1917], all users of Shell spirit will be able to obtain it from the "Shell" Marketing Company (Limited)...Users will not be compelled any longer to obtain supplies from the German-owned company.'

'Hear, hear', said the shareholders again. If their applause was somewhat tentative, it need not have been. Selling petrol to British motorists was important, not least for increasing the public's awareness of the name, but it was still only a small proportion of Shell's business. For Britain and her wartime Allies, its other oil products were infinitely more valuable – indeed, they were vital. Shortly after the Armistice in 1918, Henri Bérenger (France's Commissioner General for Petroleum) stated publicly that without Shell 'the war could not possibly have been won by the Allies'. If that was an overstatement, this is not: without Shell Transport, the Allies might very well have lost World War I.

Though it may seem extravagant, that claim is nonetheless true. Throughout the years when Marcus Samuel had tried in vain to interest the British government and its Admiralty in oil, the sticking point, from the official point of view, was security of supply. In the summer of 1914, rather than trust Shell, the government bought a controlling 51% stake in the Anglo-Persian Oil Company, whose operational area was within Britain's sphere of influence. But its production (barely 1% of the world's total) was quite inadequate for Britain's needs, let alone those of her European allies. The emergency of wartime provided Marcus and Shell with the chance to prove that their intentions had been honest and honourable all along; and with no immediately viable alternative, the government, in a fit of pragmatism, accepted.

Henri Bérenger, France's Commissioner General for Petroleum during the Great War. Without Shell, he said, 'the war could not possibly have been won by the Allies'

The result was that, from its own resources combined with those of the officially neutral Royal Dutch, Shell Transport became the principal supplier of petrol (at rates established by the British government) to the British Expeditionary Force, and until mid-1917 was the *sole* supplier of aviation spirit to Britain's Royal Flying Corps. Canning plants were built, new depots erected, and all the associated paraphernalia of distribution were put in place, usually at Shell's expense; and before long, 160,000 gallons of petrol – equivalent to nearly 3,700 barrels of oil – were being despatched every day to the Allied armed forces. (There is, incidentally, a pleasant tale about Shell and the Royal Flying Corps. All Service petrol, from whatever source, was pooled and given out in plain khaki cans; but it is said the young airmen of the RFC refused to use it unless it came freshly painted in Shell Spirit's familiar red tins.) Yet even those distinctions were eclipsed by another which, whether it originated with Mark Abrahams (as *he* said) or Marcus (as *he* said), was probably Shell's most important single contribution to the Allied war effort: the supply of TNT.

This formed a really remarkable story. By the end of 1914, barely four months into the war, Britain's armed forces were already close to running out of high explosive, and it was clear that the Admiralty's pre-war expectation of deriving sufficient toluol from coal was desperately wrong. So, during the night of 30 January 1915 (with Deterding's approval and the connivance not only of the Bataafsche but of the neutral Netherlands' government, which laid on a police guard), Shell's entire toluol refinery in Rotterdam was numbered part by part, dismantled and crated up. The following night, the freighter SS *Laertes* stole out of the port and took the

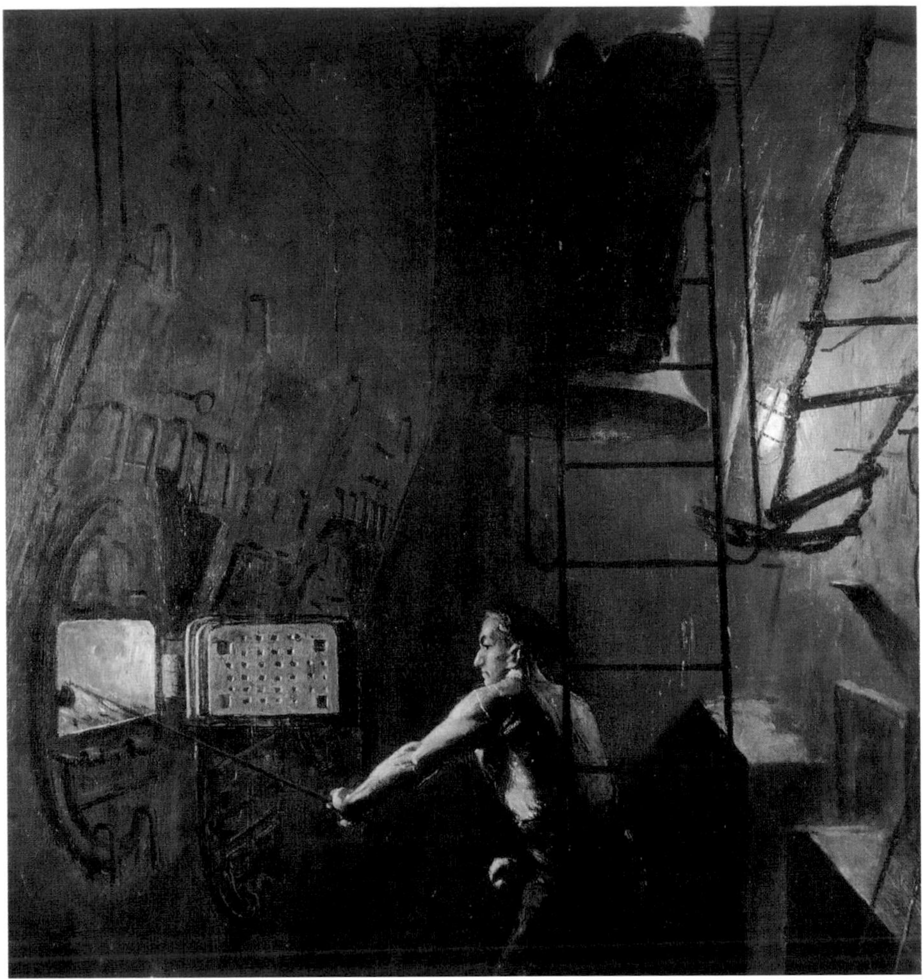

A Merchant Navy Fireman at Work *by Henry Marvell Carr. A fireman, or stoker as he was known in the Royal Navy, would feed up to five tons of coal into the furnaces each day*

components across the Channel. The Netherlands' status meant that all sailings were announced beforehand, in the peacetime way, but that of *Laertes* was given out as being 24 hours later. Her own passage to England was safe, but at her advertised departure time, by coincidence a similar-looking vessel, *Moordrecht*, left Rotterdam – and was promptly torpedoed and sunk. A prayer should be offered for those unintended victims; but *Laertes*' cargo (unloaded in London and taken a further 150 miles by rail) was reassembled at Portishead in Somerset, and within nine weeks of leaving Rotterdam, the refinery was fully functional again. Its product was passed to a new Shell nitrating plant, producing trinitrotoluene at an average 1,300 tons a month. Later that same year a duplicate pair of factories was built, and in the balance of the war the two sets of plants produced 60,000 tons of TNT. Supplied to British and French alike, this was 80% of the Allies' needs.

But whether it was petrol, aviation spirit, fuel oil or toluol, the first essential was to get the crude oil to Britain. Only thereafter could it be refined into the necessary products, packed suitably and transported to its end-user; and of course the only way to bring crude to Britain was by tanker.

At the outbreak, Shell's tanker fleet (by then the largest private fleet in the world) was immediately and voluntarily put at the British government's disposal at 'blue book' charter rates. 'Blue book' not only relieved the government of any obligation to return the vessels in their original condition of cleanliness, maintenance and so forth, but also obliged the company to accept pre-war rates fixed for the duration, however long that might be; and for a company to replace the chartered ships with others for its own needs meant chartering neutrals, at rates which were immediately double 'blue book' and soon several times higher. Shell Transport was the only oil company to volunteer its ships; all its competitors waited until their ships were requisitioned.

A requisitioned ship could command a high and rising charter fee, but any company volunteering its ships knew it would sustain a loss. Yet the fact that Shell Transport put its ships forward voluntarily is not surprising. Marcus Samuel had tried so often to assist the Admiralty, and had so often been refused, that he leapt at the opportunity when it came. When he declared at Shell Transport's 1914 AGM that 'such service, advice, or assistance as we can possibly render to the Admiralty will be freely, willingly, disinterestedly and always at their disposal', shareholders cheered. They understood that in wartime it would mean a guaranteed financial loss, but they also understood that his motive was a simple, genuine and profound patriotism, and they approved. So did Shell Transport's most dynamic director: a few days after the AGM Admiral Fisher wrote to Churchill, 'I have just received a most patriotic letter from Deterding to say he means you shan't want for oil or tankers in case of war – *Good Old Deterding*! How these Dutchmen do hate the Germans! Knight him when you get the chance.'

Tankers were vital, but terribly vulnerable. They had weak or non-existent self-defence, were comparatively slow, easily identifiable by their distinctive silhouettes, and filled with a valuable and combustible cargo. One hundred and sixteen Shell Transport fleet employees were killed by enemy action at sea – more than all its other employees in all the different branches of the armed services. Yet compared to some other companies, Shell Transport escaped fairly lightly in terms of tonnage: throughout the war only thirteen of its ships were sunk, by mine or torpedo. (Its last lost tanker was torpedoed just five weeks before the Armistice, and all her 52 officers and men died.)

But however much oil was supplied, more was always needed, and still more. Cornelis Zulver, Shell's Marine Superintendent, devised two ingenious methods to augment the flow of oil.

First, Zulver proposed that ordinary cargo ships should carry oil in their 'double bottoms', the void space normally filled with water ballast between the holds and hull. Under this scheme, introduced on 21 June 1917, 1,280 vessels nationwide were eventually adapted, giving the added capacity of at least 100 new tankers, and enabling the importation of more than a million extra tons of liquid fuel into Britain.

Zulver's second good idea was to put engines and oil tanks into eight iron-hulled sailing ships, all of which traded successfully during and after the war. These were small vessels, but there was something particularly satisfying about them: their engines were surplus ones from submarines. Powered by those, vessels with the elegant lines of an earlier maritime age helped Britain to survive the submarine, the most modern and deadly of all naval weapons.

Cornelis Zulver, Shell's Marine Superintendent, photographed in 1919...

After the United States joined the Allies in 1917 (as an 'associated power', not a full ally), arrangements were made to pool the resources of Standard Oil, the major US producer, with those of Shell Transport. The Shell Transport director Robert Waley Cohen was sent to America as part of the delegation designed to sort this out, and discovered, to his combined chagrin and amusement, that at that stage Standard viewed war merely as an extension of business by other means. After talking to their senior vice-president, he wrote:

> I really think the summing up of his argument is this: 'We are competitors, and if owing to the war we can get some advantage out of you, we want to do it...and we don't believe the American government would be interested that we should transfer from one to the other!!!'

...and MS Dolphin Shell, *one of the sailing ships converted into a tanker following Zulver's plans*

Top:
On the Deck of a German U–boat *by Claus Bergen, 1918*

Left:
Delphinula, *one of Shell's wartime losses, torpedoed 24 August 1918 by U63 in the Mediterranean while on passage from Naples*

The argument is no special indictment of Standard. Many Americans, both then and in 1939, found it difficult at first to understand why they should become actively involved in what appeared to be essentially European conflicts; but once committed, their involvement was full. This was so much so, at the company level just as at the national level, that in the years following the Great War, Shell and Standard found another bone to pick: exactly which of them had made the greater contribution to Allied victory? In terms of volume it was and remains a hard question to answer. In terms of constancy, however, there can be no doubt: Shell Transport was always there, from the first day to the last.

Robert Waley Cohen in his uniform as a Sub–Lieutenant RNVR, 1915...

Like everyone else in Britain, Shell in the war years strove to maintain its normal work while supporting the national effort in every way possible. If the scale was different for Shell, the experience was no less personal, whether in tragedy or in the rather amateurish efforts that apparently characterize British attempts at civil defence. One of the latter is worth relating. St Helen's Court, the London head office, was quite a high building, so the Admiralty decided to mount an anti-Zeppelin pom-pom gun on its roof. Selected members of staff were trained in its use by the navy, and in order to command them, Robert Waley Cohen was commissioned a sub-lieutenant in the RNVR. The experiment was an unqualified failure: on the one and only occasion when the gun was fired, a direct hit was scored on the clock at Broad Street Station. Less than a year after it had been installed, the weapon was removed.

The major difference between the wartime lives of Shell and ordinary British citizens was that Shell had a life beyond the bounds of war, in places where it was still possible for interesting, valuable and peaceful things to happen. The overall strategy was simple: firstly, to acquire producing interests in different parts of the world. This meant that over-dependence on any one area was avoided, transport costs from well to consumer were kept as low as possible, and a wide variety of grades of oil was available. Secondly, after the acquisition of producing interests, wherever practical the full panoply of an integrated industry should be established under company ownership; and thirdly, dividends should only be distributed on earnings, not on prospects of earnings. 'Our progress', said Deterding, 'must be shown by hard cash results, not by boasting'.

Through the course of the war there were several very varied items on the list of 'business as usual'. On 3 August 1914 the Panama Canal was opened to navigation; a Shell tanker became the first such ship to pass through from the Atlantic to the Pacific, and soon the company had storage facilities at both ends of the canal. In 1915, work on Shell's first US refinery, at Martinez in California, was completed and the installation came on stream; and the Trumble Refining Company was bought for a million dollars in cash. One of Shell's senior representatives in the States remarked it was the first time he had seen a cheque for anything like such a sum, but it was a wise investment. A few years earlier, Milon J. Trumble had invented an improved refining process, which Shell had begun to use under licence world-wide. Looking at the system, one of Shell's Dutch engineers, Daniel Pyzel, worked out a way to improve it still further. The problem was that under the licence terms, if Pyzel's enhancement were incorporated in the Trumble system, the enhancement would become Trumble's property, which Trumble could then license onward to whomsoever else he chose. Adding that practical loss to the royalty ($1^1/_2$ cents a barrel) payable on products of the Trumble system, it became apparent that in the long run it would be cheaper for Shell to buy Trumble's whole company.

...and members of the St Helen's Court rooftop gun-crew, legendary for their total inaccuracy of fire

In 1916, the year of the Battle of Jutland (in which three Shell employees took part), another technical advance took place, when Shell geologists began to *drill* test holes, rather than digging them by hand; and in 1917 a prolific but short-lived well (it produced 4,200 barrels a day for a year and then gave out) was found at Pangkalan Soesoe in north Sumatra.

Shell's exploration in Venezuela, which Deterding reckoned to be 'the most speculative venture of my life', began in 1913. In contrast to Mexico, where the geologists had great difficulties, Venezuela offered some very encouraging information; an American geologist, Ralph Arnold, reported:

> a well-defined zone or zones of oil sands....The structure of the beds is also practically ideal, in that they apparently form part of a great plunging anticline, a type of structure that has yielded some of the best oil fields in the world.

He was quite right. On 14 February 1914, Shell's second Venezuelan well produced a very useful 5,000-barrel-a-day strike. In 1915 oil was struck at Colon, where the company had an interest; and by 1917 the Venezuelan industry was beginning to show its future form, with the establishment of a refinery and canning factory east of Lake Maracaibo, fed by a pipeline from the Mene Grande field.

In the same year, civil strife in Mexico died down sufficiently for Shell to return after two years' absence and resume production, forming a new company, La Corona, for the purpose; and in 1918 a pipeline to the coast was completed from the Pánuco fields south-west of Tampico. But even in such distant places the influence of the war was felt: the shortage of tankers meant that Shell gradually accumulated, and had to store, two million barrels of crude in Mexico, while in Sarawak (because of the Allies' predominant demand for motor fuels) all the lighter fractions, amounting to well over three million barrels, had to be pumped back into the ground. Using the earth itself for storage – Marcus's idea – was both cheap and safe, and provided the reassuring prospect of easily recoverable reserves when the war was over. Reviewing all these out-of-war developments, Deterding observed to shareholders,

> I hope you will not think me imprudent if I declare what I have never ventured to say before...that never have prospects for after the war looked better, and never were your profits so secured for the future as they are now...and we are confident of success for many many years to come. Of this I am firmly convinced.

However, he spoke a little too soon; for in Romania and Russia, two areas of enormously valuable production, the war brought unforeseen disaster.

Perhaps the Romanian complication might have been foretold. There were at least three reasons why it could have been. For one, Romania had become the most important single source of oil in continental Europe, producing in 1913 1.9 million tons – 300,000 tons more than the Dutch East Indies. For another, Shell Transport had a long-standing interest there, predating the alliance. And for a third reason it was only necessary to look at the map of Europe: Romania was temptingly close to Germany.

The country retained neutrality as long as possible, until August 1916. Its declaration of war in that month, against Austria-Hungary, led swiftly and inevitably to war with Germany – 'we should not have been able to exist,' said General Ludendorff, 'much less to carry on the war, without Romania's corn and oil.' Its self-defence thereafter was determined but futile. In mid-November German troops burst through its mountainous frontiers and swept towards the cornfields and oil fields of the plains.

The Allies agreed only one course of action was possible: the ancient policy of 'scorched earth'. Marcus and Deterding gave their consent to the destruction of Shell's properties; as the Germans advanced, the Romanian government agreed too; and with sledgehammers, explosives and fire, the derricks, pipelines, storage tanks, refineries and all other equipment and installations were rendered useless.

'We did it thoroughly, by Jove!', said one of the participants. 'Millions of pounds' worth of property destroyed in a few days. Oil burned, wells blocked, machinery demolished, refineries put out of action. Some wreck, believe me!'

For Shell, this meant the loss, at a stroke, of 17% of its world-wide production. Yet at the next AGM in July 1917 Marcus made just one terse comment on the matter: 'In Rumania, as could only be expected under the circumstances, no revenue was obtained.'

In the same speech, Marcus had a good deal to say about Russia. Only four months previously Tsar Nicholas II had been deposed, and 'as might

Top contemporary military technology: the British-invented tank changed the face of warfare

have been contemplated, our companies have had enormous difficulties to contend with.' True enough: taxes and internal transport charges had rocketed. However, that was nothing compared to events that were soon to unfold.

From its beginning Shell Transport had been intimately connected with Russia's oil trade, initially only as a carrier. In 1911–12, however, the allied companies bought a massive stake in Russian production. Shares were used to purchase the whole of the Rothschilds' Russian organization, and though negotiations were protracted, their successful conclusion meant that overnight Shell gained control of about 20% of Russia's oil industry. This gave the company a more balanced production base: its world-wide average production was now about 63,000 barrels a day, divided roughly equally between the West (the Americas, Europe and Russia) and the Far East, with Russian sources supplying about 29% of the whole. Henri Deterding was of course closely and crucially involved in each stage of this major purchase, keenly aware both of its possibilities for the company and the responsibility that lay upon him. He wrote later:

> My own working experience has taught me that about only five per cent of people want to be saddled with any real responsibility. I don't say that the remaining ninety-five per cent object to working in moderation, but each one seeks to make his own individual burden as light as possible, and...is perfectly happy to be led by a strong man.

Having progressed through sheer ability from a very modest background to a position of considerable power in international commerce, Deterding enjoyed being seen as 'a strong man', and hitherto his judgements and decisions had been in the main correct. The decision to buy into Russia was not. The overthrow of Tsar Nicholas II in March 1917 could have been and was predicted, but not by him. The successful Bolshevik uprising eight months later surprised him even more; and when under its new Soviet masters the Russian oil industry was nationalized, he was shocked to the marrow – for Shell's interests therein were expropriated.

Nothing like that had ever happened in the world of oil, so to have conceived it would have required a prophet; but all the money, time and effort invested in Russia had been committed essentially on Deterding's decision, which now, in spectacular fashion, had been proved wrong. At the 1918 AGM Marcus Samuel hoped for 'a return to normal conditions in Russia', yet the experience of expropriation scarred Deterding deeply, changing his character for the rest of his life. He became increasingly authoritarian, increasingly right-wing and savagely anti-Communist. In 1934 he wrote with a certain bitter satisfaction:

> Today in Russia, I am the most execrated man alive. My effigy is burnt in public places. I have an amusing collection of pictures they have circulated, depicting me as a human monster in all shapes and sizes. Such ludicrously violent methods show how great is the Soviet's fear of me. But why are they afraid? Simply because they know that I see through them for what they are – a set of bluffing bullies.

Though Deterding personally never got over the expropriation, Shell shareholders found themselves less affected by the setbacks in Russia and did not hold those against him. They had, frankly, good reason to be satisfied with their company in the years of the Great War. Motivated by a patriotism which was soon to be given public recognition, Shell Transport's service to the national cause was literally vital; and though that service meant financial loss in some areas, its profits rose throughout the war (from £2 million in 1915 and 1916, to £3 million in 1917 and £4 million in 1918) and dividends remained high, at a steady 35% of par value, tax-paid each year.

Georges Clemenceau, premier of France 1906–9 and 1917–20, inspecting trenches

Oddly enough, this prosperity was to Marcus a source of embarrassment. At the war's beginning, he had announced grandly but sincerely that under no circumstances would Shell Transport profiteer from the conflict. Coming from a leader of industry this was a statement remarkable enough to prompt sceptical comment: after all, what was the company in business for, if not to make money? Nevertheless, that was how all its business with the British government was conducted – 'on a peacetime basis. We have sedulously abstained, even in cases where it would have been possible and legitimate to do it, from exacting from them more than normal prices for our products.' As with the loss-making 'blue book' charter rates, shareholders cheered this confirmation; they did not wish to make excess profits from their country's woes. However, when petrol prices rose and motorists found themselves paying more for their fuel, the correlation with Shell's growing profits seemed obvious, and outside comments intensified into criticisms which were not only unwelcome but, given that Marcus himself had taken up the question of unusually high prices with the government, unwarranted and galling. At an Extraordinary General Meeting in the summer of 1918 he laid out the facts to shareholders. Regarding profits, the bulk came not from the United Kingdom nor even from the British Empire, but from business conducted in other parts of the world; as for the retail cost of petrol, 'the price at which this article is sold is fixed by theGovernment, and if profits are made they are made by the Government, and all your company receives is a commission fixed by the Government.' With that, consumers' criticisms subsided into grumbles.

Left above: Ferdinand Foch, Marshal of France. As generalissimo of the Allied armies from March 1918, he directed the vital battles and movements which drove back the German lines and brought an end to the war

By then the end of the war was almost in sight. At 5 a.m. on 11 November 1918, in a railway carriage in a forest in France, the Armistice was signed. Six more hours elapsed until, at the eleventh hour of the eleventh day of the eleventh month of the year, it came into force and the guns fell silent. That dramatic moment was deliberately chosen; it could scarcely have been more memorable, and those who chose it did so in the faithful hope and belief that it would forever afterwards be annually marked at that precise time, so that no later generation would ignore or have to endure again the terrors which they had known.

The Great War, the war to end all wars, was over. Normal life could resume. And with the Armistice, Shell's employees found that just as the company's first thought in war had been for their welfare, so was its last. Everyone was given the day off – the first holiday opportunity in four years – and the company's Minute Book recorded that: 'a Peace Bonus was paid in cash to the Staff of the Offices and the Fleet, at the rate of one week's pay for every three months' service...in consideration of the splendid work done, and the Sacrifices made by the staff during the war.' 'We must have oil,' said Marshal Ferdinand Foch, 'or we shall lose the war.' Georges Clemenceau, Premier of France, said equally emphatically, 'Each drop of oil secured to us

saves a drop of human blood.' The ability to move men, weapons and supplies when and where you wish has always been of paramount importance in war. The word logistics (defined as 'the art of moving and quartering troops...now especially of organizing supplies') was first used in 1859, as the age of oil began, and the land aspects of World War I were characterized by one writer as 'the victory of the truck over the locomotive'. Imperial Germany dominated the railways of continental Europe, but, restricted to their 'permanent ways', these offered much less flexible mobility than wheeled or tracked motor vehicles. Moreover, unlike the tank, itself a World War I invention, railways provided no offensive capability on the battlefield. As Foreign Secretary immediately after the war, Lord Curzon provided a memorable phrase: 'The Allies floated to victory on a wave of oil.' Elaborating on 'the magnificent efforts made by the Shell company', Lord Montagu (a former member of both the War Aircraft Committee and the Mechanical Warfare Board) remarked:

> The Germans were keener on sinking vessels conveying oil than on any other ships that sailed the ocean. They realized at the very beginning of the War that the liquid fuel supply was vital to our sea supremacy then, and to our air supremacy later. Without liquid fuel you cannot fly, you cannot use submarines...and you cannot run the mechanical transport of your Army on land.

Top left:
Lord Curzon (Marquis Curzon of Kedleston). Viceroy of India at the age of 39; member of the Bitish War Cabinet from 1916; Foreign Secretary, 1919–24

Above right:
John Montagu, 1866–1929, 2nd Baron Montagu of Beaulieu. Founder and sometime editor of The Car, member of the Road Board, the War Aircraft Committee (Mar–Apr 1916) and the Mechanical Warfare Board; adviser (with the rank of major-general) on Mechanical Transport Service to the Government of India during World War I

Montagu's analysis was simple but accurate: the internal combustion engine was changing life's former habits altogether, both in war and in peace. With the return of peace in 1918, Shell Transport and Trading was a company of individuals who, from the chairman to the most junior office-boy, could feel at ease with themselves, knowing that in war each individually and all collectively had done what was needed, and more: in every respect, care had been taken. And there was a notable postscript to war. At the last wartime AGM, record profits were announced along with a dividend maintained at 35%. As one of the shareholders said to the board, 'the balance sheet which you have presented to us is a most magnificent one.' The phrase was more apt than he knew; the balance sheet of Shell's efforts in the war was also 'most magnificent'. In 1921, three years after the Armistice, the company published a book celebrating its wartime accomplishments. It included much self-congratulation (natural enough, given all they had endured) with many huge statistics of production and delivery, often printed in double-size bold type with repeated exclamation marks, and it was proudly entitled 'The Shell That Hit Germany Hardest'.

People judged the claim to be fair, and the judgement has, if anything, been strengthened with the passage of time. Even 70 years later, in dispassionate assessment of Shell's contribution, a distinguished historian of the oil industry as a whole wrote:

> the company became integral to the Allies' war effort; in effect, Shell acted as the quartermaster general for oil, acquiring and organizing supplies around the world for the British forces and the entire war effort and ensuring the delivery of the required products from Borneo, Sumatra and the United States to the railheads and airfields in France. Shell, thus, was central to Britain's prosecution of the war.

Such outstanding devotion and achievement brought deserved reward: in 1921 Henri Deterding was knighted, just as Fisher had said he should be, and Sir Marcus Samuel was ennobled as Lord Bearsted, taking his title from the village near his mansion in Kent. 'I do not know how many of the victories of this last war could be counted as decisive', wrote the brilliant wartime attorney-general, Lord Birkenhead, in 1924. 'I think that there were at least five; but if there were only three, Lord Bearsted must, in virtue of toluol, be counted as the winner of one of them.'

Marcus, said Admiral Fisher, had 'staked his all on Oil and the Oil Engine. Where would we have been in this War but for this Prime Mover? I've no doubt he is an oil millionaire by now, but that's not the point. Oil is one of the things that won us the War. And when he was Lord Mayor of London he was about the only man who publicly supported me when it was extremely unfashionable to do so.'

The ambitious East End Jewish lad had come a very long way in his 67 years. Since its nadir in 1906, the 'good name of Marcus Samuel' could not have been more thoroughly restored, and, thrilled at his ennoblement, the new Lord Bearsted expressed his delight in an engagingly innocent way. His daughters and surviving son automatically became entitled to the prefix 'The Honourable', and when writing to tell them the news, he confided to a friend: 'You can't think what pleasure it gives me to put "The Honourable" on my children's envelopes.'

Armistice Day, London
11 November 1918

G-EA.OU
G

CHAPTER SIX

A Vast and World-Wide Scale: 1919–1925

In 1920, when he was 66, Marcus announced to shareholders that the time had come for him to retire. 'No, no!', they shouted in dismay. Thirteen years on from the alliance with Royal Dutch, they credited him with much of Shell Transport's subsequent prosperity; and certainly, having survived the worst war in history, the company was in very good shape. Compared to 1918, the size of its fleet in 1919 had more than doubled, from 264,000 tons to 545,000; annual production from its East Indian wells had increased from 12 million to 14 million barrels; production from existing and new oil districts in North America was almost 5.9 million barrels; Venezuela and Mexico were looking very promising, with Marcus 'convinced' that both would 'add very largely to our profits in the future'; and profits for the year exceeded £5.8 million. Some was being donated as seedcorn for the future: Cambridge University had received £50,000 to assist scientific research into petroleum, the first of many such educational and research gifts and endowments to different institutions. All in all it had to be admitted that if Marcus must retire, he was leaving Shell Transport with a solid base. Shareholders cheered when they learned that he would be succeeded by his son Walter, but before Marcus departed in an emotional blaze of glory, he made one remark to which some should have listened more closely:

Weetman Pearson, first Lord Cowdray and founder of the Eagle Oil fleet

> I have said it over and over again, but I repeat it now – we absolutely disregard the interests of shareholders who merely go in and out of the company....We do not care what the market quotation of the shares is. It is our duty to see that this business is built on a sound and solid foundation, and we cannot take any responsibility for market fluctuations.

Shareholders, particularly ones who measured a company's strength by the size of its dividends, had become accustomed to receiving back 35% of their shares' par value each year, par value then being £1 per share. When the dividend figure was maintained in 1919 and 1920, they were

Opposite top: The Vickers Vimy (Call-sign G-EAOU – 'God Elp All Of Us') flown by Ross (later Sir Ross) and Keith Smith from Britain to Australia in 1919

Opposite Bottom: A Fokker Wolf fuelling on Shell

complacent. When it dropped to 27½% in 1921, they were disappointed, but not seriously worried. However, when it dropped again in 1922 to 22%, they began to complain – especially those who had bought into the company at the end of the war. One such had not only bought on his own behalf but, he said, had 'induced a great number of my friends, including my wife, to do the same. Since then we have seen our shares constantly fall in value, and today they only represent about half...'

Insignia of the Eagle Oil fleet

Marcus's ship designer Fortescue Flannery provided an effective answer to the complaint. Though it was still possible for speculators in the oil industry to make quick fortunes, that, as Marcus had indicated, was not Shell's goal. Running an integrated oil business, as Shell had become, involved constant high expenditure and very long-term planning, and its investors should look not for speculative rapid returns, but for solid, steady returns over as many years as they chose to keep their shares. The dividend for 1922, 22½% of par value, did not disappoint long-standing shareholders: they could recall when dividends had averaged just 1%. To laughter from the audience, Flannery said: 'I am an original shareholder, and have stuck to my shares ever since, and am not sorry for it.'

However, anyone of nervous disposition in the 1920s could have cited several reasons to be woeful. Russia ('that distracted country') was completely out of the picture; Romania's post-war government had imposed restrictive legislation on its recovering oil industry; the Egyptian fields at Hurghada appeared to be running out; and in 1922, Mexico failed in dramatic fashion.

Back in 1901, Weetman Pearson, a gifted and entrepreneurial British engineer who already had many interests in the country, decided to look for oil there. By 1910, he had acquired 1.6 million acres of oil concessions without finding anything of significance; but in that year his company, Mexican Eagle, drilled Potrero del Llano No 4 and struck 110,000 barrels a day. In the same year he was ennobled as Lord Cowdray, and (with one major rival, Pan American Petroleum) was Mexico's dominant oil producer. Potrero del Llano No 4 inaugurated a series of astonishing finds in the so-called 'Golden Lane', close to Tampico on Mexico's Caribbean coast, with wells of 70-100,000 barrels a day becoming almost commonplace. By 1914, Cowdray had also created a large modern tanker fleet, the Eagle Oil Transport Company; and when he retired in April 1919, Shell Transport took over the management of Mexican Eagle and its fleet. Mexico by then was the world's third largest oil-producing nation, and in 1921 Cowdray's Anglo-Mexican joined with Shell to create a new joint marketing company, Shell-Mex. Few areas of the world, if any, seemed more promising; but in 1922 the Golden Lane ran dry.

Calling it a 'ghastly catastrophe', an agitated Shell shareholder told Walter Samuel that 'there has been no disaster in the City which has hit investors harder'. The event must have raised spectres in Henri Deterding's memory. Back in 1897–8 the young Royal Dutch company had come close to destruction when its East Indian wells began to produce not oil but, just as now in Mexico, salt water. However, there was a very important difference: unlike the production of Royal Dutch in 1898, Shell's production was not limited to one region. As Fortescue Flannery put it,

> The strength of this company lies in the fact that not only are its products consumed the world over, but also – and this is far more important – that it draws its sources of supply from every part of the world where oil is known to exist. If one fails, the other will stand firm.

This was and is a key characteristic, underlining both the interdependence and internationalism of Shell Transport, Royal Dutch and their offspring, the operating companies. In contrast to Standard's work, which had always been domestically based, neither Shell Transport's nor Royal Dutch's ever had been; and since the Group's inception, a thoroughly

During the 1920s, following Shell's great discoveries in Venezuela and political turbulence in Mexico, the oil industry's regional focus shifted sharply away from the Golden Lane and down to Lake Maracaibo

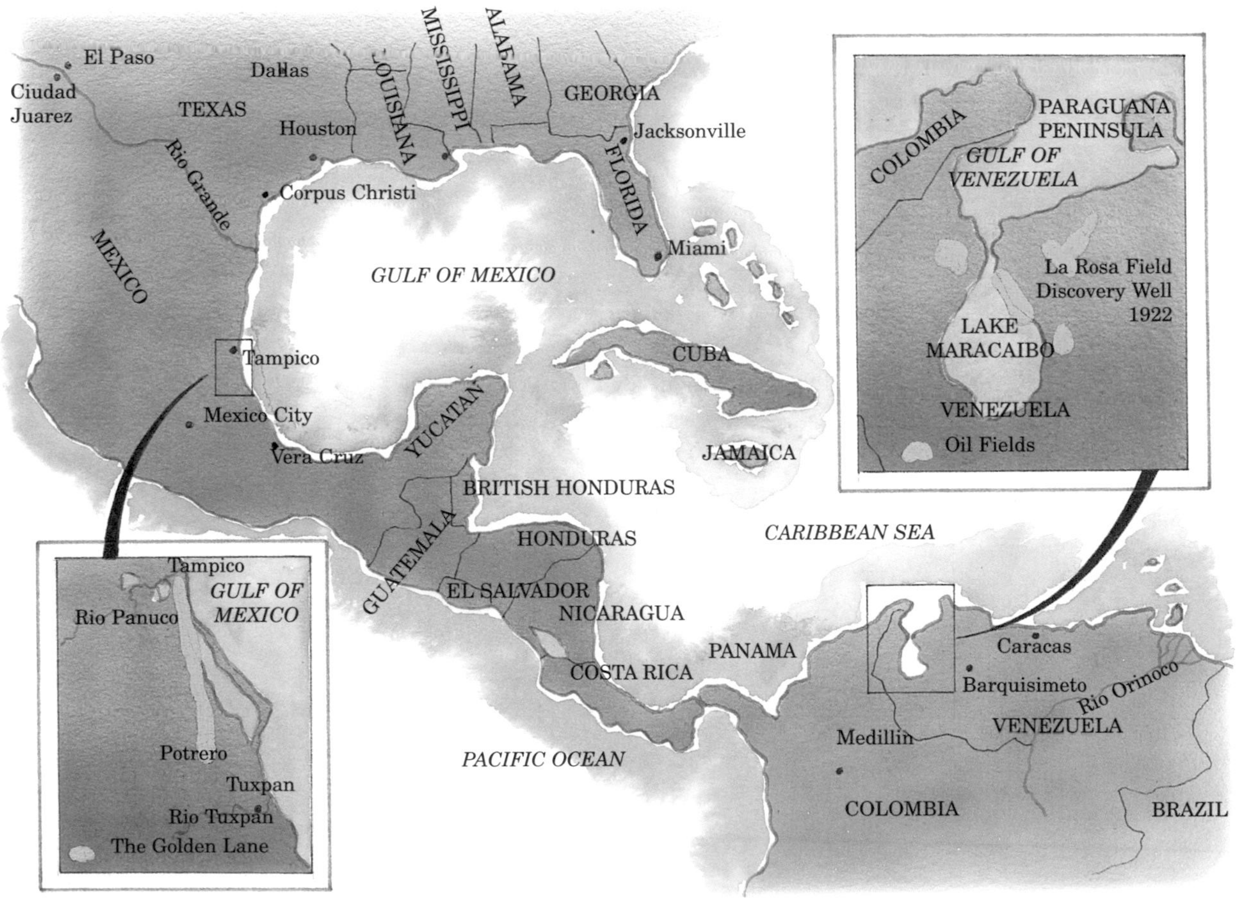

international outlook had differentiated it – perhaps more than anything else – from the other major oil companies. Despite their troubles in Egypt, Russia, Romania and Mexico, Shell companies were still drawing reliable sustenance from the East Indies; had they not been, the 22½% dividend that some thought so low would have been impossibly high. And in the western hemisphere, in North and South America, they were finding the beginnings of what would become unimaginably greater production.

Signal Hill at Long Beach in California was one of the old kind of strikes, the sort that was dreamed of with envy. Referring to the Hurghada fields, Deterding said, 'When we struck Oil in Egypt, some people said we were lucky. Apparently they forgot that we had to sink over a million pounds sterling in geological explorations before we achieved any tangible results.' Much the same was true in California. In 1864, Professor Benjamin J. Silliman, Jr., who had closely analyzed Drake's Pennsylvania oil five years earlier, visited California to judge its potential. He became very excited at what he saw, especially in Ventura County, north-west of Los Angeles. There, he said, 'The oil is struggling to the surface and is running away down the rivers for miles.' The professor was no fool, but (as a Shell historian later said) by 1920, Ventura was 'a region that had behind it half a century of disappointed oil prospectors' – among them, the Shell Company of California, which after spending 3 million dollars over five years of effort had achieved there a productive capacity of zero.

In 1918–19 a Shell geologist with (at least to the British) the improbable name of Dr W. van Holst Pellekaan suggested it might be a good idea to look elsewhere. Soon after, an American colleague called D. H. Thornburg (a native of Long Beach) remembered as a boy seeing marine fossils on Signal Hill, and returned to map his boyhood haunt. Others had already suspected that the hill, 365 feet high, could be the top of an anticline (a geological structure shaped like an elongated dome,

A diagrammatic example of oil-bearing strata

1. *Unconformity trap*
2. *Anticlinal trap*
3. *Fault trap*
4. *Fault*
5. *Seal*
6. *Pinch-out trap*
7. *Top of oil-generating zone*
8. *Immature source rock*

under which oil reservoirs might be found) and Thornburg's surface inspection confirmed this. Since the only way to find out what lay under the hill was to drill into it, the company leased 240 acres. To complicate matters, the acreage could not all be in one tract; the hill was scheduled to become a suburb of Long Beach, and much of it had been sold as domestic building lots, with some houses already erected. But at 4 a.m. on 25 June

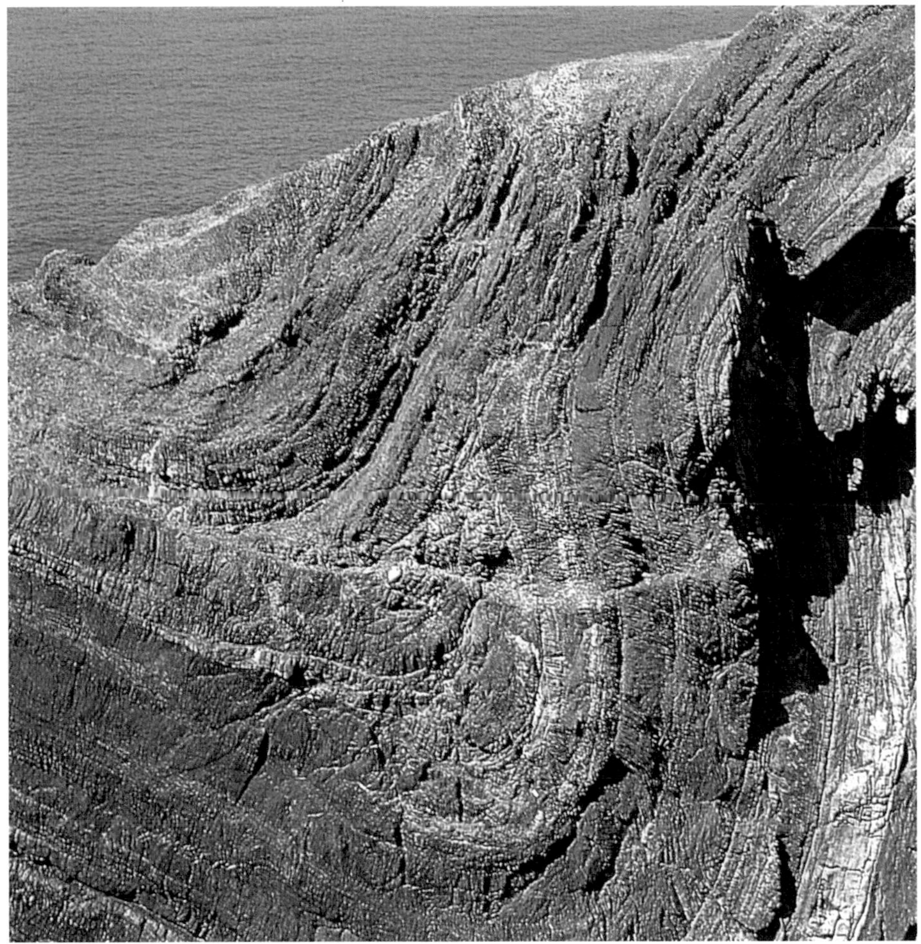

Crude oil and gas can be trapped under an impermeable dome-like structure called an anticline. If on the surface and partially exposed, an anticline can be very distinctive and easily recognized

1921, Shell's well Alamitos No 1 began flowing. Despite the early hour there were some 500 spectators, and house-building came to an abrupt halt, to be replaced by hectic bidding for land-leases.

It was just like Titusville and Spindletop in their boom years. Smaller and smaller leases were sold for higher and higher sums, derricks multiplied like mayflies, and certificates of partnership, ownership, royalty and title were hawked and auctioned. If any of the old days were good, these were some of them, at least for some people: Signal Hill was so full of oil that even holders of an interest as tiny as one three-millionth part of the lease of a single well were able to make a profit.

For the more romantic, optimistic or avaricious, this was what investing in the oil industry was all about. For Shell it was less and less so. The objective was to make money, but not in a single bonanza: its race was a marathon, not a sprint, and would be won by those who commanded not merely speed but stamina and endurance. In those conditions the untrained soon fall away, yet as any marathon runner

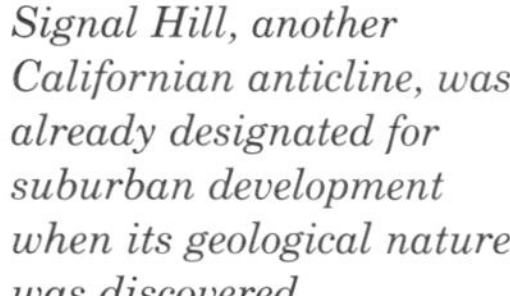

Signal Hill, another Californian anticline, was already designated for suburban development when its geological nature was discovered

knows, the surviving participants still need a regular cooling libation to see the course through. Such for Shell was Signal Hill, although rather than cold water, the refreshment it provided by the end of April 1922 came in the form of 6,000 barrels a day of crude oil. At the time the hill was producing 14,000 barrels a day through 108 functioning wells, of which 28 were Shell's. In other words, Shell was producing about 42% of the volume with 25% of the wells. Fifteen months later, the hill was host to 270 wells, 75 being Shell's; and just three months after that, on 24 October 1923, Signal Hill released its highest-ever one-day production: 259,000 barrels. Eventually Shell alone had 270 wells there, and despite 'town-lot drilling', the hill remained so rich that even 30 years later, at over 500,000 total barrels an acre, it was still the most productive oil field per acre the world had known.

The proliferation of wells was not without its hazards. Fire was the most fearful, especially with its scores of potential fuel sources; and at the beginning of September 1921, the hill's first fire ignited at a Shell well, Nesa No. 1, when a pocket of gas was penetrated. Set ablaze by the heat of drilling, it made a thundering beacon 125 feet high, visible thirty miles out at sea. Fortunately no one died, and the flames were extinguished in two days. Other operators who suffered similar accidents – and there were several – soon got into the habit of turning to Shell for help.

Meanwhile, the Shell Company of California and associated firms continued what Deterding called 'sniftering around' in the Los Angeles basin and beyond. In 1921, they brought in a 'flush field' (a high-producing but short-lived field) at Santa Fe Springs; in 1922, another in Torrance; in 1923, a few miles from Long Beach, Dominguez Hill was brought in; and in 1925 Inglewood, near where Los Angeles' airport would be built. Two wells there, Rindge No 1 (drilled May 1925) and No 9 (July 1925), produced over 1.8 million barrels by Christmas – enough to pay Shell's Californian expenses for the whole year, with more than a million dollars in the bank besides.

In 1923 the Los Angeles basin alone produced the amazing figure of a full 20% of the world's entire oil – increased, if the whole of California was included, to the yet more staggering level of 25%. Inevitably, even within a region of such abundance, not every attempt was successful and not every decision was correct; but altogether, from a base of about 16,500 barrels a day in 1920, production by the Shell Company of California more than tripled by 1925 to about 53,500 barrels a day. And there was more.

In the same period in the mid-continent, Roxana (Shell's first producing company in the United States) was doing almost as well in percentage terms: 8,200 barrels a day in 1920 rose to nearly 21,400 in 1925, an increase of 260%. But its method of achieving this was completely different; instead of searching for new fields, it bought proven or semi-proven ones.

The first of these acquisitions was in Burbank, Oklahoma, where in February 1921, for a million dollars' cash and a half-million on account, the company bought a half-interest in and took over the management of three proven 160-acre leases. The vendor, Ernest W. Marland (who only two years earlier had paid just $15,000 for the leases), then travelled to London, met Deterding and persuaded him that Roxana should join in a venture at Tonkawa, 35 miles west of Burbank. Two million dollars was the agreed price (a million in cash and another when necessary for development) with the deal signed in September 1921. Neither party ever regretted it: Tonkawa crude was of such quality that by simple distillation it produced 46% gasoline, and of such quantity that by the end of the decade the $2 million investment had been returned more than fifteen times over.

In 1922 a young Englishman named Frederick Godber came to Roxana's new headquarters in St Louis. After starting with Shell as an office boy in 1904, he had been told off by Marcus for his illegible handwriting. Without a neat and readable hand, Marcus declared, Godber would never get on in life. Godber must have paid attention, for he was joining Roxana as its new president; and when he retired in 1961 it was as Lord Godber, chairman of Shell Transport and Trading. Under his presidency Roxana continued to bloom, with other properties either wholly- or partly-owned in Louisiana, Texas, Kansas and Arkansas; the last included the Smackover fields, where for a while the oil had to be stored in earth-walled reservoirs so big that the only way to gauge their content was to go out in a rowing-boat and take soundings of the depth.

Roxana and Shell of California were not the only members of the family in the United States. A third (and completely new) one was the Shell Union Oil Corporation, created from the 1922 amalgamation (in the ratio 72:28) of some of Shell's US interests with those of the Union Oil Company of Delaware, which after a brief, colourful and financially unsound history needed to be salvaged.

The arrangement brought Shell several more Californian properties, as well as some extra ones in the mid-continent. Between 1920 and 1925, when the US oil industry as a whole expanded crude oil production by 72%, Shell's companies there increased theirs overall by 275%. More importantly for the still-expanding automobile industry, they enhanced their gasoline production by 660%. In those few years Shell was transformed from a peripheral challenger in America's oil business to a permanent, prominent presence in the continent; but very little of that would have been possible without a variety of important technical developments.

Wildcatting (choosing a drilling site more or less at random) was a feature of the oil industry for many years. It could be very successful, but was generally one of the surer ways of losing large sums of money. Consequently, all oilmen were eager to find cheaper methods of working out what lay under the ground. When Ernest Marland was looking for a reliable partner to develop his Tonkawa fields, he studied Shell's exploration and development techniques before committing himself to a contract with them, and came to the conclusion that 'their scientific advancement was way ahead of any American oil company.' Shell's surface inspection of Signal Hill by D. H. Thornburg was actually a fairly primitive effort, but its key (noting the presence of marine fossils) was sophisticated. There was still, then, argument about the way oil was formed: according to one widely held theory it was the rotted remains of dinosaurs, and one American oil company even used a picture of a brontosaurus as its logo. But marine fossils were a feature observed as common to most oil-bearing regions and strata, and as optical instruments improved, so it became possible to study ever-smaller fossils, including foraminifera – 'hole-bearing' shells as tiny as half a millimetre in length.

These could either occur in other rock or (because there were often millions of them in one place) could together form the rock. In either manner they gave a firm indication of the age of the rock in which they appeared, and Shell geologists began using them as valuable signposts towards oil-bearing strata in the early 1920s.

However, to read these signposts they first had to be dug out cheaply and systematically. This became possible with the introduction, again in the early 1920s, of diamond core drilling. By this worm's-eye view of underground materials, geologists and palaeontologists could investigate core samples of rock, extracted at intervals of 10–16 feet, and recommend either further drilling or the abandonment of a site. And in order to decide where the core drilling should take place, three further techniques were developed, all at around the same time: the use of torsion balances, refraction seismographs and magnetometers.

The torsion balance measured local variations in gravity created by the presence of particular underground rock formations. Torsion balances were not easy things to use: their central component was a wire scarcely thicker than a human hair, they took several hours to put up and use, and a large number were needed for a single survey. But they did provide an idea, at least, of the shape of things underground. So too did refraction seismographs, which were faster, less delicate and a lot more fun.

Here the principle was that, just as light refracts through water, so energy waves transmitted into the earth refract through layers of rock, travelling more slowly through soft layers (such as shale) which absorb them, than through harder layers such as limestone. The practice – the creation of these earth-bound energy waves – was to make a small earthquake. This was done by planting explosives in the earth and setting them off: a small boy's dream, and one which could make money in the oil industry. The speed of the echoes reverberating through the ground was recorded electronically, and the results were drawn. Roxana first used these gadgets in 1924–5, and with them quite quickly became able to make maps of underground structures; and though they were inexact, the maps could be rendered in a three-dimensional fashion.

Magnetometers were roughly equivalent to a large-area version of the torsion balance, except that rather than measuring local variations in gravity, they tracked variations in magnetism, and so provided hints on the substructure of a given locality. This was broad-brush prospecting, best done from an aeroplane for the preliminary covering of wide tracts of desert, forest or jungle; but in such terrain it was incomparably faster than any possible surface geological exploration, and it helped both in North America and South to select or eliminate areas considered for closer examination. Thus aeroplanes, dependent on petroleum for their every movement, became the prime transporters of a machinery which

could locate possible sources of their own liquid fuel – a strange magic indeed.

Technical developments of the time were also making life safer at the wellhead. When drilling began in a likely field, the chances were still high that no oil would be located, or not enough for a paying operation; but if something was found, then the moment of its discovery could be very dangerous. It is easy, but erroneous, to imagine the oil or gas which was the target of the search as lying in some murky cavern. In fact, it was exceedingly rare for the reservoir to be a single great subterranean space; a more accurate image of the reservoir is that of a gigantic rigid sponge – a layer or layers of porous rock, with innumerable tiny interconnecting holes. As oil and gas were created, they rose to the highest possible place, displacing water in the process, until they either emerged at the surface as natural seepages, or became trapped beneath an impermeable layer of rock.

When the pressure on the trapped oil was released by a drill piercing the impermeable rock, the results could be uncontrollable – hence the legendary gushers of oil's early days. Those may have been welcome sights to early oilmen, but whether by mere blow-out or consequential fire, they could kill, and were always wasteful.

The problem of containing and regulating a well's pressure was solved largely by blow-out preventers (essentially a series of large powerful valves) and by mud. Somewhat unexpectedly in an industry of such complex technology, mud was a very useful substance. At first it seemed that any mud would do, but it turned out to be a much more complicated affair than anyone would have thought – to the extent that a whole new branch of engineering developed around it, with any test well being attended by a mud engineer, who would check anything up to seventeen different characteristics in the mud. This was designer mud, a careful blend of several

Signal Hill, 1923. In this dramatic picture, dominated by a gusher, the curve of the anticline – the hill itself – is clearly visible. Gushers occurred following an uncontrolled breakage of the crust above a reservoir, and could all too easily be initiated by...

components each with a specific purpose, pumped down the drill-pipe and back up the bore-hole in a continuous operation. A variation in any single component could alter the conditions in the hole for better or worse (hence the need for a specialist engineer) but whatever its exact composition, all drilling mud had to fulfil four basic criteria. First, it had to be dense enough to lift out cuttings and, if drilling was stopped, to hold them suspended so they did not fall back and clog the drill-bit; second, it had to be naturally lubricating and non-abrasive; third, it had to be heavy enough to prevent fluids from entering the bore-hole; and fourth, it had to maintain its characteristics over a wide range of temperatures, since the heat at the bottom of a hole could be greater than the boiling point of water on the surface. Of course, it also had to be easy to pump, and such was the ingenuity of the mud engineers that they soon evolved a mud which was more than twice as dense as water and yet just as fluid.

Studying the subject systematically in their Californian laboratories, Shell scientists quickly found that a light mud of low viscosity left a smooth plastering coat on the wall of a hole, while a heavier mud (including very finely ground haematite and barytes) enabled safe drilling into high-pressure reservoirs. Before these heavy muds were introduced, accidental blow-outs and subsequent fires were common to every oil company in any part of the world; but once the new muds appeared, events like these became progressively rarer.

Unfortunately such muds were not available in Venezuela in 1922. Managed and part-owned by Shell, Venezuelan Oil Concessions Ltd (VOC)

...the activities of a 'torpedo man'. If a well-shaft became clogged during drilling, such people were called in to lower a 'torpedo' – an explosive charge – down the shaft. The work of a torpedo man was an extremely inexact science, fraught with risk

Another uncapped blow-out – the historic Barroso No. 2 well at La Rosa, Venezuela, 1922. The oil-workers appear quite unconcerned, but fire could easily follow...

had been poking around on the shores of Lake Maracaibo without much success since its discovery of a 5,000-barrel-a-day well in 1914. The region was plagued with exploitative bugs and insects. A lucky victim would only get malaria; an unlucky one could come away with liver disorders, intestinal problems, and insect eggs growing underneath the skin. From experience with the indigenous people, it was deemed wise to avoid going into the jungle unless absolutely necessary, and to keep undergrowth cut back to beyond the length of an arrow-shot. For these and other reasons (expense, and the physical difficulties of movement and transport compared to anywhere in North America) oil prospectors were not crowding

Venezuela's ports of entry; but set against Shell's comparatively trivial returns, a review of its costly efforts worried Standard. As an investigating agent noted, 'The fact that they have spent millions there leads us to suspect there is considerable oil in this country.'

Stressing later that the Venezuelan investment had appeared to be 'a very big hazard', even a 'sheer gamble', Deterding liked to say that he had had only what he called his sixth sense ('that blind instinct which I believe every successful Oil-operator must have') to justify authorizing this expenditure. Certainly, as always in oil exploration, a degree of faith was involved; but as he admitted in passing, he also had a very favourable geological report on the region. On 14 December 1922 both the report and his

...and did, at the same place three years later, when more wells blew out of control

judgement were vindicated: near the lakeshore, Barroso No 2 blew wild. It did not ignite, but for nine days, until it was brought under control, the well gushed out an astonishing 100,000 barrels a day. 'Venezuelan Oil exploration and development thenceforward made the whole world industry sit up', Deterding said smugly. That was true enough: Barroso was the beginning of Venezuela's history as a major oil-producing nation. From a total of 2 million barrels in the year 1922 (all produced by Shell at an average 5,500 barrels a day), its national production hurtled up exponentially to 9 million barrels in the year 1924, 37 million in 1926 and 106 million in 1928, making it second only to the US.

The prodigious rise of Venezuela was closely connected with – indeed mirrored by – Mexico's second decline, but for political rather than geological reasons. After the Golden Lane dried up, production shrank throughout Mexico, but despite the fears of shareholders Shell's own production there actually increased in 1923 by 75%. Nevertheless, in 1921, when Marcus Samuel's son Walter took his first AGM as chairman of Shell Transport, he sounded at once a warning note:

> It is greatly to be regretted that oil, once a mere article of commerce, has now become a political bone of contention and a cause of international jealousy. We were, if I may say so, quite happy to and quite capable of looking after ourselves in commercial rivalry. Today, however, every Government in the world seems to want to meddle in the oil business.

So it has remained ever since. Most people dislike government interference in their lives, and oil companies perhaps most of all: the business has quite enough inherent difficulty, and becomes nearly impossible, if not entirely so, when it is the subject of international political dispute. Domestic political instability is almost as bad, but the costs of setting up in any given nation are so great that circumstances have to become very hard indeed for an oil company to wish to leave. In 1924, Shell reached that point in Mexico, as its operations were increasingly plagued by what Walter called 'the incalculable manifestations of political unrest' – in a word, revolution.

When pressed by Mexico's earlier civil turbulence, Shell had already vacated the nation once. Now all the oil companies present wanted out; and following Shell's strike at Barroso, Venezuela offered an avenue which, at Mexico's expense, was eagerly followed. This was not just because Venezuela was now a nearby and proven source of oil at least as valuable as Mexico. No less importantly, Venezuela had already had its revolution (or rather, a series of them) and from 1908 until 1935 its new president, Juan Vincente Gomez, provided an era of political stability. In an industry where decisions are expensive and take many years to bear fruit, stability is attractive. But stable though it was, the regime also created a severe moral quandary for Shell, because with the use of secret police, sudden arrests and torture, Gomez ruled as an absolute tyrant. For any business with the primary responsibility of making money for its shareholders, there could be no complete way of resolving such an ethical dilemma. But a partial resolution was found, at least. Rather than Venezuela itself, Shell chose the Dutch-owned island of Curaçao, 50 miles off the coast, as the site of a new large refinery for Venezuelan oil; and the Curaçao refinery grew steadily into a very important regional focus.

At the end of this period, in 1924, Walter Samuel remarked, 'We have not neglected the technical side of our activities'. It was already a modest understatement, and much more will be heard of the techniques Shell invented or developed to find, extract and refine crude oil. Walter then proceeded to give shareholders a neat thumb-nail sketch of the giant that Shell had become:

> You must remember that we are producers, refiners, distributors of oil; we make candles, road material, lubricants, medicinal oils, and a host of other by-products. We are even our own bankers, for we do not borrow a penny. All our enormous stocks of oil, tin-plates, boring materials, etc., are bought and paid for out of our own funds. We are amongst the largest ship-owners in the world, with a fleet of over 1,300,000 tons...

Right:
Taking its name from St Helen's Court and Finsbury Circus, the Lensbury Club on the banks of the Thames has provided a social and sporting haven for Shell's staff since 1921

Above:
George Engle in naval uniform, 1916. In 1919 he conceived the idea for the Lensbury Club. No mean sportsman himself, The Shell Magazine recorded in 1934 that 'On May 30th, at Bushey Hall Golf Club, Mr. G.S. Engle did the 6th hole (198 yards) in one stroke'

He went on to say, quite correctly, that the company was now 'working on a vast and world-wide scale'. Yet even with its global scope, which would have been inconceivable twenty years earlier, it still did not forget to look after the people who made it function – the staff. During the Great War, both for the recreation and the emergency training of non-combatants, a rifle range was installed in the basement of the London head office at St Helen's Court. The site has long since moved, but in the present premises the facility is still available. However, its popularity there has long been overtaken by another much less martial amenity: the Lensbury Club.

Taking its name from a contraction of the head office at St Helen's Court and a subsidiary office at Finsbury Circus, the idea for the club was first submitted to the board of management on 19 March 1919 by Mr George Engle. Always known by his initials as GSE, the plump and genial Engle had no particular title in the company and no precisely designated function, but over the years had taken on the work of what might now be called a senior manager of personnel. He did this rather well and was a popular man, gently lampooned by his juniors with cartoons and little inside jokes in company publications. What he had in mind in 1919 was that the company should buy a piece of land, somewhere not too far from central London, and establish a staff sports club. The board liked his idea and told him to investigate. On 2 April GSE came back with details of a plot of land at Romford. This was rejected as being too distant, so on 9 July 1919 he suggested instead Ham, on the north bank of the Thames. This seemed rather more suitable. After a little more investigation the final choice was made, and just across the river from Ham, a site was bought at

The 1st Scammell, 6-wheeled road-tank wagons at Fulham, 1924

Teddington, on the south bank. Two years later the Lensbury Club was opened, membership being available to every permanent employee world-wide. At first its buildings consisted of three adjacent houses with their combined gardens. The site itself was gradually extended, and in 1933–5 the original houses were demolished and replaced by much larger, purpose-built premises. With a long river frontage, pleasant terraces, wide lawns, and sports courts and pitches of every sort, as well as bars, dining rooms and overnight accommodation, Shell's own comfortable and imposing country club has arguably affected staff's lives more to the good (certainly more pleasantly and much more visibly) than any figures of production or acts of high international drama; and GSE, the man who thought of it, became its life president.

1

CHAPTER SEVEN

Roaring Along: 1925–1929

By the middle of the 'Roaring Twenties' Shell Transport and Trading, in partnership with Royal Dutch, was a global presence. In addition to their traditional working areas in the Far East and their newer ventures in the Americas and West Indies, Shell companies were active (whether in exploration, production, transportation, refining or marketing) in most parts of Europe, as well as in Aden, Australia, the Sudan and Turkey. The tanker fleet, growing steadily towards two million deadweight tons, was becoming the largest private fleet anywhere, representing 10% of the world's total tanker tonnage; and the Group's companies produced on average 11% of all the world's crude oil, every year.

In 1926 *The Economist* called Shell 'one of the most successful oil organizations in the world'. It was rather more than that; at the time, only one other oil company remotely approached its stature. The result of the legal break-up of Standard Oil in 1911 had been much as Deterding predicted: for a decade or so Standard's former constituents did not really compete with one another, but continued to sell under one name and respected each other's traditional geographical territories. This was partly because the individual companies' structures meant that none was a properly integrated company: one would be mainly production-based, another marketing-based, and so on. However, as the reality of their separation gradually took hold, competition between them also became real, and by the early 1920s one was emerging (or re-emerging) as the dominant force in the US market. Led by Walter Teagle, this was Standard Oil of New Jersey, known later as Esso (from its first two initials, S. O.) or subsequently Exxon. Other former Standard companies prospered too, to almost comparable degrees – in particular, the Standard Oil Company of New York, known by the acronym of Socony, which later became Mobil. But in the middle 1920s Shell, with its attitude of co-operative internationalism, was the oil world's predominant body.

It remained so until the outbreak of World War II in 1939, when Esso took pole position. Since then, either Shell or Esso has topped the small select league of oil majors. The situation has become so customary that it is easy to forget how astonishing, in the mid-1920s, Shell's achievement actually was. Barely thirty years earlier, it had not existed; now it was the world leader in an industry of central importance to the whole of modern society.

Its founder, Marcus Samuel, was by then approaching the end of his life. When he died in 1927 (less than 24 hours after his wife Fanny), it was in the knowledge that he was a wealthy man: once, in 1919, he worked out that the family holding in Shell Transport was worth £7 million – in the money of the time, a very considerable fortune. More important to him, though, was the social recognition which he had won; for in 1925 he, the son of an East End shell merchant, had been raised further in the English peerage, from the rank of baron to viscount.

Sir Harry Ricardo, FRS (1885–1974). Prolific inventor and highly gifted mechanical engineer. His work on the problems of ignition, combustion and detonation in the internal combustion engine led to the use of octane numbers as the measure of a fuel's anti-knock capability, and his improved combustion chamber has been universally adopted

Succeeding him as second Viscount Bearsted, his son Walter was already second chairman of Shell Transport and Trading. Some people reckoned that with its Dutch ally, the company's wealth and responsibilities were equal to those of a minor European state. The comparison is probably accurate; yet none of it could have been achieved without Marcus, Sir Henri Deterding and Robert Waley Cohen – or Sir Robert, as he was from 1920, 'for services in connection with the war'. These men, the first oil knights, formed a mutually indispensable triumvirate. Sir Marcus had the imagination and audacity to create Shell Transport and Trading; Sir Henri had the single-minded brilliance to sustain it; and Sir Robert had the businesslike vision to extend it to the supply of ever more sophisticated oil-based products. Prime amongst these was the evolution of the best possible fuel for internal combustion engines; and to an increasing degree there was also the advancement of chemicals derived from oil.

In the earliest days, Shell Transport's petrol was made simply by distilling crude oil; Marcus Samuel would boast that Shell's 'motor spirit', unlike those of some other companies, was of such naturally high quality that nothing needed to be added to it. But as engines improved, fuels had to improve too – and as fuels improved, it was possible to improve engines further. Shell's continuing work in this field began in 1917, led by the great mechanical engineer Harry (later Sir Harry) Ricardo, whose brilliance later earned him Fellowship of the Royal Society – a rare distinction for an engineer.

More Shell in the air: Sam Samuel presides over a dinner held on 26 September 1922 to celebrate the first King's Cup Air Race, won by Captain F. L. Barnard (far left) on Shell...

In 1917, Ricardo was working for the War Office on the development of engines for tanks. These novel weapons, which it was hoped would overcome the deadly stalemate of trench warfare, produced many technical problems. They had to be strong enough and heavy enough to withstand enemy fire, yet light enough in handling to be manoeuvred by one driver; they also had to be able to go into a trench and climb out again at an angle of up to 35^0. For this they needed engines which for the time were relatively large, producing (on Ricardo's designs) 150 and later 225 horsepower. The engines also had to be simple to maintain, and free from both tell-tale exhaust smoke and the danger of stalling on the battlefield.

In combating this, Ricardo concentrated on the phenomenon of 'pinking', or 'knock'. This occurred when the air/fuel mixture in an engine's cylinder ignited (or as he put it, detonated) prematurely, resulting in a loss of power – a common phenomenon with early tanks, mainly because of the poor quality of fuel provided for them. Service fuel supplies were allocated by a War Office committee, so one day Ricardo presented himself before it to ask for better fuel. But he asked in vain:

> I was given to understand that the best quality petrol was earmarked for aviation; the next best for high-speed staff cars, and the lowest grade for tractors and heavy vehicles, and that the Tanks, which only waddled along at walking pace, would have to be content with the dregs of the barrels.

The committee, composed mainly of senior naval and military officers, was chaired by a civilian, 'a huge and formidable looking fellow' – none other than Robert Waley Cohen, by then one of Shell Transport's directors. Since the committee would not, or could not, supply him with better petrol, Ricardo asked instead for benzole, or benzene. He explained that it was more stable and less inclined to detonate than a kerosene-based fuel, so with it, he would be able to increase his engine's compression ratio, which in turn would provide enhanced power, economy and range. But the committee was still not interested – 'they would have none of it.' It was not

Shell on land: Ferdinand Porsche (right) watches a mechanic adjusting the engine of a Porsche-designed Daimler Benz, Monza, Italy, 1922...

...Enzo Ferrari at the wheel of the Alfa Romeo in which he won the Coppa Acerbo at Pescara, Italy, 1924...

until after the meeting that the chairman took Ricardo aside and said, 'What's all this stuff about benzole and detonation? I would like to hear more about it.'

Waley Cohen had read chemistry at Cambridge, and though he had been unable to sway the committee's Service personnel, he had sensed there must be something in what Ricardo said. Over dinner the two men agreed that Shell should send some samples of its different petrols to Ricardo for evaluation. Ricardo found that one of these, from Borneo, could give an engine at least 20% more power – 'far and away better than all the

...Count Brilli-Perri in the winning Alfa Romeo P2 after the 1925 World Championship. It was following this victory that Alfa Romeo added the laurel leaves to their insignia...

...and the team of Mariononi and Behuist in their Alfa Romeo, first overall in the 24-hour race at Spa, Belgium, 1929.

others', he noted. This was so astonishing that he repeated all the tests on another sample of the same fuel, and obtained identical results. Yet what had Shell been doing with it? Burning it as waste, by scores of thousands of barrels, because its specific gravity was deemed too high to be of commercial use. Armed with this 'shocking revelation', Waley Cohen promptly cabled Borneo to stop the waste and took Ricardo on as a consultant.

So began Shell's continuing programme of matching fuels to engines, employing the best brains; and some of Ricardo's other investigations showed that brilliance can promote both heroism and harmless pleasure, as well as sound businesslike products. On his suggestion, Shell's Borneo petrol was used to make a super aviation fuel. Doing this was costly and quite wasteful, and the product was too scarce for normal aviation use; but it did provide 10% more power from 12–15% less fuel – and it was this fuel which Alcock and Brown used in their pioneering non-stop transatlantic flight. As Ricardo remarked,

> They said that even a small increase in power or fuel economy might make all the difference between success and disaster for their enterprise, for it was touch and go whether they could take off with enough fuel for the crossing...

It is easy to imagine the thrill and satisfaction derived from being involved with the success of that historic crossing on Shell fuel; and just as easy to understand the enjoyment that came from another of Ricardo's suggestions. Thinking 'it might be amusing to concoct a special fuel mixture for racing cars and motor-cycles', he put the idea to Robert Waley Cohen, 'who had no objection; in fact, he too thought it would be rather fun.'

From this whim, 'Shell Racing Spirit' was developed. It was a blend of ethyl alcohol (supplied by the Distillers Company), benzole, acetone, water and 2% castor oil. The formula was kept confidential, partly because Ricardo wished to patent it and partly because the 'secret composition' was a splendid marketing device. But applying for a patent produced a hitch: any competitor could simply analyse the fuel, work out its components and their proportions, and by adjusting those proportions very slightly could make an equally efficient fuel without breaking patent law. Shell Transport's chief chemist, James Kewley, decided 'to find some complex organic substance which would both defy analysis and give to the exhaust a peculiar and characteristic smell.' The answer was to add to the mixture – of all things – a small pinch of finely-powdered bone meal.

It certainly gave the exhaust a distinctive ('and,' said Ricardo, 'I am afraid, a rather repulsive') odour. It also provoked enormous speculation among competitors, who mostly reckoned a secret wartime development, probably a high explosive, must be involved. Speculation and theories increased when, with Shell's agreement, the Distillers Company began marketing 'Discol R', a 'rival' fuel – which was in fact exactly the same thing, but in a differently coloured can. The invented rivalry proved another good marketing device, and Ricardo much enjoyed overhearing riders extol the virtues of one fuel against 'the other'. The deception was an innocent one, because whatever the can's colour, the fuel worked outstandingly well: it could produce as much as 30% more power than anything else available. Indeed, with motor-cycles it was so outrageously efficient that after one season of its use, in which a Shell-sponsored rider took every prize that was going, the authorities at Brooklands racetrack banned it. It cannot be often that a company produces something which is just too good.

Waley Cohen's recognition of Ricardo's talents laid the basis for all Shell's subsequent work towards producing an ideal petrol, and some of the merit for initiating Shell's work in chemicals must go to Waley Cohen as well. Back in 1901 he had asked an old Cambridge friend, Humphrey Jones, to analyse Shell Transport's Borneo oil. 'This', he said later, 'was, I think, the first occasion on which a scientific analysis was made of a petroleum distillate.' Jones, who died young in 1916, was another exceptionally brilliant man – First Class passes in Parts I and II of Natural Sciences, with distinction in chemistry (an achievement no one else repeated before his death), led him to be the Royal Society's youngest Fellow – and from the oil sample given him by Waley Cohen, he extracted no fewer than 350 separate pure chemical substances.

One might suppose this had irresistibly exciting business potential – and indeed the time would come when, if put together, Shell's chemical companies would form one of the world's largest such companies, with unequalled petrochemical production. But there was no serious Group attempt to investigate the possible value of its chemicals until the late 1920s. In 1901 Waley Cohen was too junior a member of Shell Transport to have much influence: he did not become a director until 1905. Thereafter he remained for many years the only board member to be interested in chemicals; all the others regarded the production and sale of fuels as quite enough. Eventually, in 1919, they agreed he could establish the London-based Central Laboratory. (Its first chemist, James Kewley, who later thought of adding bone meal to Shell Racing Spirit, was another Cambridge graduate personally selected by Waley Cohen.)

But despite the laboratory's grandiose title, it had nothing to do with chemical research; from a descriptive memorandum, Godber learned that 'the scope of the work was very limited, and practically confined to the straightforward examination of cargo samples.' In 1928, however, this situation changed suddenly and dramatically.

According to one account it was only when Deterding had 'a violent personal clash with an arrogant German professor' that he 'became suddenly converted' to the idea of pursuing a commercial chemical interest. The story may be true, but he was not normally a man to sink large sums of money in a long-term, untested project merely through a fit of pique. Good commercial reasons are far more likely; and at that very time there were at least three such reasons. Firstly, some important potential customers were emerging: the British chemical group ICI had been formed in 1926, followed in 1927 by the similar I. G. Farben group in Germany. Secondly, Standard Oil of New Jersey was experimenting with chemicals in the United States. Any significant success could bring a large business advantage to Shell's main competitor. Thirdly, a world-wide surplus of crude oil was developing, with a correspondingly severe fall in its price and that of existing products. Fully alert to these alarming changes, Deterding was ready to listen to proposals; and in 1928 J. B. A. Kessler junior (son and namesake of Royal Dutch's former chairman) joined Shell Transport's board. In him, Waley Cohen found a welcome ally, another supporter of a venture into chemicals. With the changed times, Kessler was the catalyst of action.

J.B.A. Kessler junior joined the board of Shell Transport in 1928 and gave great support both to Sir Robert Waley Cohen and James Kewley in the expansion of Shell's chemical interests

There were indications that producing chemicals from petroleum could be a commercial success; yet doing so was still almost uncharted territory. For Shell to start work in such a new area would be expensive, with unpredictable results. Kessler challenged these worries directly, saying

> We should have confidence, energy and courage enough to develop this new part of our business, even if it does not yield profits to start with. Do not let us be frightened by some preliminary cost calculations which show that we may not make money in the beginning. I do not expect to make money; on the contrary, I expect it will cost a good deal of money before we can actually produce on a fairly large scale. If we can work ourselves out of the routine in our rather simple oil business, we shall certainly be in a position to develop our industry on a very complete scale.

Our rather simple oil business? One wonders how Deterding reacted to that remark. Kessler also pointed out that in addition to crude oil, other raw materials were abundantly available and going to waste – the gases from thermally 'cracked' crude.

Thermal 'cracking' was recognized around 1909, and rapidly became an important part of the refining of crude oil. Though dangerous in execution, the concept was simple: by subjecting crude to heat under pressure, the long hydrocarbon molecules of its heavy fractions were cracked into smaller, lighter ones with more uses and greater commercial value. For example, gas oil (the fuel for diesel engines and central heating systems) could be 'cracked' to make a petrol with a better anti-knock value than ordinary petrol. Furthermore, the method meant that a single 'cracked'

The Central Laboratories, Fulham, in 1931

barrel of crude produced more than double the amount of petrol obtainable from an 'uncracked' barrel. The method also produced about ten cubic metres of gas from every barrel, only some of which could be used as refinery fuel. The rest was simply burned – yet the gases too were made of hydrocarbon molecules. In Kessler's view, it seemed:

> logical that we should turn all this energy that is going to waste at present into something that we can put into packages and sell. If we had a lot of gas in a very thickly populated area we might make electricity and sell it, but what can we do with the waste gas we have in the United States, the East Indies, Venezuela, Romania and so on? The only thing we can do is to make something we can ship.

But what? Even he could only surmise that 'the range of products...may be found after some time to be much wider than we know it at present'. Though realistic, that uncertainty was not entirely reassuring. However, Deterding and his fellow directors were probably most influenced by another of Kessler's points in favour of chemicals: 'the larger the number of products we make, either in the form of oil or in some other form, say artificial manure, the safer we are against any drop in prices of some of these products.'

Once the strategic decision had been made, early in 1928, it was rapidly implemented. A budget of ten million dollars was set for a dual programme of chemical production and fundamental research – for as one of the staff said later, 'You could not buy a book and look up how to build an acetone plant. You had to find out for yourself.' Three main parts of the Group were affected. In London, Shell Transport's Central Laboratory was relocated from Bishopsgate to Fulham and expanded; in Amsterdam, Royal Dutch's Central Laboratory, previously used mainly for quality control, was provided with a chemical research department; and in the United States, a refining company called Simplex (established by Shell in 1915) was renamed the Shell Development Company, with its headquarters relocated from New York to California. Given the task of finding it a research director, Waley Cohen turned once again to his academic contacts. From University College, London, he recruited Dr Clifford Williams, who at only 35 was already dean of the college's science faculty and in charge of the country's first chemical engineering curriculum. Williams designed Shell Development's brand-new laboratories at Emeryville, near San Francisco, completed in October 1928. Just a few months later, in February 1929, the Shell Chemical Company was chartered in the US; and the following September, at Ymuiden in the Netherlands, Mekog (in full, Maatschappij tot Exploitatie van Kooksovengassen) became the first chemical company in the Royal Dutch/Shell Group to commence production, extracting twenty tons of nitrogen from the air each day and reacting it with hydrogen to make synthetic ammonia as a fertilizer – the 'artificial manure' that Kessler had imagined.

A Romanian oilfield, c. 1927

Thus Shell invested swiftly and determinedly in an entirely new area of the petroleum industry. Thinking in 1931 of the results of Humphrey Jones' experiments at the turn of the century, Robert Waley Cohen grumbled, 'When you reflect that that was something like thirty years ago, we cannot congratulate ourselves very much on the use we have made of those 350 pure chemical substances.' But better late than never, and Kessler's hopeful forecast was proved correct – the range of discovered products was indeed 'much wider than we know it at present'. Shell chemists soon found that from refinery gases which had previously been burned as waste, they could make feedstocks for all kinds of things: to name just a few, insecticides, resins, disinfectant, detergents, solvents, anti-freeze, fertilizers and glycerine – and even a base for synthetic perfumes.

Calouste Gulbenkian, 'Mr Five Per Cent'

Shell's entry into chemicals was undertaken partly to avoid being left at the post by others, and partly in order to extend and diversify the product base in the face of a forecast global surplus of oil. In 1928–9, the idea that the world might soon have too much oil was new. A decade earlier, just after the Great War, many well-informed people had been sure that the world's known sources of oil were close to exhaustion. All the large oil companies had then put great efforts into exploration, which for Shell involved two widely separated but simultaneous sets of negotiations: one called 'the Red Line', focused on the Middle East; and one in Russia, called the 'Front Uni'.

The term 'Red Line' was invented by one of the legends of oil, Calouste Gulbenkian, Mr Five Per Cent. He had been associated with Shell Transport and with Royal Dutch since the very early 1900s, acting both before and after their alliance as an intermediary with other companies in acquisitions or the raising of capital. He gained his nickname when in 1914 he held a key position in arrangements concerning the Turkish Petroleum Company (TPC) made between Shell, the Deutsche Bank, and the British government's oil company Anglo-Persian. The outcome, just before the Great War, gave Anglo-Persian 50% of TPC, Shell and the Deutsche Bank 22.5% each, and Gulbenkian his famous 5%.

At that time – the last days of the decaying Turkish Ottoman Empire – the concession areas belonging to TPC included lands which are now independent republics. By 1919, when the Great War was over and the Ottoman Empire had collapsed, it was widely believed that the world's last untapped reserves of oil lay within those lands, especially in the former Ottoman province of Mesopotamia, now Iraq. Another widespread opinion was that if oil were discovered there, defeated Germany deserved no beneficial part of it. The Deutsche Bank's share was accordingly transferred to French interests; but the Americans, deeply fearful that their own resources were drying up, insisted (because of their part in the Allied victory over Germany) on having a share as well, and with their government's permission a syndicate of US companies was formed to press the case, under the leadership of Walter Teagle from Standard Oil of New Jersey.

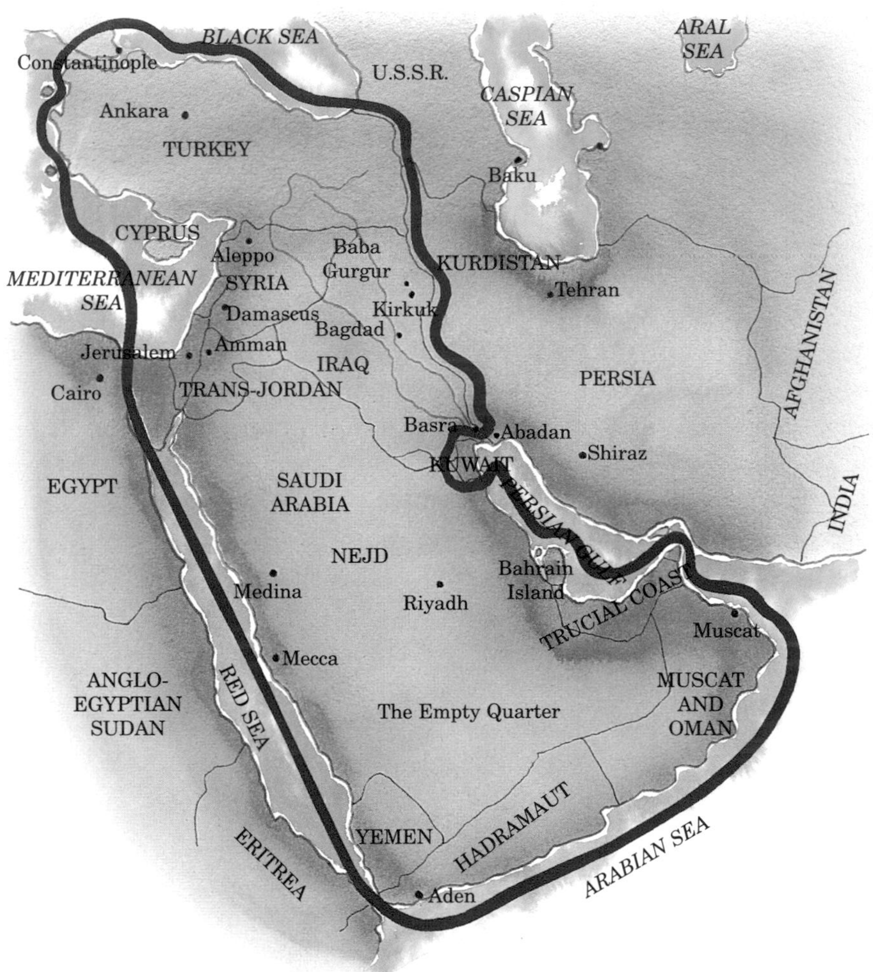

The area of the 'Red Line' agreement of 1 July 1928

Negotiations continued for years. So did the TPC concessionaires' joint exploration; and on 27 October 1927, at a place in Iraq called Baba Gurgur, north-west of Kirkuk, oil was struck in spectacular fashion. Improved techniques had made gushers largely things of the past, but the fountain at Baba Gurgur could be seen twelve miles away and took eight days to bring under control, during which time it flowed at a rate of 95,000 barrels a day, with the oil forming itself into a river.

Negotiations now became compelling, but were not concluded until 31 July 1928. Shell, Anglo-Persian, the French and the American syndicate would each receive 23.75% of the oil, making 95%; Gulbenkian would receive 5%. All involved agreed they would only work within the region on a joint basis, the region being the old Ottoman Empire, as defined by 'the Red Line'. Drawn by Gulbenkian in red pencil on a map, it stretched from Constantinople (modern Istanbul) in the north to Aden in the south, covering almost the whole Arabian peninsula. Kuwait was excluded, as was the whole of Persia (modern Iran); but otherwise, as further discoveries would reveal, the Red Line embraced all the major oil fields of the Middle East.

As the same time as Iraq was being explored, the post-Great War fear of an oil famine was also spurring Western oil companies to reach a settlement with Soviet Russia. For years after the Bolshevik expropriation of Western oil properties, Shell and all its former competitors there regarded themselves as the victims of outright theft, and in the summer of 1922 Henri Deterding invited his rival Walter Teagle to London, together with representatives of fourteen other oil companies. The meeting arose from one of Deterding's favourite Dutch proverbs: *Eendracht maacht macht*, or 'co-operation gives strength'. The idea was that instead of continuing to pursue, in vain, some form of compensation separately, all should join together. They did so, in the *Front Uni;* but though the French phrase emphasized the internationalism of the grouping, it could not overcome the companies' intrinsic mutual distrust and competitiveness.

The story of the *Front* was brief and bitter. It had taken the Soviets some time to reactivate the expropriated properties, but in 1923 oil – priced more cheaply than most other crude oil in international trade – began to flow from them into Europe. Despite the unified intent of the *Front*, its member companies, including Shell, could not resist the bargains and began secret purchases, undermining their joint efforts to win compensation. In 1924, as it became clear the *Front* was not working, Deterding and Teagle agreed to establish a joint company for the purpose of buying Russian oil. Both the Western tycoons found the business distasteful; neither could overcome the sense that they were having to buy back their own stolen goods, and Deterding was further influenced by a personal reason – he had fallen in love with, and soon married (as his second wife), a White Russian lady: an aristocrat in exile, whose family fortunes, like those of Shell Transport and Royal Dutch, had suffered at the hands of the revolutionaries. 'The idea of trying to be on friendly terms with the man who burglarizes your house or steals your property has never appealed to me', said Teagle. Deterding agreed writing to Teagle at the beginning of 1927, I feel that everybody will regret at some time that he had anything to do with these robbers.'

Regret came much sooner than Deterding imagined – and in an extraordinary manner. Within weeks he learned that other former Standard companies were busy buying cheap Russian oil and selling it in India and other parts of Asia. This enraged him not so much because they were undercutting Shell's prices, but because they were purchasing from the Communists at all. Promising that he would prevent anyone from buying Soviet oil, in an untypical reaction, he opened a fierce price war in the subcontinent. It was the very tactic which the old Standard Oil would have used, and which he and Marcus Samuel had jointly and separately so often decried. The former Standard companies responded in kind, and like a forest fire started by a carelessly thrown match, the war quickly spread to other markets.

By then, Deterding was sixty years old. He had been in effective charge of Shell Transport and Trading for twenty years, and of Royal Dutch for twenty-seven, and hitherto few people could have faulted his management of either. He had been Shell Transport's saviour, if not explicitly acknowledged as such. Just the year before, Sam Samuel had described him as 'indefatigable; he is always at work, and I might say that he has absolutely a genius for money-making', which brought laughter and cheers from shareholders. But opening the price war was a great mistake.

This was not immediately apparent to shareholders. On the contrary, as far as they were concerned, all appeared well in the late 1920s. Their investments were bringing satisfactory returns, and Shell Transport's profits rose from just over £5 million in 1925 to £6.72 million in 1929. In the same period, shareholders also learned with pleasure that the allied companies' leading technological edge was being maintained: in 1927, at Ventura, the Shell Company of California drilled a well to a record-breaking depth of over 7,000 feet – 'the deepest commercially producing well in the world' – soon followed by another at 7,503 feet. In that year too a thousand miles of pipelines were planned and completed in various parts of the world, three new American refineries were built, and distributing facilities were 'enormously multiplied'. Year on year the tanker fleet was being modernized and extended. In land transport, 'motor tankwagons' now brought liquid fuel and lubricating oils in bulk to factories, and with the recent introduction of petrol pumps, 'petrol is today brought practically to the actual consuming car'. Shell petrol stations of a standardized, distinctive design (known in the United States as 'crackerbox') were being erected everywhere to serve the motorist direct, and Shell's pump attendants wore a special uniform.

As Walter Samuel observed to shareholders, these were costly changes, forcing 'a large increase in the capital necessary to bring each gallon of oil from the well to the consumer.' His audience understood that to increase profits in the modern world, a great deal of extra expenditure was required; and profits were being increased. But from the annual reports, the assiduous reader could calculate another set of figures. The rise in profits represented an apparently healthy 34.5% in five years, and Shell Transport had maintained its percentage share of annual world production – yet in the same period its actual production had risen by nearly 80%, from an average of about 273,000 barrels a day in 1925 to 486,000 a day in 1929.

If an 80% greater production brought only 34% greater profits, profitability had shrunk very badly indeed; it had become much harder to make money out of oil.

Seminole, Oklahoma, USA – an oil-boom town in around 1927

The trouble was two-fold. Firstly, in their post-war efforts to avoid an oil famine, explorers around the world had achieved too much success. Baba Gurgur was the largest discovery, but elsewhere other companies made other big finds. Added to these, the tide of cheap Russian oil meant that in the industry as a whole, production accelerated from an average 2.9 million barrels a day in 1925 to over 4 million in 1929; and, with the more efficient processing that thermal cracking permitted, this in turn meant that much more oil was for sale than anyone wanted to buy.

Having too much oil to sell is almost worse than having too little, for prices are bound to slip; but at least a surplus is unintentional. In contrast, the second cause for the slide in profitability was wholly deliberate – the price war unleashed by Henri Deterding in 1927. Combined with the surplus, its effect was dramatic: a barrel of oil which cost $1.85 in 1926 could command only about one dollar in 1930.

As recently as 1926 Walter Samuel had emphasized that modern competition depended upon the quality of service offered to the customer, and had spoken of price-cutting as 'the crude form of competition', with 'certain well-defined limitations'. And just a few years later, in 1934, Deterding himself said in his memoirs that price-cutting was not competition at all.

> You may call it a refined form of throat-cutting, a stranglehold, a dog-fight [but] annihilation, not competition, is then the right word. You can't compete with a man, nor he with you, if all the while you are both bent on squeezing each other to death. Quality and service are the only sure foundations on which competition can survive.

Yet squeezing the competition to death was precisely what he was trying to do – even though he knew, as did his co-directors in Shell Transport, that it could not possibly succeed and could only bring harm.

This act, to which we shall return, was uncharacteristically irrational of Deterding. It stands in striking contrast to Shell's major decision of 1928 – the risky, daring but wise and justifiable entry into chemicals, taken following the advice of J. B. A. Kessler junior, the new boy on the Shell Transport board. But going into chemicals was not Shell's only important decision in 1928. Another, of which Kessler disapproved, was taken when Sir Henri chose to have a very unusual summer holiday.

As the world's forecast production compared to actual demand moved progressively from famine to flood, a new mood was arising in the oil industry: something had to be done to stabilize both production and prices. The Americans were the first to react publicly, taking steps to limit their production – a practically unheard-of idea for any oilman. In June 1929 Walter Samuel applauded this as 'the first step towards the active rationalization of the oil trade.' Adding that 'we can never hope to get away from human fallibility, and it is not to be expected that any perfect scheme can be evolved', he also hinted gently to shareholders 'there is very much that can be done, and I believe we are slowly but surely getting on to the right road.' As well as the public American initiative, a secret one had already taken place, on Deterding's instigation.

The chairman only had to report to shareholders once a year, and then a year in arrears; and Walter Samuel's hint to shareholders in June 1929 referred to an event which had occurred in the August of 1928, that busy year. Officially, it was simply a private vacation in Scotland, a group of friends and business associates getting together for a fortnight's shooting and fishing. Unofficially, in the remoteness and privacy of a Scottish castle called Achnacarry, it was a congress of the world's oil leaders.

Sir Henri and the new Lady Deterding were host and hostess. Among their guests were Walter Teagle of Standard Oil of New Jersey; Sir John Cadman, head of Anglo-Persian; William Mellon of Gulf; and a crowd of office staff and advisers, sufficiently numerous to need a separate house for their accommodation. Press speculation was considerable, but no one connected with the household would say anything to reporters. Many years earlier, at a time in World War I when petrol prices were rising, Marcus Samuel had attempted to explain the reasons to the press, emphasizing how small a profit margin was involved. The sums were clear, but the message was not well received – probably because in exasperation he ended with the remark, 'The price of an article is exactly what it will fetch.' It would be hard to summarize the principle of free market competition in fewer words, yet consumers felt suspicious. They viewed the industry more as a service than a commercial supplier, and, enjoying but wary of their own dependence on oil products, were highly sensitive to fears of monopoly,

Walter Teagle, 'The Boss', chief of Standard Oil of New Jersey (later Exxon) and grandson of John D. Rockefeller's first partner

cartel and exploitation. In 1928, had word of the talks at Achnacarry leaked out, there might well have been public uproar; because a global agreement – or as consumers would have seen it, a global cartel – was just what the oil chiefs intended to establish.

The dictionary definition of cartel ('an association of business houses for the regulation of output, prices, etc.') is neutral, yet the word is an emotive one, associated with exploitative price-fixing. However, the purpose of the houseparty was nothing so unsophisticated. Instead, the participants wished to end their mutually destructive price warfare and bring stability to the industry – for the benefit of their companies, to be sure, but also (as they saw it) for the benefit of their customers.

Their discussions were summarized in a document which they termed 'the Pool Association', more generally known as 'the Achnacarry Agreement' or 'As-Is'. That phrase indicated one of the agreement's most important aspects: the competing companies consented to maintain their existing market shares 'as is'. To stabilize revenue, prices would be based on the prevailing US Gulf Coast price plus the current freight rate from there to the marketplace, even if the area of production was closer to the marketplace. (This was one part which consumers would have criticized sharply.) In addition, in order to keep costs down, the oil chiefs agreed to share and not to duplicate facilities, adding to them 'only...as necessary to supply the public with its increased requirements in the most efficient manner'. They also agreed to share the supply of markets from the nearest geographical source, and either to shut in wells if their production was greater than the local market could support, or else to offer the surplus for sale elsewhere without undercutting the price elsewhere. Finally, the document's seventh paragraph stated:

> The best interests of the public as well as the petroleum industry will be served through the discouragement of the adoption of such measures the effect of which would be materially to increase costs, with the consequent reduction in consumption.

Although the agreement was obviously something very different from normal free-market competition, it did not seek to create an artificial captive market vulnerable to exploitation. Rather, it was designed to break the cycle of boom and bust, to render the industry as efficient as possible, and to reduce the risks involved in its inevitably long-term investments. Profits would be maintained, to the benefit of the companies and their shareholders; resources would be conserved and their waste minimized, to the benefit of the consumer.

At any rate, that was what Henri Deterding and the other architects of 'As-Is' believed and hoped. In their view, the only alternative would be chronic instability, from which all would suffer, producers and consumers

Murex *discharges 1st bulk gasoline cargo at Miramar, New Zealand, in around 1925*

alike. Yet at least one Shell Transport director was deeply sceptical about the arrangement – J. B. A. Kessler. On 13 September 1928, with the Achnacarry conference barely over, he wrote in somewhat idiosyncratic English to Teagle. 'The figures we had before us', he said,

> showed that, of the potential world production, a large part is controlled by companies which are not controlled either by you or us or any of the few other large oil companies. From this [it] followed that the present balance in the world's production cannot possibly be maintained by you and we alone.

As the prime mover of Shell's major new strategy – its vigorous and full-blooded entry into the chemicals business – Kessler had already shown himself to be one of the most inquiring, forward-looking and imaginative members of Shell Transport's board. Shell was ending the 1920s as a world-class concern, selling fuels of very high quality, with the beginnings of a wider product base than ever before and with the assurance of prolific fields in the Middle East; but the price war had been a gratuitous and self-inflicted wound whose full effects were yet to be seen, and in Kessler's analysis, Shell could not pretend to govern the world's oil production, even in covert agreement with other oil majors. If he was correct, Achnacarry was an idea whose time was already past, and there would be no easy solution.

CHICHESTER CHOSE
SHELL
OILS & SPIRIT
G-AAKK
CHICHESTER'S PLANE in which he flew from LONDON to SYDNEY
AVIATION
£4000 ART UNION

CHAPTER EIGHT

Surviving the Flood: 1930–1934

1930

The great majority of the world's oil still came from the USA, and late in 1929, in a talk given to the American Petroleum Institute, Henri Deterding posed the central question:

> Are the United States going on producing all they can – more than they can consume – with the result that they are exporting today at a low price what they are likely to be importing later on at a higher price?

Indeed they were. In a time of overproduction and falling prices, the last thing any major oil company wanted was another big strike. On the other hand, any small operator – and there were thousands in America – wanted a big strike more than anything else; and on 3 October 1930, a 70-year-old wildcatter called Columbus Joiner struck lucky. The location (Rusk County in East Texas, 120 miles or so south-east of Dallas) was a place where no one else believed oil could exist – but perhaps 'lucky' was not the word to describe Joiner's find. At the time, crude oil was selling for about a dollar a barrel (already little more than half its price in 1926) and Joiner's discovery introduced a field which put every other one in the United States into shadow. Within four months, in February 1931, the region was producing 25,000 barrels daily; two months later, 340,000; and just four months after that, in August 1931, a full million barrels a day – one-third of the United States' entire production for the year, and over one-quarter of the whole world's production. At earlier rates this would have meant fortunes for all; but when the existing surplus of oil turned suddenly to an actual glut, the market simply could not absorb it, and the price collapsed almost completely, to fifteen cents – six cents – even two cents a barrel. A physical barrel was worth more than its contents, which could scarcely be given away.

Opposite top:
A bustle of activity on the Seria field in the Brunei in the 1930s

Opposite below:
Gipsy Moth

World production in 1930 averaged 3.86 million barrels a day, of which Shell produced about 466,500 barrels daily, or about 12% of the whole. This maintained the very respectable proportion of the preceding five years or so; but the price curtailment was reflected in Shell Transport's profits and dividends. The former fell from £6.72 million to £5.13 million, and the latter from their steady 25% of par – still, then, the way of expressing dividends – to 17.5%.

Walter Samuel felt this merited some comment at the AGM for 1930, held on 24 June 1931. The new East Texan region had yet to reach its devastating top level of production, but was already sensational. After outlining the generally sound and stable situations of Shell Transport and the Group, he referred to the American discoveries. In some fields, he explained, 'a very sensible drilling policy has prevailed', with confirmed discoveries being treated as reserves and kept in the ground. But 'in many other fields the industry has been unable to prevent almost reckless drilling campaigns, and as a consequence these fields have been not only over-developed but ruthless drainage has taken place...'

This seemed to him not only economic nonsense, but immoral as well. He and his colleagues felt that oilmen owed a debt to the future; none of them could guarantee that production would continue more than a few years, and at present rates, he said, 'some day posterity will have to pay for this unwarranted prodigality of natural resources.' This real moral concern was heightened by the worry that wasteful overproduction could bring the day of reckoning forward into the lifetimes of those present (indeed, because of the price collapse, the American-based Shell Union Oil Corporation was obliged to pass its dividend for 1930) and so Walter was relieved to be able to say that US State Governments seemed likely to enact legislation which would curb production to 'within the safe limits of actual consumption'.

Hotel Cecil

He also had some positive news. Dividends to Shell Transport from operations elsewhere, including Mexico, were satisfactory; a new South American enterprise was on stream in Argentina, complete with a functioning refinery in Buenos Aires; and a large London property had been bought to house Shell-Mex. Formed with a staff of 780 in 1921 (two years after Shell took over the management of Mexican Eagle), this company handled the combined UK marketing of Shell products and Eagle products, and now had a staff of 1,319. Its original accommodation at Shell Corner on Kingsway had become too small; overflowing into an adjacent building had been inefficient; and so the large Hotel Cecil had been bought freehold. The hotel was to be demolished and its site redeveloped. Running in a prime location from the Thames Embankment right through to the Strand, the new building, Shell-Mex House, would be much bigger than necessary, so the surplus space would be rented out to other businesses; and 'our agents have already received applications from very many would-be tenants of good standing.'

Shell-Mex House under construction

Yet behind these items of good news, as his listeners knew well, another spectre stood alongside the oil collapse: the general slackening of the world's economy, the depression sparked by the crash of the New York Stock Exchange in 1929. Walter recorded 'with the greatest regret' that many employees had had to be dismissed, and though they at least were given lump sums from the Provident Fund, he warned shareholders that 'taking all the adverse factors I have enumerated, together with the extreme improbability of their sudden removal, I cannot hold out rosy prospects for the current year.'

His audience was extremely sympathetic: their spokesman declared that a 17.5% dividend was very good in the circumstances, and added, to general laughter, 'that if the majority of companies in London today could produce balance-sheets half as good as theirs, there would be little anxiety for the future of business in this country; and incidentally, the Chancellor of the Exchequer would be able to look forward to next year's Budget not only with equanimity but with optimism.'

It was as well they took the chance of laughter while they could, for their revenue was about to plunge. The £5.15 million profit of 1930 tumbled in 1931 to just £2.8 million; crawled up to £2.85 million in 1932; and limped on to £2.99 million in 1933. In each of those three years the dividend on par was a mere 7.5%. Not until 1934 did profits and dividend alike stage a partial recovery, respectively to £4.19 million and 12.5%.

If the shooting and fishing at Achnacarry, back in 1928, had not been unusually poor, then perhaps Sir Henri and Lady Deterding's houseguests might have simply enjoyed themselves. As it was, their attempt to rationalize the industry had been almost a complete waste of time. With the effects of East Texas magnified by world-wide economic depression, Kessler's analysis was proved correct: 'the present balance in the world's production cannot possibly be maintained by you and we alone.' No one in the industry, not even Shell and Standard in secret agreement, was big enough to govern it, or even, in present circumstances, to influence it very much. But as Walter Samuel remarked, the industry was concerned with 'a vital requirement of civilization and must therefore continue to exist'; and he added optimistically that

> it will be one of the first to revive when world conditions improve. And in the oil industry there is no unit better equipped technically and financially to see through the bad times, and be the first to take advantage of the good, than the Royal Dutch-Shell Group.

Garage in Westerham, Kent, with Shell sealed pump, 1929

In 1931, Britain's abandonment of the gold standard devalued the pound sterling by 30% overnight. On 20 September 1931, £1 could buy \$4.86; on 21 September, \$3.50. For Shell Transport, the most unpleasant immediate effect of this was that one of the Group's main operating companies, the Dutch-based Bataafsche, had to write off £4 million. 'This is, I hope and believe, a non-recurring loss', said Walter. 'Of course, you will appreciate that if it had not been for this enormous and unavoidable depreciation your dividend might have been considerably greater.' But as exchange rates fluctuated unpredictably and as governments imposed currency controls, it was difficult to forecast how much revenue from outside the sterling area could be transferred into it, or how much it would be worth.

These were indeed 'troublous times', as a shareholder said. With international trade and commerce becoming weaker every month, governments around the world sought to protect their enfeebled domestic economies by increasing taxation. Tax in general, but especially on oil and its products, was a topic guaranteed to rouse Walter Samuel's ire. In 1931, when the average weekly wage in Britain was about two pounds sixteen shillings (in modern terminology, £2.80), duty on petrol rose from fourpence to sixpence a gallon, and then to eightpence, bringing the cost to the motorist of a single gallon to one shilling and fivepence (7p); and a penny a gallon was levied on lubricating oil. Special petroleum taxes in the UK had already brought the Exchequer £44 million in 1930; in 1934 that rose to more than £73 million. Some users changed to other fuels, but the majority could not, or did not wish to, and had to pay the government-created higher price. 'The tax', said Walter, 'is therefore a definite handicap on industry and is paid, for the most part, by the most progressive and most efficient section of industry – by that section, in other words, which deserves the greatest encouragement rather than penalties.' Even in 1932, when Shell Transport's profits were only £2.85 million, the Group paid about £50 million in taxes world-wide, rising, with the modest increase in profits in 1934, to £57.7 million. Walter was scathing:

> I doubt whether any prime commodity is, or ever has been, so mulcted. There is a production tax or royalty on your crude; there is an import duty on your products, and, if there should remain any profit, there is an income tax on that.

'Iron maiden' petrol pump from 1920s after restoration

Much needed to be done to reorganize the business in the face of adversity, and much that was done was, in the short term, very painful. Since 1921 the Lensbury Club in London and its Dutch equivalent, Te Werve, had taken turns in hosting the Group's annual international sports day. In-house magazines carried all the results, and everyone looked forward to this enjoyable event; competition was intense but friendly, because no one really minded whether the Dutch or British won. But in 1930 the need for economy was so urgent that suddenly, even the sports day had to be cancelled indefinitely. It was a sad loss, but small. Far worse sacrifices were essential – including at Shell Transport's very heart, the tanker fleet. For lack of carrying trade, five tankers belonging to Anglo-Saxon had to be laid up. The same fate befell 5–10% of the Mexican Eagle fleet, manned and managed by Shell employees; and though one might say that to contrive employment at such a time for 90–5% of the fleet should have been seen as a considerable triumph, it did not appear so then. These were, in the directors' view, 'the worst years', ones which they hoped never to see again. They felt so because of the all too visible human cost involved. When a ship was laid up, her crew became redundant, and in the 1930s it was commonplace for experienced merchant navy captains to volunteer for employment as third officers, or even stokers. Ashore as well as afloat, redundancies were rife. Those who were lucky enough to remain employed shared the pain: everyone in Shell Transport, from directors downward, received a substantial cut in pay. As for the unlucky ones who lost their jobs altogether, nobody can measure what it meant, in financial and emotional terms; but a hint may be gathered from the cost to the corporation. The Provident Fund then totalled about £25 million. In the years 1929–32 alone, £10 million had to be paid out.

Yet distressing as these events were for those who endured them, they may have been something of a blessing for Shell Transport. The economies brought changes which, though forced, were logical; and some proved beneficial for the long term as well as the short.

INTERNATIONAL SPORTS MEETING "DETERDING" CHALLENGE CUP

Saturday 23 June 1934 was 'a red-letter day in the annals of the Lensbury Club': the international sports meeting, suspended in 1930, was revived. Most spectators found seats where they could...

Lensburian: 'Yes, we have our own bar now. Let's order something while they're slack-what?'

...but senior personnel (Sir Henri Deterding on the right, with George Engle, president of the Club since its foundation in 1921, next to him) stood apart

...and apparently did not care for interlopers

Competitions included lawn tennis (singles, doubles and mixed doubles), fencing (ladies' foil, men's foil, sabre and epee) and ten different athletics events

LENSBURY AND BRITANNIC HOUSE	23 POINTS
REST OF THE CONTINENT	9 POINTS
TE WERVE (HOLLAND)	8 POINTS

The mile

Relay race

High jump

Miss Mary Burgess presenting a bouquet to Lady Deterding

Mr G.S. Engle after receiving the Cup from Lady Deterding

These took three main forms: on the one hand, the creation of a number of joint operations, in the place of competitive ones; on the other hand, great progress in fuel and chemical research and development; and in between, a spectacular explosion of brilliant advertising.

New joint marketing ventures commenced in 1931. Since 1919, Shell-Mex had worked as the UK marketing arm of Shell and Mexican Eagle combined. Simultaneously, British Petroleum (or BP, the marketing company which in World War I had been German-controlled) had done the same for Anglo-Persian; and now Shell-Mex and BP were united. If their joint name – Shell-Mex and BP Ltd – was pedestrian, nonetheless the marriage proved successful and durable, lasting 45 years.

At the same time, joint marketing operations were established far away as well. In India, the Burmah-Shell Oil Storage and Distribution Company appeared, while in the Near East, east Africa and southern Africa the new Consolidated Petroleum Company Ltd put Shell's marketing operations together with those of Anglo-Persian; and in America the Shell Union Corporation linked with one of its old rivals, Standard of California, to buy a company called Union Oil Products (UOP), which held valuable patents for the 'Dubbs' thermal cracking system.

Back in the British Isles, by 1932 Shell-Mex and BP Ltd were covering distribution in England, Wales and Northern Ireland. Scottish Oils and Shell-Mex Ltd covered Scotland; Irish Shell Ltd (40% Royal Dutch/Shell, 40% Anglo-Persian and 20% Canadian Eagle) served the Republic of Ireland similarly; and in the same year the economics of refining were improved as well, when Shell Refineries Ltd was created by the take-over (with full effect from 1936) of the UK's three main refineries, at Shell Haven on the Thames, Stanlow near Liverpool, and Ardrossan in Scotland. In a time of world-wide recession, this was altogether an impressive degree of active realignment; and it was brought to the attention of the buying public in a long-running series of advertisements which remain classics of their kind.

Before the 1930s, Shell's advertising was often quite ponderous. As late as 1924, seeking to emphasize its Britishness, there was a poster which showed Britannia filling her flagon at a lion-headed fountain. The informative but cumbersome caption read: 'SHELL *distributes more petrol refined from crude oils* PRODUCED WITHIN THE BRITISH EMPIRE *than all other petrol distributing companies in Great Britain combined.' But fashions changed as publicists recognised* that fast, clear, memorable communication required uncluttered images and briefer texts. From about the mid-1920s, and coming to full flower in the 1930s, Shell's advertisements were a wonderful outpouring of imaginative talent. Implicitly or explicitly, the themes were power, purity, reliability, modernity, quality of service, and a mildly romantic escapism – getting away from it all.

In addition to the press, posters formed the main medium of advertising. These were not on roadside hoardings; there were none in Britain in those days, and many people, including Shell Transport as a company, believed they would deface town and country alike, and sought to prevent their introduction. Instead, the message was brought to consumers by billboards on the sides of lorries. On these the passer-by might see, above a perfect summer countryside, five high-flying biplanes bearing one letter each: S - H - E - L - L. With special fuels for different seasons, Summer Shell (available from May to October) was depicted by cricketers, tennis-players and sailors, while in the cold months a motorist

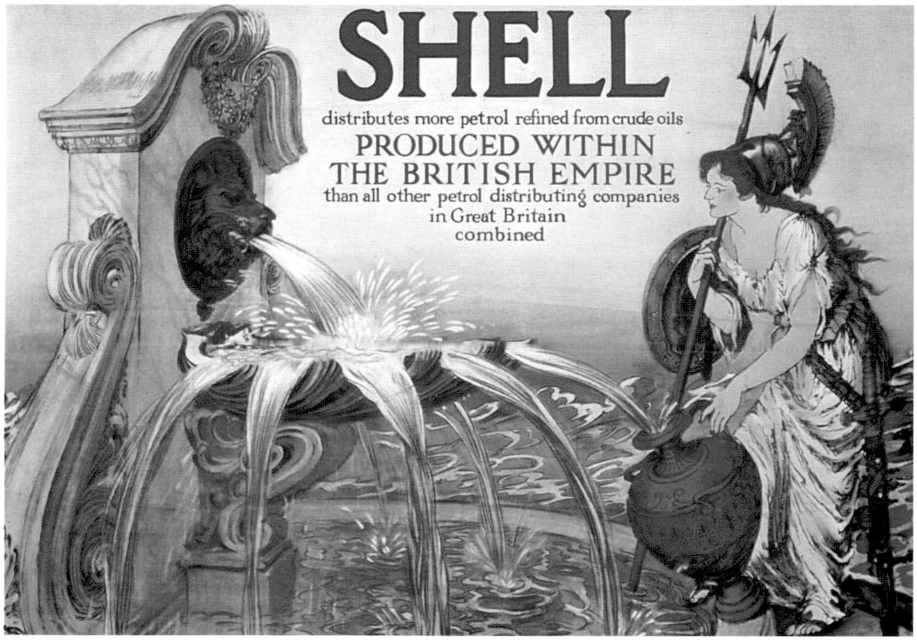

whose car had broken down in a snowdrift was told, 'Don't use bad language – Use Winter Shell'; and year round, Shell oil and Shell petrol became 'The Quick-Starting Pair', symbolized by grasshoppers, greyhounds, horses, birds and even newsvendors.

Sometimes it seemed the advertising copywriters were competing for brevity. 'No Essential Missing' brought the message down to three words; 'Unadulterated' to just one. And one outstanding advertisement was an image of such simplicity that it was nearly abstract: power and reliability together were expressed by a thick, taut chain, with nothing else in the picture.

In solid three-dimensional contrast, the introduction of petrol pumps brought the opportunity for pumps whose outsides were locked by a tamper-proof seal, as visible confirmation to the motorist that only Shell petrol could be inside. Lorry-bills excited the curiosity of those who had not

seen the 'sealed cabinets'. One of the early 'spelling' advertisements had shown five of the familiar red cans, each pouring out a letter of the name; but with the novelty of pumps, new advertisements pointed out that there was 'No Waste – No Mess', 'No Tin To Pay For', and that nowadays 'Every Drop Counts'. And when they went to have a look and fill up, buyers discovered there were also glass domes on top of the pumps, so they could appreciate the fuel's transparent purity as it flowed into their tanks.

Humour was frequently used. To begin with it usually took the form of rather stilted puns, such as 'the Spirit of Many Triumphs' and, for aviation, 'the Spirit of the Air'. Around 1912, as a pronounced maritime theme emerged, a castaway sailor had the sense to clutch a red can and was 'Buoyed Up by the Best of Spirits'. Another popular advertisement, linking Shell with the Royal Navy, showed a naive midshipman looking innocently down the barrel of an enormous gun and enquiring 'Where's that Shell?' A third, crossing over the boundary of sea and land, suggested that a motorist who had run out of petrol near a seashore need only look around to find a salty fisherman holding up a welcome red can. The motorist would then naturally feel impelled to exclaim, 'He sells *the* Shell – Saved!'

Quality was repeatedly emphasized, with 'the Knock-Less Monster' (a creature of friendly appearance) reminding drivers of the fuel's high anti-knock capability. That particular picture was allied with another caption which was a lasting favourite. The visual accompaniment was always an apparently two-headed stationary person or animal with a blur of speed in front. The left-facing head saw an approaching vehicle; seconds later, the right-facing head saw the vehicle disappear into the distance. There could be no doubting the identity of such a powerful fuel: 'That's Shell – That Was!'

Advertising campaigns outside the UK were usually quite differently devised. For example, Shell's newspaper advertisements and billboards in France used an artist's mannequin as their main feature, accompanied by long descriptions of the virtues of Shell's fuel. In Great Britain, the home of Shell Transport, the buyers of its products were seen more and more clearly as people who responded to a specific set of images; and from that perception sprang two brilliant advertising series.

Their themes were 'To Visit Britain's Landmarks' and 'Everywhere You Go', either phrase being followed by the same enduring slogan: 'You can be sure of Shell'. Coined in 1931, it sounded good; it rolled off the tongue; and in just six words it said everything the company could wish, evoking with magical economy a sense of total stability, unerring quality and proven reliability. It was unbeatable.

Another long, delightful series allied with 'You can be sure' was called 'the Conchophiles' – the Shell-lovers. The idea was that, given the choice, anyone and everyone would prefer to use Shell; so picture after picture appeared – 'Farmers prefer Shell', 'Seamen prefer Shell', 'Footballers prefer Shell', 'Theatre-goers prefer Shell'. As one of the artists said, 'There is no reason why we should be limited to a fixed viewpoint once we have discovered that in life itself there is an unlimited number of points of view.' Well-known artists and the most promising young artists of the day were selected: Ben Nicholson, Paul Nash, Rex Whistler, McKnight Kauffer and Edward Bawden amongst many others, some of whom were brought to public prominence for the first time by their commissions from Shell. The main credit for this must go to Jack Beddington, advertising manager from 1932 to 1939, who had been given the job after saying the current advertising was unsatisfactory. He looked more like a bank manager than an artist, but Great Britain's artistic heritage owes much to him and to this exceptionally enlightened period of Shell's history.

A fundamental part of the advertising policy was that using art of such high quality would itself enhance the public's impression of the quality of Shell's products. Extending this, Beddington soon found another successful medium of publicity: a brand-new kind of cinema, the documentary film. With the age of television only just beginning, cinema-going was at its peak as a pastime in Britain, and in 1934 the Shell Film Unit was formed. Its first film produced for public screening was *Airport*, depicting a day in the life of Croydon Aerodrome, London's main airport at the time. The film lasted only seventeen minutes, but nothing could quite compare with aircraft, and everything associated with them, for excitement. Many people had never seen an aeroplane (or if they had, only at a great distance) yet everyone recognized the exotic glamour of flight. *Airport* informed, entertained and educated while simultaneously indicating Shell's own position in the vanguard of modernity, and it was very well received by the public and professional critics. 'An admirable example', said one. 'Concise, consistent in its interest, and good to look upon and listen to.' At first glance the subjects of some of Shell's films do not sound very promising – for example, *Transfer of Power*, whose theme was 'the principle of the lever'. But the films were made so skilfully, often with mechanical animation, that another critic described *Transfer of Power* as 'most exciting and beautiful...a short but dazzling demonstration of the human genius for invention.' Since those days, the technology has changed and the range of topics has widened, but the same self-set standards and objectives – the vivid, clear, simple exposition of complicated subjects – remain unaltered; and in the Shell Film and Video Unit, descendant of Beddington's brain-child, the showcases glitter with dozens of silver and gold international awards.

Jack Beddington, Shell advertising manager 1932–9...

...and in caricature

Even the most highly creative marketing drive could not succeed for long if it were not backed by products of authentic quality. In the 1930s, Shell's efforts to survive the world-wide flood of oil continued, with expanding and fruitful programmes of research and development in fuels and, building on the strategic decision of 1928, in chemicals too. Of all the wide variety, perhaps the most striking was the work (conducted by Shell companies in the USA) on anti-knock additives.

An effective additive was already well known. Shortly after the Great War, two scientists working for the American corporation General Motors discovered that putting tetra-ethyl lead into petrol made engines run much more smoothly, and in 1924 Standard Oil of New Jersey began marketing the first leaded petrol. But the lead had drawbacks: only a tiny amount was needed, and too much could damage the engines it was supposed to help. Much more importantly, the lead was toxic. Fifty or sixty years later, this became an important argument in the growing

John Grierson, father of the British documentary film industry, advised Shell to set up its film unit

environmental desire to replace leaded fuels with unleaded, and even in the 1920s it was enough to make many people view Standard's leaded petrol with concern. In the factory where it was made, several employees contracted lead poisoning, and four died; and in 1925 it was withdrawn for a year.

Shell was reluctant to use tetra-ethyl lead, partly because of its toxicity, and partly because its use would involve paying patent royalties to a rival. Different approaches were tried, with some success, and different premium fuels were marketed to face the challenge of leaded fuel; but it gradually became apparent that the General Motors scientists had been quite extraordinarily lucky. They could have chosen any of thousands of different additives, yet on almost their first attempt, they had identified the best available one. In 1931 the commercial imperative obliged Shell to begin selling leaded fuel, but research continued, and shortly afterwards the Shell Chemical Corporation of the USA made an important discovery: the synthesis of the hydrocarbon called iso-octane.

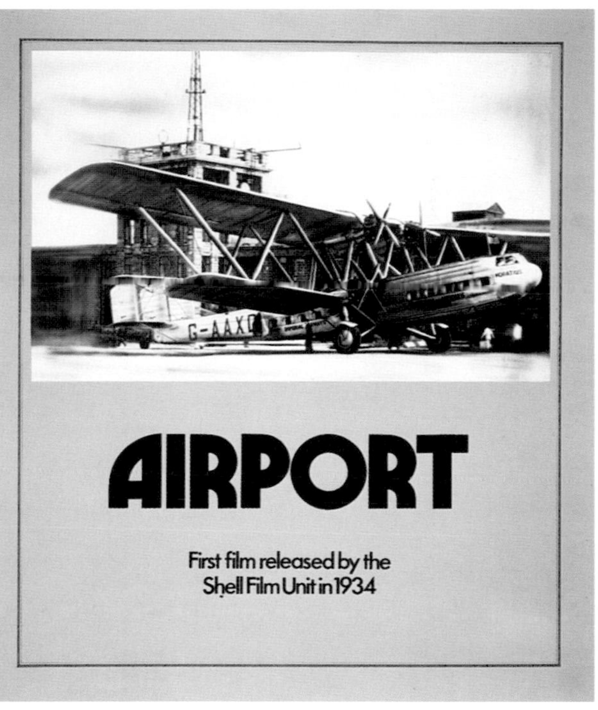

AIRPORT
First film released by the Shell Film Unit in 1934

Although this was not, unfortunately, a replacement for lead additives, it was a marked step towards a permanent solution of the problem of knock. Since the early 1920s, iso-octane had given its name to the octane rating of fuel, expressed as a percentage of anti-knock capability. Iso-octane rated 100 – the perfect anti-knock additive – but it had always been very expensive to produce: typically, in the 1920s, \$20 a gallon. The discovery that it could be synthesized brought two implications. First, its production cost could drop dramatically; second (although because of the volumes that would be required it could not possibly replace petrol as a fuel for ordinary car engines), it could – and did – become a vital part of the fuel for another more specialized form of internal combustion engine: the aero-engine. Early in 1934, the US Army Air Corps ordered a thousand gallons of synthesized iso-octane from Shell for aviation spirit. The delivery was made in April, and in the balance of the year a further 18,750 gallons were sold. The average price obtained was 71 cents a gallon. If that sounds extremely cheap compared to the prices of just a few years earlier, it was; yet the price of production was only about 35 cents a gallon. To Shell, twenty thousand gallons or so was a small business, but at that level of profit it was a very good business; and just a few years later, it became

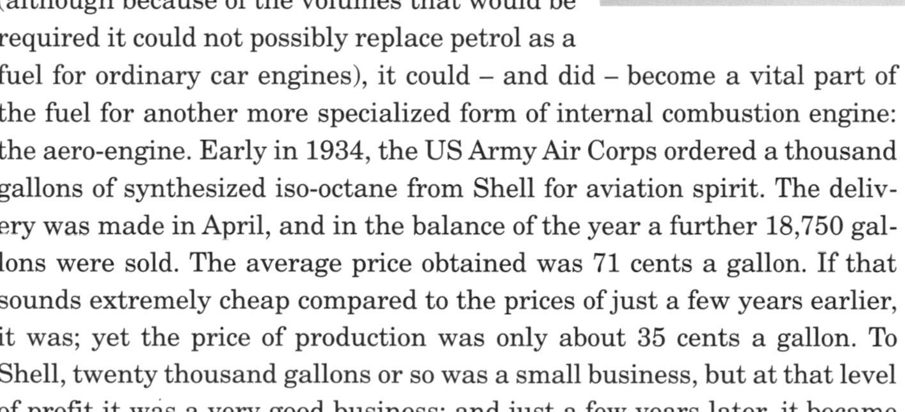

even more so – not only for Shell, but for all the Allied air forces, as once again the world was plunged into the chaos of global war.

Before those cataclysmic events, Shell Transport and Trading initiated the international carriage of yet another fuel: liquefied petroleum gas, or LPG. This is not to be confused with LNG, or liquefied natural gas. LPG refers to several different gases (including butane and propane) which are extracted from crude. In contrast, the first important distinction of LNG, which may be found in association with crude oil or by itself, is that it is usually composed of 80–95% methane. The special values of methane as a fuel (it has a high calorific value, is non-toxic and virtually pollution-free)

US Army Aircraft

were already well known, and in this period an interesting incidental event took place in Venezuela. Finding large quantities of methane in the Maracaibo basin, Shell faced the practical problem of what to do with it. At the time, methane could only be transported by pipeline. The gas could therefore either be flared off as waste, or supplied to local consumers. Choosing to do the latter, Shell gave the town of Maracaibo a natural gas system of heating and lighting, and supplied the fuel free. If this was a small gesture for a corporation of such size, it was nonetheless a practical demonstration of its best characteristics. As well as being a generous act towards the local population, it was also an early indication of Shell's dislike of the wasteful use of natural resources.

The reason for the contemporary difficulty of transporting methane over long distances was the second important distinction between LPG and LNG – namely, the way the gases are turned into liquid form. In its normal state, any gas obviously occupies an extremely large volume; yet if it can be liquefied and kept liquid, then, taking up much less space, it can be shipped in a suitable vessel. LPG becomes liquid either by being compressed or slightly chilled: butane liquefies at only -2^{0}C., and propane at -42^0. However, to liquefy methane, it must be chilled to -165^{0}C.; and whether with LPG or LNG, the conditions must be maintained, or else the liquid will boil and revert to vapour. If one considers the implications for a voyage of some hundreds, perhaps thousands, of miles, the vast technological difference is clear. LPG can be shipped in fairly simple pressure tanks; LNG must be pressurized and deeply refrigerated as well.

Decades would pass before dedicated LNG tankers became a regular feature of ocean traffic, but carrying LPG across the seas was the fulfilment of Kessler's foresight. In 1927, wondering what could be done with the waste gases from petroleum refineries in remote places, he had said, 'The only thing we can do is to make something we can ship.' To do so required a new application of technology; but since 1892, when *Murex* inaugurated the bulk transport of kerosene, Shell ships had been at the forefront of tanker technology. So in 1931, it was wholly appropriate that Shell Transport and Trading built and launched the 4,700dwt *Agnita*, the world's first purpose-built LPG carrier. She sailed as a member of the Anglo-Saxon fleet, and was a remarkable sight for her time, with her domed pressure tanks protruding above the deck; but as the world's LPG trade grew in volume and value, she and other later, similar ships became increasingly familiar in ports around the world.

The year 1934 was sensed, even then, as a time of transition for Shell Transport. The world economic depression appeared to be easing; global overproduction of oil was slackening; and the pain they had produced – the laying off of staff, the economizing in every sphere – was lessening. Profits had increased by more than a million pounds over the previous year, from £2.99 million to £4.19 million; for the first time in three years the dividend rose from 7.5% of par to 12.5%, and with cautious optimism Kessler remarked: 'Even the most serious crisis must come to an end, and I think we are approaching the end of the present one.'

By exerting tight economy, exploiting brilliant marketing and expanding its product range, Shell Transport and Trading appeared to have survived the flood. When Walter Samuel spoke of 'the recent most difficult years' and the severe tests they had imposed, shareholders responded warmly, praising 'the good work that you have done and are continuing to do in regaining the prosperity of pre-depression years'. But a slow return to prosperity was not the only visible transition. In 1934 the board of directors lost a highly regarded member when, on 23 October, aged 79, Sam Samuel died at his home in Hay Hill.

Walter, speaking both as his chairman and his nephew, described Sam as 'not only a loyal and hard-working colleague but also a most loveable and benevolent man. His kindnesses were as many as they were unostentatious.' This was not platitudinous; Sam had no children to

Agnita, *built 1931, was the world's first purpose-built LPG carrier*

remember him, yet even sixty years after his death, people who had known him in their childhood still spoke affectionately of him.

Henry Neville Benjamin died just a few months after Sam. Compared to the Samuel brothers, Benjamin was a background figure in Shell Transport: he had brought capital into it in its earliest days, but he had never been a full partner with them in the business. However, he played a part in early negotiations – most notably the alliance with Royal Dutch. He proved so obdurate a bargainer that Henri Deterding described him as 'worse than the late Shylock!!!' (Written to the Royal Dutch chairman, who was himself Jewish.) Benjamin was the last of Shell Transport's original directors. He was also, as it happened, Marcus Samuel's brother-in-law. His death and Sam's marked a new stage in Shell Transport's history. In thirty-seven years they had seen it grow from a small family-based concern into something which was so large, it was unrecognizably different – and yet which still maintained its original values. Back in 1899, when Shell Transport was independent, Marcus had said firmly, 'it is not part of our policy to inflict either loss or annoyance upon friendly competitors' – a view now echoed equally firmly by Henri Deterding. 'Always', he wrote in 1934, 'our Royal Dutch-Shell policy has been to create goodwill...To crush a rival is to make an enemy.'

To fill the vacancies on the board and continue Shell Transport's established business traditions, two new directors were appointed:

Andrew Agnew and Frederick Godber. Agnew was former chairman of Venezuelan Oil Concessions, part-owned by Shell; Godber was former president of Shell's wholly-owned American company Roxana. Both men brought valuable new international experience to the board as it moved into a new era; and both, as the years went by, would earn distinction for themselves and the company for which they worked.

Samuel Samuel

WORKS.
UTOMOBILE
NGINEERS.
EPHONE-CLEVEDON.81.
VULCANIZING.
CARS & VANS
FOR HIRE.
CHAR-A-BANC

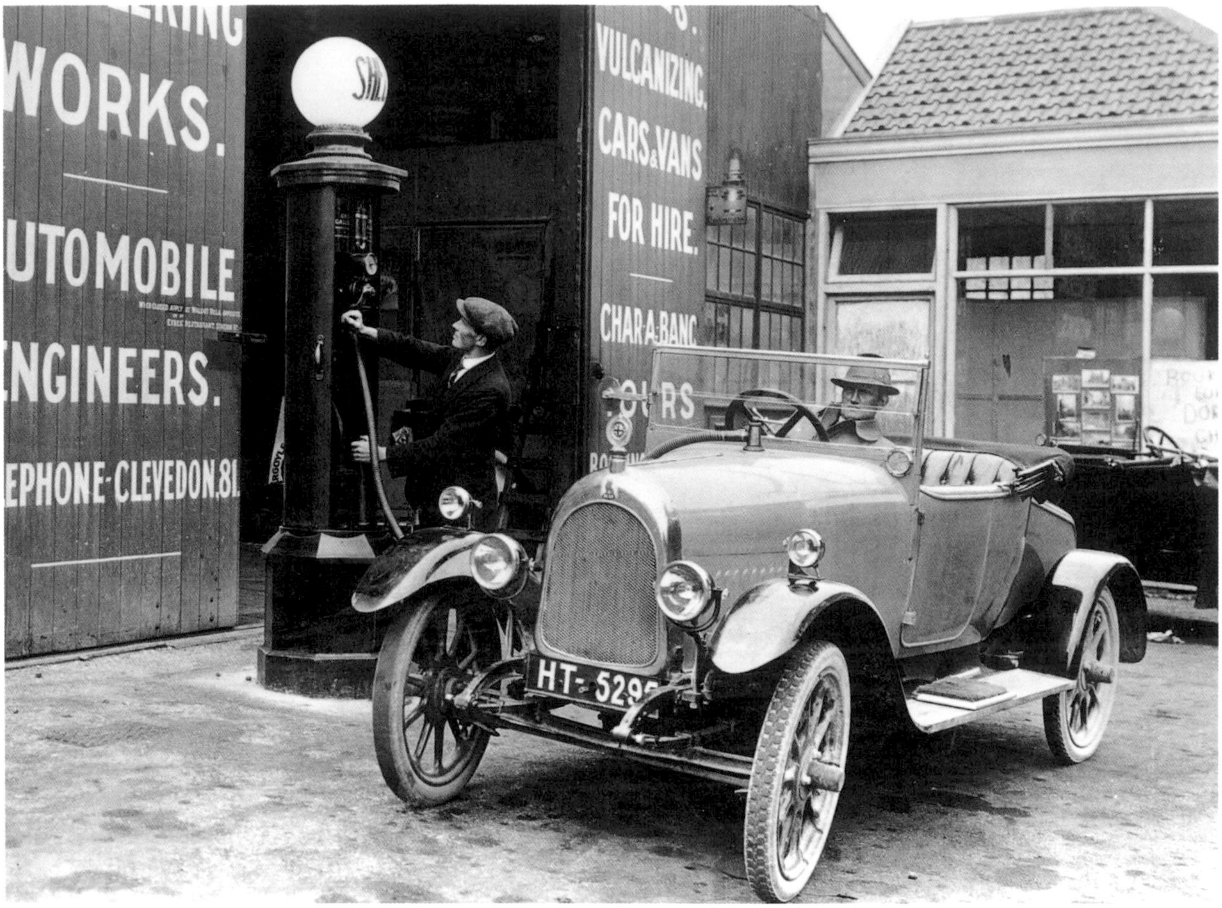

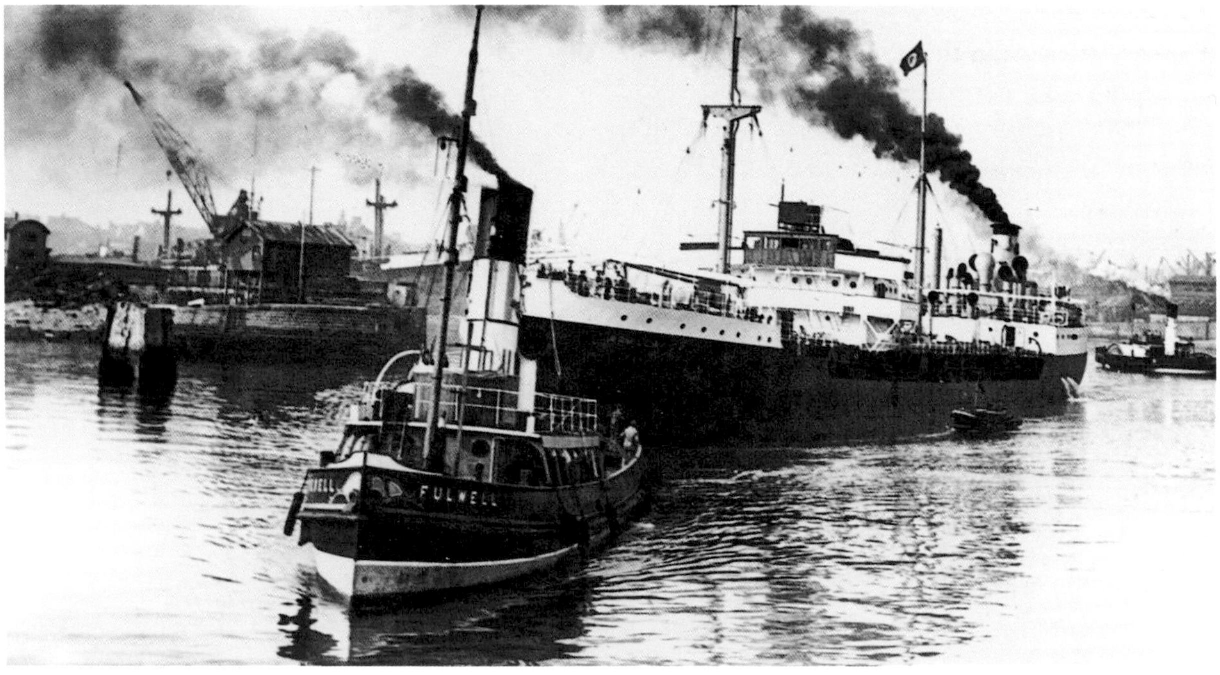

FULWELL

CHAPTER NINE

On the Roller-Coaster: 1935–1939

By 1935 Shell Transport and Trading had been, in its alliance with Royal Dutch, a significant member of the world community for over a quarter of a century; but it had never been able to roll along comfortably for any significant length of time. Every corner turned had brought another into view. In the middle 1930s Shell hoped for a return to prosperity and some years of tranquillity, and measured by profits and dividends, prosperity did return; yet rather than tranquillity, the latter half of the decade included some of the more trying experiences in its history. Of these, one of the most important was, as we shall see, the increasingly challenging attitude of Mexico's government; and that was set against an ever more ominous international background, as unfolding political events brought the rising probability of a renewed war in Europe, and perhaps beyond. During the later 1930s, memoranda by Shell Transport directors were just as likely to be lodged in the archives of the Foreign Office as in those of Shell Transport itself.

Germany had been one of Shell's markets both for kerosene and petrol since 1902, and in 1932 also became one of its producers of crude oil, when Shell and Standard Oil of New Jersey jointly began to drill in a number of small fields in the north-west. Italy then had no known indigenous oil in commercially viable quantities, and Japan only a little, but both were valuable and well-established Shell markets – indeed, trade in and banking interests with Japan had been one of Shell Transport's foundation stones. However, on approaching the middle 1930s, each country displayed an accelerating momentum towards extreme right-wing militarism.

The rise of the Nazis in Germany, the Fascists in Italy and the political power of the Japanese army are too well known to need repetition here. At first, the European aspects of these deeply disturbing events affected Shell Transport more personally than commercially; in particular, Robert Waley Cohen made strenuous efforts to expedite the evacuation of as many Jews as possible from Germany to Britain. Yet 'in the distressed and distressing state of world affairs', as Walter Samuel said, Japan's

Opposite top:
1930s service station

Opposite bottom:
Helix *refitted as* Armilla, *1938*

The Foreign Office, London

occupation of Manchuria and subsequent war with China quickly caused business problems. Early in 1936 the Japanese government enacted a new Petroleum Law, designed, said Walter, 'with a view to obtaining from the oil-trading community what amounts to a very substantial contribution towards the defence measures which are considered advisable by the Government.' For Shell Transport, which acted as a purely commercial, non-political enterprise, this was most unwelcome. Negotiations were under way with Japan; but while emphasizing that 'it must in the long run be to the advantage of the public whom it is our duty to serve that the company should keep itself entirely clear of political issues', Walter added that doing so was very difficult – 'Unfortunately petroleum has become an article in which governments all over the world have become specially interested.'

This was as true of the left wing as of the right. Before the outbreak of World War II, Germany, Italy and Japan were ominous and unpleasant as customers, but still just possible to deal with in business. However, in the same years Mexico shifted politically very much to the left, and made itself an impossible business partner.

A letter concerning Mexico, published in *The Times* in 1932, contained a good indication of Shell Transport's method of doing business. 'The key to real success', said the writer, 'must be the honour and integrity of the men at the top. I hope Sir Robert [Waley Cohen] will forgive me if I quote from a cable which he once sent to me...when I was representing the Shell

Company in Mexico. We were negotiating for the purchase of certain oil properties, and the message ran as follows: "If you have passed your word on our behalf, whether it is in writing or not, you must stick to it, and we will support you." '

Charged with negotiating on the company's behalf, the man on the spot was expected to deal fairly and honestly, and if as a company representative he gave his word of honour to a fair and honest deal, he could expect company support, no matter what difficulties arose. Yet few could have foretold the difficulties in Mexico.

Japanese filling station

It had never been a very easy country to work in. As we have seen, its instability after the 1913 revolution had led to its temporary abandonment by Shell and the Shell-managed Mexican Eagle as well as other oil companies, and the invasion by salt water of the 'Golden Lane' oil wells in 1922 had caused high concern. Overall, though, it had been worth the effort. But it was still politically volatile, which was not an encouragement to further costly exploration, and compared to nearby Venezuela, its oil was expensive to produce; so during the period of global glut, less and less was produced. From a peak of over 499,000 barrels a day in 1922, Mexican production plunged in ten years to less than 90,000, and a vicious circle began as the government, seeking to maintain its revenues, raised oil taxes. Then in 1934 a new president was elected: the former War Minister, left-wing General Lázaro Cárdenas, intent on challenging the legality of all foreign-held oil concessions. Within months, Mexican Eagle reported that 'politically the country is quite Red' – an accurate statement, and one which immediately rang shrill alarm bells in the mind of Sir Henri Deterding, still smarting at the memory of the Soviets' expropriation of the oil industry in 1918. His reaction took Eagle's resident manager aback: Deterding, it appeared, could not imagine Mexico 'as anything but a Colonial Government to which you simply dictated orders', and when the manager tried to explain the reality of things, then far from gaining support, he was accused of being 'half a Bolshevik' himself. Deterding would not attempt to understand the mood of Mexico. Had he done so, he could have learned that although Shell produced about 65% of Mexico's oil, the nationalism sweeping the country was not directed at Shell in particular but at foreign-owned industries in general, and especially at American-owned ones. A more conciliatory man might have used this to the advantage of his business. Instead, he sided with the American oil companies in Mexico, and Shell was sent down the road of maximum resistance. Insulted and exasperated, Eagle's resident manager wrote some prophetic lines:

> The sooner that these big international companies learn that in the world of today, if they want the oil they have got to pay the price demanded,however unreasonable, the better it will be for them and their shareholders.

Lázaro Cárdenas (1895–1970), Mexico's nineteenth and youngest president, taking the oath of office, 12 February 1934

An epidemic of labour strikes hit the Mexican oil fields. With their government's support, workers called for more pay, shorter hours, longer holidays and large early pensions. The government itself called for the replacement of all foreign technical staff by Mexican ones within two years. As the situation deteriorated, the wives and children of Eagle employees were evacuated. Eventually, in 1937, the Mexican Labour Board ordered the oil companies collectively to increase their workers' wages. The total sum involved was 26 million pesos; the companies offered 22 million – nearly 85% of the demand. But President Cárdenas personally added another condition: management and administrative functions alike must be transferred to Mexicans. This was refused; and on 18 March 1938, Cárdenas nationalized the entire industry.

For Shell Transport and Trading, the expropriation of its Mexican assets was the single most shocking business event of the period, a nightmare repetition, almost to the day, of the expropriation in Russia twenty years earlier. A small part of the unpleasantness was the circulation of rumours (which Walter Samuel vigorously denied) that the oil companies together had funded and fomented attempts at revolution against Cárdenas. But bad as the loss was, as things turned out it could have been still worse, for although despite legal appeal the expropriation was not revoked, Mexico (unlike Soviet Russia) did eventually pay some compensation: in 1943, $30 million to the dispossessed American companies, and in 1947 a disproportionately high $130 million to Shell and Mexican Eagle.

The reason both for the difference and the delay was that whereas the American companies had little support from their own government and had to settle for what they could get, Shell Transport was strongly supported by the British government; and the reason for *that* lay in Europe. Germany's march into the Czechoslovakian Sudetenland on 1 October 1938 brought a sense of horror throughout Britain. With the scent of war in the air, Shell and the British government shared more secret fears. What would happen if other Latin American oil producers followed the Mexican example? And what would happen if Germany chose to annex the Netherlands?

Both scenarios were all too plausible, and 'worst case' discussions were apocalyptic in tone, as shown by a letter copied to Godber. If a German annexation of the Netherlands occurred, Royal Dutch 'would go forthwith

Curaçao in the Netherlands West Indies, c.1930, soon to become the site of the largest refinery in the region – and, through World War II, the legal headquarters of Royal Dutch

under German domination. As was evidenced in the annexation of Austria by Hitler, all things of value in German-acquired territory become German at once: the liabilities of the annexed country are the only things not taken over.' Thus, all liabilities jointly held by Shell Transport and Royal Dutch would fall upon Shell Transport,

> which company would be absolutely at the mercy of the Germans...The question of legality will never arise, for the Führer has stated his intention of abandoning Roman Law in the Reich and formulating a new law entirely disassociated from international law as it is observed today. It will be useless appealing to him in the name of common justice; he only admits the rights of Germans...

Various precautions were considered, including the idea of moving everything to Canada; but a decision on that could be deferred. (Actually, when the Netherlands were invaded, Royal Dutch shifted its headquarters and its entire legal basis to Curaçao.) Just before the war, the most immediately urgent consideration was the need to avoid any other Latin American expropriation. The world held only eight main oil-producing regions: the United States, the East Indies, Soviet Russia, Romania, Iraq, Iran, Mexico and Venezuela. Added to the last was a smaller level of production from Peru, Colombia and Argentina. In a European war, US neutrality legislation would cut off supplies, exports from Russia would almost certainly cease, and geography would hinder if not entirely prevent supplies from Iraq, Iran, Romania and the East Indies. Latin America, which already provided nearly 40% of Britain's oil, might become its only remaining source.

High national policy thus coincided with Shell's commercial policy, which was to encourage Latin American producers to maintain the status quo and avoid the road taken by Mexico. Frederick Godber travelled to Venezuela in the summer of 1938 to try and establish its national mood. On his return to Britain, while advising the Foreign Office that the Venezuelan government did not appear to view expropriation as a practical policy, he also warned that:

> the conviction that the Nation is not getting an adequate participation in the exploitation of its natural wealth will result in continually increasing demands and burdensome legislation...At the moment it is not thought likely that any such drastic action will be attempted. But the degree of success in other countries in a policy of confiscation and repudiation must be expected to have a definite influence on the future policy of the Venezuelan government in this connection.

As with the Soviet expropriation twenty years earlier, all the oil companies involved in Mexico regarded themselves as the victims of outright theft: at the stroke of a pen the millions they had invested were removed. The British government demanded that Shell's property be returned; the Mexican government in response cut off diplomatic relations; the companies collectively embargoed Mexican oil; and Pemex, left-wing Mexico's new national petroleum company, found that ironically it had very few customers apart from the nations of the far right. But if the lesson on one side appeared to be that the producing nations needed the established oil majors, the reverse was also true.

Shell understood this quickly, and reacted accordingly when, as feared, Venezuelan sabres were rattled. Following the death in 1935 of the dictator Gomez, there had already been a six-week oil strike there, and in the autumn of 1938 the new president, Lopez Contreras, threatened nationalization if higher royalties and taxes were not forthcoming. Unlike Cárdenas in Mexico, however, Contreras had no ideological leaning towards dispossession, and offered in return new agreements to last for forty years. The reactions of Shell and the resident American companies were noticeably different: while the latter talked truculently of the 'sanctity of contracts', Shell's management recognized the need for compromise and the value of stability, even if it came at a high price, so a far-reaching agreement was soon attained, involving more money for the Venezuelan government and better pay and improved conditions for the workers – although these were actually already good by the standards then prevailing in most of Venezuela.

Because of its long-standing preference for co-operation rather than confrontation, Shell's recognition of the industry's new political realities was comparatively readily achieved, and by its agreement with the Venezuelan government, a large proportion of its own production and Britain's needs were ensured. Furthermore, for Shell as a group that was not the only important result of the Mexican debacle. At least as valuable, if not more so, was an increased emphasis on what one might call the localization of expertise: that is, the recruiting and training of members of the local population for key jobs – drillers, geologists, engineers and managers – thereby strengthening ties with the local community and earning from it much enhanced loyalty and commitment. Shell was already tending strongly towards this policy, which would have emerged eventually, even without the Mexican experience; but to derive such a positive outcome from so apparently disastrous a situation was a real sign of maturity.

Amid these shocks and throes there were some brighter aspects – not many, but that was all the more reason to mark them. By April 1937 all the laid-up tonnage in the tanker fleet had resumed work. During the year, the fleet not only worked to full capacity but had that capacity increased by about 10% to 2.2 million deadweight tons. Further very large expansion was anticipated for 1938–9: 282,000 dwt more were to be added in 1938 and another 264,000 in 1939, almost all of which was built in Great Britain, to the great advantage of the British ship-building industry; and, at the beginning of 1938, a new benefit for staff was introduced – a pension fund, to run in conjunction with the existing Provident Fund.

Both the fleet's increase and the new pension arrangements became possible because of the policy followed during the Depression years, namely, reducing all indebtedness. Shell Transport had had a very bad fright in the early 1930s, uncomfortably reminiscent of the situation in 1906 when its very survival demanded alliance with Royal Dutch. Any company might run into financial difficulty once; to do so twice would show only that nothing had been learned from the past. The Depression created a correspondingly cautious and conservative frame of mind in the board of management. If it had less vim and sparkle than in Marcus Samuel's day, nevertheless it was right for the time, and it worked: Shell Transport's

By the mid-1930s Shell could ensure supplies to customers all round the world

From refineries such as the great one at Petit-Couronne in France (seen here in 1935)...

... fuels, lubricants and other products might be moved by river barges like the Alexandre...

...by tankers like Auris, *launched in Trieste in 1935 (and sunk by enemy action in the Atlantic only six years later)...*

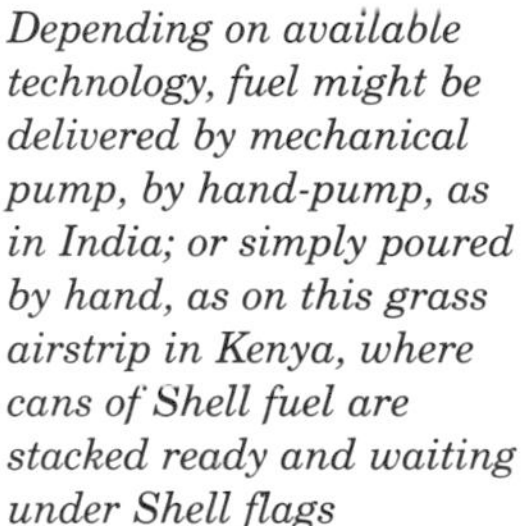

...or by road tankers, whether in the Philippines, South Africa or Malaya

Depending on available technology, fuel might be delivered by mechanical pump, by hand-pump, as in India; or simply poured by hand, as on this grass airstrip in Kenya, where cans of Shell fuel are stacked ready and waiting under Shell flags

assets in 1938 totalled £53 million, against which it listed debts of a mere £230,363. Shareholders approved: they had been through the Depression too, had seen its dire effects on neighbours and nation, and from it had learned something which Marcus in his later years had often reiterated – that it was better to be sound and solid than to gamble on quick high returns and to risk losing all. This policy, a wariness of making large borrowings, therefore continued.

Shell's touring guide to Tasmania and New South Wales

The year 1938 brought another pleasure – reflected but real for shareholders, and for its recipient no doubt quite considerable. Andrew Agnew, who had joined the board along with Frederick Godber in 1935, was knighted in the Birthday Honours for his services to the petroleum industry. In the coming few years he would earn the recognition all over again, because his services, and those of Shell Transport as a whole, were soon to be called upon to an unprecedented degree.

Ships, funds and public honours were not the only positive lights in that otherwise dismal decade. Global production of oil was at last easing and demand was rising. New exploration was undertaken, in the then British colony of Nigeria, and in Ecuador and Colombia – locations which if successful could make good the lost Mexican production. In the United States, Shell Union returned to profitability in 1935, and Roxana's transport costs were greatly reduced by the construction of a 450-mile pipeline from its Wood River refinery in Illinois to its marketing company in Ohio; and on the chemicals front, three intriguing and rather complicated advances, called co-polymerization, alkylation and isomerization, were made in the production of 100-octane aviation fuel. All brought about marked cost reductions and production rises, and stimulated aero-engine research. Even with existing engines designed for 75-octane fuel, 100-octane gave 15–30% more power, and when in 1936 the US Army learned that an experimental engine designed for 100-octane also brought fuel savings of at least 15%, the order was given that with effect from 1 January 1938 all its aero-engines should be so designed. It was a most timely decision.

THE IDEAL PAIR

Shell had been active in aviation matters since almost the earliest days of heavier-than-air flight. Its fuel had powered Blériot across the Channel in 1909, and in 1919, when Alcock and Brown flew the Atlantic with the super aviation fuel formulated by Harry Ricardo, Shell Aviation Services was established. Over the next ten years, almost every pioneering flight anywhere in the world was achieved on Shell fuel and lubricants. The habit of being first in the air continued in the 1930s, including (in

December 1931) the first air mail delivery from Australia to Great Britain. Piloted by Air Commodore Kingsford Smith, the *Southern Sun* carried out 'a memorable flight', as Shell's magazine *The Pipeline* recorded. Bringing Christmas cards and packages half-way round the world, the journey was 'accomplished entirely on Shell products.' By 1935 Walter Samuel was able to tell shareholders that aircraft could be supplied by Shell at 'all important aerodromes throughout the world'. He could add too that 'Happily we foresaw some years ago that aviation would become a normal means of transport', and that in contrast to the pioneering days when special supply dumps had to be made for any given flight, Shell's network was so well established that 'it is now possible for any aviator to leave Europe unannounced and fly to the East or to the Cape with the full knowledge that he will be able to obtain supplies of Shell petrol and oil at all the landing grounds *en route*.'

Walter did not publicly foretell that aviation would also become a normal means of warfare, but at least one Shell man was convinced it would be, and made it part of his job to convince others. When James H. Doolittle joined Shell Petroleum in St Louis as manager of its aviation department on 15 January 1930, he was a lieutenant in the US Army Air Force Reserve. When he retired from Shell Oil as a director in 1959, he was also a lieutenant-general and a national hero: for in the dark days of April 1942, with America still reeling from the shock of Japan's surprise attack on Pearl Harbor, it was he who led a flight of sixteen B-25s in a largely symbolic but morale-boosting attack on Tokyo.

It was also he who in 1934 succeeded in selling Shell's first batch of 100-octane to the US Army, and who persuaded his own senior management to invest heavily in a business which did not as yet properly exist. Shell had on several occasions spent very large amounts of money on developing aviation fuel, without looking for a short-term return, because in any branch of the aviation business technical prestige was vital; and on this occasion two million dollars were spent in little more than a year, on what was realistically a risky and speculative venture. The risk was emphasized early on: it took two years for the Army General Staff to be persuaded that 100-octane was the aviation fuel of the future, during which time Shell built iso-octane refineries capable of producing 14.5 million gallons of 100-octane a year. For a while it looked unnervingly as though this was over-catering, for Doolittle calculated that the US Army and Navy combined would only need about ten million gallons a year. But when war came to the United States, they required rather more – in fact, they needed the staggering figure of *twenty million gallons every single day*. No single company in the world could possibly meet this gargantuan demand, but with the necessary information made freely available, the industry as a whole was able to, with Shell providing on

*James H. Doolittle,
the Tokyo air-raid leader,
in 1969*

average about 14% of the total; and the fact that it could was a tribute both to Doolittle's powers of persuasion and his senior managers' faith in an untried business.

The 1930s had proved a difficult and unpredictable decade for Shell Transport and Trading – the Depression, the successful move into chemicals, the increasing politicization of oil as governments of both extremes came to power. Yet even if none of that had occurred, it would still have been a climactic time, for on 17 November 1936 Sir Henri Deterding retired. He was then a few months over 70 years old. His forty years in the oil business included twenty-nine as an executive director of Shell Transport and Trading (in modern terminology, a Group Managing Director) and thirty-six as General Manager (that is, president) of Royal

Dutch. He had been a decisive, governing influence in Shell Transport, and in almost complete charge of Royal Dutch, for more than half his life: he had become a dominant force throughout the world-wide industry, earning the respect of almost everyone who knew him, and often their affection too. Naturally, therefore, his departure engendered a considerable sense of loss; and yet it was not entirely unwelcome, for as he had grown older he had become rather an embarrassment to his colleagues.

Given all his achievements, this is an unhappy story, and one which has caused lasting distress within Shell Transport and Royal Dutch; but it is as much a part of the history as the more glorious days, and enough time has passed for it to be seen in some perspective.

Briefly, Deterding had become increasingly right-wing, bordering, some said, on the megalomaniac. His memoirs, published in 1934, were a masterpiece of vanity and egocentricity, reading as the self-portrait of an autocrat. For example, there was his talk with Mussolini – 'a man who, regard him as you may, has shown a driving force almost unparalleled in running a country'. Deterding decided that this conversation:

> proved that there were several points on which we saw eye to eye. We both agreed that the coping-stone of Education is a sense of discipline and a respect for prestige, lacking which no youth can be considered to have been properly educated at all... To people unacquainted with the Italian character his manner in public may seem at times to be a trifle theatrical, but what chiefly interested me at our meeting was that he seemed so direct. One felt that, if faced with a difficulty, he would get out his sledge-hammer and strike straight at its root.

So too would the ageing Sir Henri. When he wrote that, he was 68. Many people, as they grow older and see the world changing around them, become more conservative, with a hankering for 'the good old days' and a growing belief that things are not what they were. With Sir Henri the process was becoming somewhat marked. In the same text, he wrote this memorable sentence:

> If I were dictator of the world – and please, Mr. Printer, set this in larger type – I WOULD SHOOT ALL IDLERS AT SIGHT.

But in a world where millions of working men and women were idle through no fault or desire of their own, Deterding's colleagues (particularly in The Hague) were very sensitive to the public display of such sentiments, and still more so to his open admiration of what he perceived as the firm government which had recently been elected in Germany.

Back in 1914, just before the outbreak of the Great War, Britain's Admiral Fisher had written to Winston Churchill: 'I have just received a most patriotic letter from Deterding to say he means you shan't want for oil or tankers in case of war – *Good Old Deterding!* How these Dutchmen do hate the Germans!' The roots of this antipathy may be summarized as the reaction of an economically strong yet militarily weak nation towards a neighbour whose qualities were the exact opposite but whose language – the cardinal identifier of nationhood – was similar. The strength of feeling among 'these Dutchmen' was such that in 1934 (the same year as Deterding's memoirs appeared) the Netherlands Ministry of Education decreed that English translations of documents originating in Holland should henceforth use only the adjective 'Netherlands'; the use of 'Dutch' was banned, since it could too easily be confused with 'Deutsch'.

Once upon a time, seeking to tell Deterding how much he admired his financial genius, Marcus Samuel said to him: 'I am going to pay you the highest possible compliment. You ought to have been born a Jew.' There is no record of how Deterding received the intended accolade, nor is there any indication that he was particularly pro- or anti-Semitic. But he was violently anti-Communist, and after his retirement he lived neither in his native country, Holland, nor his adopted country, Britain. Instead he bought an estate in Germany, settled there with his third wife, and at his death in February 1939 was buried there too.

The new Lady Deterding was German. In a striking lack of imagination on Sir Henri's part, she was also his former secretary; and because the Nazi regime was visibly restoring order to her country's chaotic economy, she was very much in favour of it. So was Sir Henri, who saw the disciplined economic aspects of Nazism as the world's most powerful weapon against Communism. The Nazis, eager even after his death to exploit the publicly-avowed support of this world-famous individual, virtually hijacked his funeral: Field Marshal Goering, chief of the German air force, sent a wreath; so did Hitler himself; and, even Germanizing his name, the functionary who represented them said as he laid the wreaths: 'In the name and on the instructions of the Führer, I greet thee, Heinrich Deterding, the great friend of the Germans.'

To his former colleagues both in Shell Transport and Royal Dutch, these events were intensely painful and hard to come to terms with. Recalling his irrational and damaging price war in 1927 against buyers of Soviet oil, and his high-handed 'colonial' treatment of the left-wing Mexican government in 1934, some wondered privately if he might have been going mad. Probably he had not; rather, traits that he had always possessed – simplicity of outlook, clarity of goals, strength of character and forcefulness of speech – had become accentuated by old age. By then, their expression was crude and humiliating. In his youth and middle age, though, the same traits had been priceless business assets. Using them, he had rescued Shell Transport from virtually certain extinction, and had built its fortunes, together with those of Royal Dutch, to an level which simply would not have been credible when he began; so both as a friend and an inspiring leader, his passing was genuinely mourned. At his memorial service in London, which took place at the same time as the Nazi-dominated funeral in Germany, Robert Waley Cohen – Jewish to the core – read the lesson. Walter Samuel had wanted to, but could not: he was attending the Palestine Conference, one of many contemporary efforts to win Arab approval for the establishment of a Jewish national homeland. Tainted by his late and brief association with the Nazis, Deterding left the saddest possible memory for his former colleagues, whether Jewish or Gentile, Dutch or British; but the man they all liked and admired had, in truth, died several years before.

Sir Henri Deterding, KBE, 1866–1939

The Bystander, July 19, 1939

GERMANY

No country is more easily seen. From Schleswig to Carinthia, Silesia to the Rhine, great arterial roads supplement the railways.

Above all, a German holiday is good, simple fun. It is jolly to sit at a lakeside restaurant, listen to the band and watch tow-headed babies splashing in the blue water. It is amusing to sample Rhine wine in the vineyard where it grew. It is fun to paddle down the Isar in the deepening twilight and hear the Wandervögel singing. To learn the Tyrolean part songs for yourself—perhaps a step or two of the folk dances—to wear lederhosen or a Dirndl dress and discuss life with mine host of the "Hirsch" over a stein of Schwartzbrau, to enjoy the good humour of the throng which crowds Salzburg, Bayreuth, Munich and many other cities during the festival season.

There are the little things that linger in the mind—the goodness of the "Kuchen," the girl you danced with in the Platz, the fragrance of a moonlit night in the Black Forest, the stillness of a Tyrolean lake, black glass, mirroring the stars. Of such is a German holiday.

60% REDUCTION IN RAIL FARES
40% to 50% SAVING BY PURCHASE OF TRAVEL MARKS

Full information and literature from:—

GERMAN RAILWAYS INFORMATION BUREAU

19 REGENT STREET, S.W.1 and 140 WEST GEORGE STREET, GLASGOW, C.2, OR ANY TOURIST AGENCY

CHAPTER TEN

Fuelling Democracy: 1939–1945

There was no more modern way of holidaying at reasonable cost than to go on a motoring tour, and in the summer of 1937 one Shell employee (a Dutchman named de Maat) did just that, taking his wife on an eighteen-day, 1,800-mile tour of Holland, Belgium, France, Italy, Switzerland and Germany. Naturally he used Shell oil and petrol throughout. Recording that he needed one gallon of oil and 98 of petrol for his journey (which meant, incidentally, that his 'trusty Ford' gave barely eighteen miles to the gallon), he advised colleagues who might wish to do the same that while choice of hotel would be a factor, a comparable holiday – with food, accommodation, transport and all – would probably cost £25.

In Germany, on the last leg of his tour, de Maat was impressed by 'the new *Reichsautobahnen,* the great new motor highways, which aroused our admiration for modern German enterprise.' Neither Britain nor Holland had any such roads; they were wide, straight and easy to drive on, and their building had brought much employment. Two years after de Maat's holiday, in July 1939, tourism in Germany was still being warmly encouraged, with the autobahns as a particular attraction. 'Seeing is believing', said a travel advertisement of the time. 'Come and see Germany – no country is more easily seen. From Schleswig to Carinthia, Silesia to the Rhine, great arterial roads supplement the railways...' Driving on similar ones in Stalinist Russia in August 1939, on his way to the signing of the Nazi-Soviet Non-Aggression Pact, Hitler's interpreter Paul Schmidt thought to himself, 'Dictators seem to delight in the magnificence of broad roads.' The Romans had done the same, yet not for vanity or the convenience of holidaymakers but rapid military communications; and as the German air force took to the sky on 1 September 1939, tanks, trucks and soldiers of the German army streamed down the 'great new motor highways' and stormed into Poland.

FOR HIGH PERFORMANCE
HAWKER HURRICANES
ROBERT BUHLER
LUBRICATION BY SHELL

Shell Transport entered World War II in uncompromising mood. Its chairman, Walter Samuel, had been decorated in World War I; his brother Gerald had been one of the 109 Shell men killed. If the survivors had faced only a repeat of 1914–18 it would have been quite bad enough, but rather than imperial Germany, the enemy was now Nazi Germany. In a company like Shell Transport, founded and in 1939 still to an important extent directed by people who were proud to be both British and Jewish, rumours and reports of Jewish persecution by the Nazis – often discounted as incredible, but far short of the forthcoming truth – gave added urgency and weight to decisions and actions. Today these are matters of history, and Deutsche Shell has long been a valued and respected member of the Group; but in 1939, terror remembered brought the determination to resist, to the utmost, terror renewed. It was perfectly obvious that oil would once more play a central role in conflict; so after the declaration of war, one of Shell Transport's first actions was to provide the British government and the Royal Air Force with all available data on its refineries and oil fields in Germany. Though easily seen as a correct decision, it was nevertheless depressing, because it was a clear invitation for the bombing of Shell installations and Shell personnel; yet the alternative was too awful to contemplate. The human result is uncertain. But as discovered after the war, the physical result was that, out of all the refineries and other oil establishments in Germany, those belonging to Shell were the most severely attacked and the most comprehensively destroyed.

At the outbreak of hostilities, the Shell Film Unit was similarly put at the complete disposal of the government, and over the following six years made 47 films (some for secret instruction and training, some for the boosting of public morale) for the Admiralty, the Ministry of Home Security and the Ministry of War Transport. Shell Transport's role in the campaign against Nazism was a matter of action, invention, decision and deed, of total involvement and total commitment throughout the period and in every part of the world where Shell oil was produced, transported, refined or needed. It was correspondingly complex, often with the simultaneous occurrence of vital events far apart, or the gradual development of processes and products no less vital. Describing these can at best be only partial reflections from a fragmented mirror; but so that the reflections should not be too confusing, let us look separately (while remembering that many took place simultaneously) at some of Shell Transport's crucial actions in the war years.

Great Britain was in many ways under-prepared for war in 1939, but at least one lesson of 1914–18 had been well learned: every possible effort must be made to ensure sufficient oil, wherever and whenever it was needed. Under Shell Transport's leadership, the creation of the Petroleum

Given over to the British government service shortly after the outbreak of war, productions from the Shell Film Unit were supervised by Edgar Anstey...

...and Arthur (later Sir Arthur) Elton

Board, or 'the Pool', symbolized the nature of the British wartime oil industry. Planned during 1938, it was activated at 'time zero', the last second of Britain's first day at war: midnight on the night of 3–4 September 1939. Following the plan, the Pool was ready to continue its work for as long as the war continued. VJ Day, the day of victory over Japan, arrived on 15 August 1945, and for the entire intervening six years, the Pool ran every part of Britain's network for the importation, storage and distribution of oil. At its core were the four largest distributing companies (Shell-Mex and BP, Anglo-American, National Benzole, and Trinidad Leaseholds) with another 94 smaller companies attached. All had been pre-war competitors, but at 'time zero' they suspended their competition for the duration of hostilities, and instead worked together as one. To American politicians, accustomed to anti-trust fences around every oil deal, this was a real eye-opener:

> The oil men of Great Britain have fashioned an effective 'monopoly' of oil for the purposes of prosecuting the war, but they also, they feel sure, have arranged to ensure the return of a truly free and competitive oil industry when the war is concluded. Perhaps the most striking feature of this 'monopolistic' arrangement and the plans for return to a free industry is the most informal way in which

the legal side has been handled, not only as between the oil men but with their government.

Shell Transport's director Sir Andrew Agnew was chairman of the Petroleum Board throughout its existence, and in that capacity sat on the Oil Control Board, a sub-committee of the War Cabinet. The organizational challenge facing him and his Pool colleagues was formidable, not least because in suspending competition the former marketing rivals effectively merged all their financial interests in the UK. Their joint assets and capital there totalled about £50 million, and as a later commentator said, this was 'the most dramatic, the quickest and probably the most important merger in history.'

Charged with total co-ordination of Britain's wartime petroleum supplies and led by Andrew Agnew of Shell Transport, the Petroleum Board was headquartered in Shell-Mex House. The ARP (Air-Raid Patrol) control room; the main telephone switchboard and the teleprinter room, which never closed, day or night, throughout the war

The directorates of the Pool's component companies collectively held an unrivalled store of experience and expertise, enabling them to plan a working administrative structure very quickly. Committees handled Overseas Supply, Tanker Tonnage and general Management; individual departments handled legal matters, insurance, inland distribution, chemicals, storage, pipelines and so forth; and – although this planning had to be done in secret and without staff training before 'time zero' – the Board companies' personnel soon got over the surprise of no longer being competitors, so the planned structure was able to function smoothly with very little modification.

It was also able to function more efficiently than in normal competitive life. For example, where before the war there had been several different grades of petrol, now there were just three, all sold without distinction of trade-mark as 'Pool' petrol from pumps and road-tankers painted a uniform drab olive green. Although those accustomed to choosing Shell furrowed their brows in dismay, marketing was made infinitely simpler.

So was storage: rather than three neighbouring depots each handling a variety of products, one might be given over exclusively to petrol, another to kerosene and the third to diesel. Similarly, incoming supplies could be directed to the most suitable destination, whether by pipeline, rail or road, with the Pool's 6,000 drivers. The Pool was not a rationing authority – it was up to the government to say who should receive what – but it did

Fire damage: the pump-house at Thames Haven (above left), a storage tank at Purfleet (centre), and Stanlow

have the responsibility of ensuring that all those entitled to a rationed supply actually received their entitlement. *Shell Magazine* took a wry view of this: 'An ABC of Wartime Conditions' included '*Bicycle*. The successor to the motor car in taking the place of the horse. ... *Petrol ration*. Enables you to run a cigarette lighter and drive your car to the nearest dump.' Private motorists' mobility for pleasure was certainly much restricted, but the Petroleum Board supplied everyone – agriculture, hospitals, the fire service, all three armed services and the whole of industry – and the judgement of an American colonel was nearer the mark: 'If the Petroleum Board says it's difficult, it's in the bag; if they say it's impossible, it'll take a little time.'

The Board's headquarters were located, fittingly enough, in Shell-Mex House. Its construction had been completed – so people thought – in 1933, with its official opening taking place on 25 January. But it was not long undisturbed: in 1938 the workmen were back, constructing air raid shelters in its basements, and the summer of 1939 the electrical engineers were there as well, putting in the Pool's central communications facilities. Duplicate telephone systems were installed in the shelters, with their lines re-routed: some ran to the Strand and others to the Embankment, so that if the building was bombed, one set or the other would probably survive. To further increase their chances of survival if external telephone exchanges were hit, the lines were also divided to go through not just one but four exchanges. Added to these duplicated telephone systems were duplicate terminals of the then marvellous new instrument, the teleprinter. Although superseded to a large extent in recent years by fax machines, teleprinters had one enormous strategic advantage: they required their own dedicated lines, separate from the telephones. They thereby offered enhanced security, as well as redoubling the communication system's possibility of survival. The nationwide teleprinter net focused on Shell-Mex House was even linked to those of the three armed

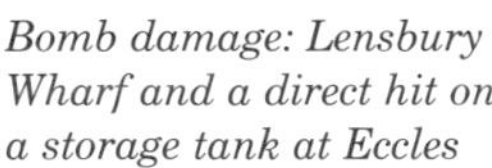
Bomb damage: Lensbury Wharf and a direct hit on a storage tank at Eccles

services; it could (and did) handle as many as five thousand messages a day, and it never closed. Nor did the telephone switchboard; but even in the height of the Blitz, when bomb damage meant that telephone lines could be out of action for twenty-hours at a stretch, the teleprinter system survived unscathed.

So, to a very large extent, did the building. During June 1940 the possibility of invasion seemed frighteningly real, and in an atmosphere of apprehension the Local Defence Volunteers (later named the Home Guard) were formed, with national membership eventually totalling two million. Shell-Mex House, the 'Headquarters of the Petroleum Industry' had its own Company – D Company – with 216 men from the petroleum industry. These, like all the others in 'Dad's Army', were men whose age, health or occupation prevented them from joining the active forces, and like all such companies they were initially ill-equipped: at their first training session, D Company's total weaponry consisted of two double-barrelled sporting guns without ammunition, and six 'formidable pieces of gas-piping' for use as truncheons. No doubt they would have resisted valiantly if invasion had taken place, but happily they were never put to that test. Instead the crux of their duties, especially during the Blitz, became 'fire-watching': climbing up to the roof, protected by nothing but tin helmets and armed with little but binoculars, to give early warning of the approach of incendiary bombs.

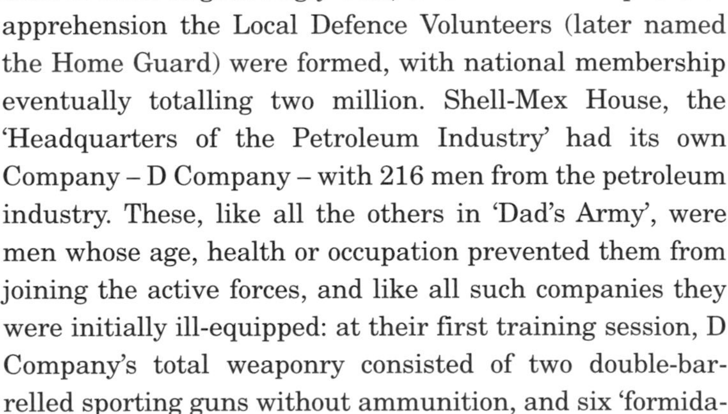

"QUICKER TURN ROUND" – INDEED!
LOOK WHAT'S HAPPENED NOW!

There was a small but useful economic factor to this task: having watchers on the roof meant that staff in the rest of the building did not have to evacuate to the shelters as soon as an air-raid warning was sounded, but could remain at work until bombs began to approach the building. Yet being a fire-watcher was hardly a soft option. It tested the nerves to be up there in the darkness, with search-lights stabbing and sweeping in search of the bombers as they droned overhead, and during a raid in September 1940, Shell-Mex House received a direct hit from a 500-kilo bomb. Given the elaborate precautions surrounding the laying of telephone and teleprinter cables, it was a little ironic that the bomb hit neither side of the building, but instead the tower in the very middle of its roof, where the lift machinery was housed. Fortunately damage was limited and no lives were lost, and Shell-Mex House was never hit by an incendiary; but some nights the fire-watchers could count as many as twenty-seven separate large fires ringing the building, and on at least one occasion, after a night of bombing by parachute mines, dawn revealed to them a mine dangling just a hundred yards or so away, entangled in the ironwork of Hungerford Bridge.

'Righto then – we'll go on.' Although drawn as a cartoon (by A.R. Midgeley, one of the Petroleum Board's driver staff), this sketch shows a real incident. On 18 December 1940, three loads of benzole were en route for the far side of Manchester when a particularly violent air raid began. The drivers briefly conferred and decided to proceed with their 'normal' blitz routine

Before the Petroleum Board could distribute anything, the oil first had to reach Great Britain, and there was only one way for it do so. The imaginative foresight and detailed practical planning that enabled the Board to start work from the outbreak of hostilities was mirrored at sea by the Royal Navy's immediate introduction of the convoy system. Long established and proven as the safest means of bringing imports into Britain, the delay in its introduction in the Great War had brought defeat very close. In 1939, the mistake was not repeated.

Just a few nights after the bombing of Shell-Mex House, Shell Transport's head offices at St Helen's Court were hit, again by a 500-kilo bomb. On that occasion one man died (the first victim of enemy action in the City of London) but fortunately, since the City had already been deemed vulnerable to air attack, most of Shell Transport's other staff had been evacuated: a thousand or so did their wartime work at the Lensbury Club, while tanker management was transferred to Plymouth, the great seaport and naval base at the western end of the Channel. They did not stay there long: port and base were so heavily bombed that after a year it was decided London would be safer after all.

Shell's tankers served in every part of the world. In the latter part of the war they acted as oilers for the British Pacific Fleet; they brought fuel to Malta and, through the Murmansk convoys, to Russia; they were present on 6 June 1944 at D-Day, the Allied invasion of France, not only carrying millions of gallons of fuel but also distilling and providing fresh water, that other essential for armies on the move; and throughout the war they brought oil to Britain on the core transatlantic routes. Most seamen regarded the Russian convoys as the worst, with the natural dangers of ice, fog and storms compounding the threats of surface, submarine and air attack. Shell's ships were extremely lucky in that theatre: though many were damaged there, none was lost. But elsewhere, their losses were severe. In the Group overall, 66 vessels were sunk and 1,434 officers and ratings were killed, with Anglo-Saxon's fleet (the main one) bearing the brunt: 41 of its vessels (about 40% of its tonnage) went down with the loss of 1,112 lives, most being victims of the unremitting U-boat warfare that characterized the Battle of the Atlantic.

Some of the ships and their brave crews became famous, with their actions being turned into morale-boosting films – for example, the tiny *Africa Shell*, captured and sunk by the surface raider *Admiral Graf Spee*; or Eagle Oil's *San Demetrio*, blown in half in mid-Atlantic by the pocket battleship *Admiral Scheer*, abandoned ablaze, then reboarded by her own sailors and nursed all the way to Britain; or the Shell-manned *Ohio*, which

Simnia, *attacked and sunk by the pocket battleship* Gneisenau, *15 March 1941, en route from Stanlow to Curaçao*

Erodona, *torpedoed the same day while travelling in the opposite direction, broke in half. Her stern section sank; her forward section, still containing 3,000 tons of cargo, was towed to Iceland, onwards to the UK, and eventually rebuilt*

in convoy to Malta was repeatedly hit by torpedoes and bombs and yet managed to bring 80,000 barrels of fuel to the besieged island.

Ohio would never have made it to Malta without an ingenious new life-saving device – compressed air. This was a marvellously versatile

The Shell-manned Ohio struggling towards Valetta harbour, Malta, with less than two feet of freeboard left after repeated attacks

system. When forced into ruptured compartments, the compressed air simultaneously helped to expel seawater and prevent the ingress of more water. It also provided buoyancy, and could provide power for pumps, emergency steering gear and fire-fighting. By mid-1941 the system was a standard fitting in all Allied tankers; and it was only one of five inventions from the combined staff of Shell Transport's operating company Anglo-Saxon Petroleum and Eagle Oil, managed and run by Shell. The other four were the 'Eagle hood', a special soap, fire-proof lifeboats and MAC ships. These inventions, though not well known at the time, were literally vital – without them, many more Allied tankers would have been sunk, and their crews lost – so it is worth having a closer look at them all.

The Eagle hood

If a ship could not be saved, the next priority was to enable her men to save themselves. But a sailor plunging into an oil-covered sea could all too easily find that he would surface to find the sea itself on fire, his lifeboat ablaze and himself blinded by oil. Eagle's technicians therefore came up with the 'Eagle hood'. Covering the head completely, it had a valve which kept sea-water out while allowing a man to breathe, and a series of removable optical lenses through which he could glimpse a route to safety. Having reached his life-boat or raft, he could then wash himself with the special soap which, developed from ester salts, worked either in fresh water or salt. This was particularly valuable if he had not had a hood, for then the oil would probably have filled his ears, eyes and nose. Both these inventions did much to alleviate the misery of shipwrecked sailors; and their chances of survival were further enhanced by Shell's fourth life-saving brainwave – the fire-proof lifeboat, devised (and personally tested in temperatures above 2,400 degrees F.) by Anglo-Saxon's Marine

Superintendent, John Lamb. This, like the other inventions, was approved by the Ministry of War Transport and put into general service.

But the best way to keep ships afloat, sailors alive, and oil flowing in to British ports was to keep the enemy away from the tankers altogether; and the fifth of Shell's maritime inventions helped accomplish that very thing. The idea came simultaneously to John Lamb and his colleagues, and to the Admiralty. Both knew that for Britain's essential transatlantic convoys, the most dangerous area lay in mid-ocean. Naval cruisers and destroyers could escort a convoy for the first thousand miles or so either way, but when crossing the central 'Black Gap', a convoy's main protection against U-boats was the wide emptiness of the ocean itself. What was needed was air escort, because by scanning and threatening dozens of square miles of sea at once, even a single aircraft could keep U-boats submerged and impotent. Later in the war, VLR (very-long-range) aircraft became available and American shipyards turned out scores of small escort carriers, but at first there was nothing – until the Merchant Aircraft Carriers, or MAC ships.

Pondering the possibility of putting flight decks on merchantmen, John Lamb concluded that oil tankers could do the job very well, with little loss

Top:
One of the Coastal Command's 'Liberator' aircraft escorting an oil tanker, part of a convoy...

Middle:
...and one of the Shell tankers reconfigured for convoy protection: the MAC ship or merchant aircraft carrier Amastra, *here seen undergoing conversion...*

Right:
...after conversion...

of cargo capacity. At the same time, the Admiralty decided that grain ships were the best, and (shades of Marcus Samuel's experience!) dismissed Lamb's suggestion as too dangerous. But once again, time and necessity told. Not only grain ships but also a total of nine Shell tankers were converted for this unusual duty, and proved their worth. Still able to carry 90% of their pre-conversion cargo, each of Shell's MAC ships carried three aircraft. In the two years of their service, they made 323 Atlantic crossings; their aircraft flew 2,177 sorties, escorting 217 separate convoys; and only one of those convoys was successfully attacked.

Top: ...launching a Swordfish on patrol...

...and safely landing a Swordfish back on board again

The sheer quantities of oil that were involved required additional methods of shore transport: whether in the United States or the United Kingdom, existing systems simply could not cope. In the UK, great use was made of the railways, and rather than the accustomed random numbers of tank-cars, it was soon found that the vagaries of shipping delivery (and, worse, of supply to HM warships, which could turn up en masse thirsty but unexpected) were better handled by establishing standardized tank-trains: this many tank-cars could drain one tanker, that many could fuel one destroyer, that many more one cruiser, and so on. For smaller quantities, barges were used as well, whether on river or canal: not for the first time British people had cause to thank the engineering genius of their Victorian forebears. And both there and in the United States they had cause to thank their contemporaries, who devised pipelines that had never been seen before. These were of two entirely opposite kinds – portable, and huge fixed ones.

In the US, Shell participated in the construction of the 'Big Inch' and 'Little Big Inch' pipelines, running respectively from Texas and the American south-west to the East Coast tanker ports. Big Inch, completed at the end of 1943, was 24 inches in diameter and 1,254 miles in length - the largest crude-carrying pipeline in the contemporary world, handling half of all crude oil for the East Coast. Little Big Inch, which carried petrol and other products, was narrower (20 inches diameter) but, at 1,475 miles, still longer. Both these wonders of American engineering were fixed pipelines, and Britain saw their smaller brothers, in the shape of a 1,000-mile network of permanent pipes constructed by Shell engineers, from the Bristol Channel to Walton-on-Thames, the Mersey, East Anglia, the Solent and the Isle of Wight. These supplied RAF and USAAF bases throughout the country, as well as invasion forces on the south coast. Chemicals further enhance their performance: the Shell Corrosion Inhibitor process (introduced in the US and described as 'a liquid "go-devil" preventing internal corrosion of pipelines') brought about a direct 15% increase in pipeline throughput wherever it was used.

Big Inch, Little Big Inch and the new UK pipeline network were all novelties by virtue of size; but it was in Britain that Shell developed a total novelty – the portable, flexible pipeline. Made in pre-fabricated sections, it was light, strong and mobile; it proved equally functional whether in transporting fuel or drinking water, and was used not just in the UK but in North Africa, Sicily, Italy and France.

Where pipelines of any sort were inappropriate, Shell Transport's subsidiary companies still ensured the effective transport of its products. The old-fashioned standby of the tin proved a more or less eternally practical form of packaging, whether for petrol, diesel, kerosene or water, and huge numbers of them were made – for example, over 100 million in Egypt alone in the four years 1941–4. And if there were neither roads nor airstrips available, well then, there soon would be with Shell's 'mix-in-place' system of surface construction. This enabled local wet sand to be com-

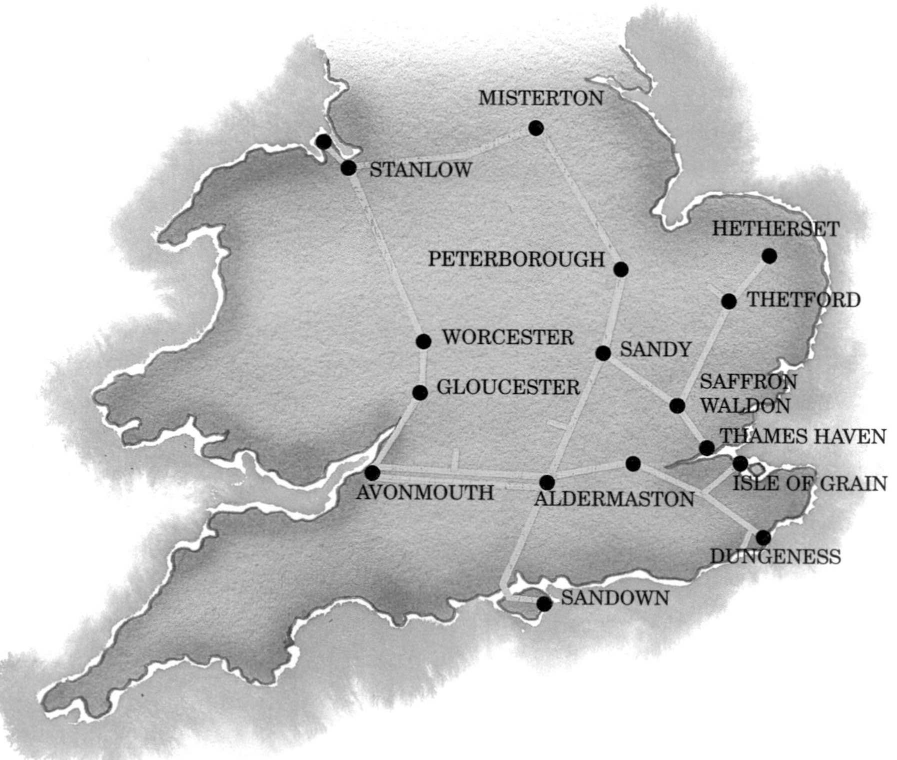

Routes of the UK pipeline system

FOR HIGH PERFORMANCE
VICKERS WELLESLEYS
JAMES GARDNER
LUBRICATION BY SHELL

bined with soil-stabilizing bituminous binders, providing a surface of exceptionally high load-bearing capacity. To take Egypt as an example again, in the six years 1939–45 Shell's asphalt divisions built 60 aerodromes and 250 miles of roadway in and around Cairo alone.

By these and many other means the Allies kept the oil flowing to British ports; and on land the ingenuity in putting it to good use was no less, with Shell's laboratories developing between twenty and thirty different categories of new products or improved processes, ranging from a system for the large-scale manufacture of the wonderful new antibiotic, penicillin, to the extraction of vanadium – a soft, silver-grey metal which adds strength, toughness and heat-resistance to steel, and which occurs in minute quantities in fuel oil. At the urgent request of the Ministry of Supply, Shell provided enough of this material in the course of the war to treat 85,000 tons of steel.

Shell Transport's leadership of the UK's wartime distribution of petrol and petroleum products was a crucial part, but only a part, of its efforts for the Allied nations' cause. Its lead was equally evident, and its efforts equally essential, in the wartime developments of the petroleum industry's chemical aspects.

Even a summary of these is an impressive indication of the effort that Shell Transport invested on the Allies' behalf. Its range of solvents grew enormously, with one of the most important new additions being acetone: this was used not only as a stripper of paints, varnishes and lacquers, but also in the manufacture of perspex (from which the canopies of bombers and fighter aircraft were made) and of smokeless explosives. In World War I

Shell's Borneo oil had been used as a prolific source of toluene, essential for the high explosive TNT; in World War II, even greater quantities were produced by a new process (first discovered in Shell's Amsterdam laboratory and developed in California by Shell Development Company) called dehydrogenation. Another development, evolved in co-operation with other companies, was petro-gel, used both in flame-throwers on land and to create camouflaging smoke-screens at sea. A similar substance was used in a product known as FIDO. Fog, whether at the time of take-off or landing, could be a serious, sometimes insuperable handicap for the RAF; but as everyone knew who had ever watched the sun burn away a morning mist, it could be dispersed by heat. Hence FIDO, the RAF acronym for 'Fog, Intensive Dispersal Of'; and by day or by night (particularly on the flat, low lands of the East Coast counties, where fogs from the North Sea could be a daily occurrence) many an English aerodrome was illuminated with FIDO's spectral glare, as long troughs of blazing gelatinous liquid cleared the air.

Anti-corrosive lubricating oils were also developed both for marine turbine engines and aircraft engines, along with greases which worked at low temperature (and hence at high altitude) for aircraft controls, de-icing fluids, mineral and non-mineral hydraulic fluids, fluids for cleaning, decarbonizing and degreasing aircraft engines, and easily applied camouflage paints based on bitumen and petroleum resins. From 1940, a great deal of Shell research into aero-engines and lubricating oils was conducted at its new laboratories established for the purpose at Thornton on the Mersey – a place of which more will be heard.

Numerous synthetic chemicals were invented or brought to mature production in the war years. These were valuable for one of two reasons: either because they were cheaper than the natural equivalent, or (if there was no natural equivalent) because their invention enabled the manufacture of completely new products. Compounds discovered and developed by Shell were widely used in synthetic resins, including new plastics, laminates and coatings, and formed the essential basis of synthetic glycerine, from which such things as celluloid were made. Similarly, Shell chemists were responsible for the invention of two synthetic fuel additives, of which one made the manufacture of 100-octane fuel cheaper, while the other increased the power output of aviation spirit. But perhaps the most strategically timely of Shell's wartime synthetic products emerged from Houston, Texas, where in the summer of 1941, a plant was completed for the manufacture of butadiene. This, the single most important starting material for synthetic rubber, was more resilient over a wider temperature range than the natural product, with greater resistance to ageing and weathering. Just six months later, Japan entered the war and swiftly overran the oil- and rubber-producing islands of Malaysia and the Netherlands East Indies. The loss to the Allies of those oil sources was a serious blow; combined with the loss of the rubber, needed for aircraft and motor vehicle tyres, it could have been fatal – except that butadiene immediately began to be produced in very large quantities.

Water, often as much an enemy as a friend, came in for close scrutiny. Engines and other such devices can founder or fail if it is present, and during the war Shell developed WD (Water Displacing) compounds, which would either keep water off or remove it if it was present. Likewise, de-icing greases were evolved for use at sea: by preventing ice from bonding onto metal they helped keep ships' decks clear. Even by-products were found to be useful, such as naphthanic acids which made canvas rot-proof; but undoubtedly the most spectacular use of Shell water-proofing materials occurred on D-Day.

D-Day was the largest amphibious operation ever conducted. Nearly three million Allied officers and men were directly involved, along with more than 3,500 major naval combat vessels, landing craft and special vessels, supported by ships from the Merchant Navy – including, of course, members of Shell's fleet. And for the assault to succeed there was another, not exactly negligible factor: over 150,000 motor vehicles had to be got safely ashore from the invading fleet – which meant they had to be driven off the ramps of landing craft and through the sea up onto the beaches.

On 18 January 1943, the Ministry of Supply asked Shell in London if a suitable water-proofing material could be made. The challenge was uniquely daunting. First and foremost, the material had to provide 100% water-proofing efficiency, to 'enable a vehicle to wade in sea-water' to a depth of three feet with 18-inch waves on top. It had to be easy to use, so that comparatively unskilled personnel could apply it with a high degree of reliability. It had to be a good insulator; it had to be rigid up to 200^0 Fahrenheit, and to survive exposure to that temperature for long periods without either sagging or hardening and cracking, and without showing any oil separation whatever; it had to smear easily and 'take' on slightly greasy surfaces, without being tacky or clinging to the operators' hands; and it had to be something which could be provided soon, in enormous quantities.

The Ministry reasoned that if anyone could meet this alarming specification, Shell could; and without dismay Shell did, very quickly indeed. After discussions, tests, modifications and more tests, the first 28-pound batch of Compound 219 (codenamed after the number of the Shell-Mex office in which the request was first made) was ready for trials on 24 February – just 37 days after work had begun. An engine was covered in the mixture, placed in a water tank, warmed up, immersed in salt water, and (to quote the first public post-war account) 'Eureka!' It worked, even when submerged a full eighteen inches. Better, it worked several feet under the sea, whether in the chilly Channel or the Mediterranean. Over 24 million pounds (more than 10,700 tons) of Compound 219 were used at D-Day, and when the operation with its 150,000 invading vehicles – tanks, trucks and jeeps – had become history, Sir James Griggs (Secretary of State for War) remarked with satisfaction that 'despite the fact that many of them went ashore through five feet of water in heavy seas, less than two out of every 1,000 of the vehicles were "drowned" off the beaches.'

There was of course another aspect to Shell Transport's war: the effect of the conflict on its friend and partner, Royal Dutch. On 10 May 1940, in the face of the Nazi invasion of the Netherlands, the sister company's headquarters (and those of its operating companies, such as the Bataafsche) were legally transferred to Curaçao, while the Bataafsche's laboratory staff evacuated to the UK and joined their British colleagues in setting up the new Thornton research laboratories. Underlining the tragic circumstances, this was Royal Dutch's 50th birthday year.

In the Netherlands East Indies (although British and Dutch personnel alike were involved) it was mainly Dutch personnel who sustained and endured the greater part of the devastation wrought after Japan joined the war. As in Germany and Italy, several Japanese companies since the war have become important partners with or customers of Shell; co-operation has long replaced conflict, a most welcome change for all concerned. But it does not alter those four years in history, from December 1941 to September 1945, when (as Walter Samuel said sadly) amid the physical destruction of wells, terminals and refineries,

> The Japanese over-running of the Far East resulted in the loss of many of the Group's senior experienced officials in those areas. Large numbers of the staff were able to get away in time, but in some cases managers and senior officers, who remained until the last minute, lost their lives, whilst others were taken prisoner by the Japanese. ...We have them in mind and shall welcome them back at the end of the war.

It would be wrong not to mention these events; yet by the same token, this is not the place to describe them in detail. To make such a distinction is at best legally hazy. Ever since the alliance in 1907 of Shell Transport with Royal Dutch, and their simultaneous assumption of mutual responsibility, it has been very difficult to differentiate one from the other. Moreover, this is not just a legal but an emotional matter: whichever country they may come from, and whichever they may work in, colleagues within the Royal

Dutch/Shell Group see each other as members of a trans-national family. Nevertheless (though this may sound parochial), Japan's invasion of the Netherlands East Indies impacted far more directly on Shell's *Dutch* personnel than on their British friends and partners; and so their story at that time, which was both terrifying and enthralling, belongs more correctly in the histories of Royal Dutch and of the Group as a whole. If here, in the history of Shell Transport and Trading, they are accorded only a brief salute, it is given in the same spirit as when Walter Samuel addressed Shell Transport shareholders just after the invasion of the Netherlands in 1940:

> I cannot refrain from referring to all those members of our Group who now find themselves in enemy-occupied territory, particularly our Netherlands friends. The happy relationship which has always existed between the Royal Dutch and your company makes the present unhappy position of the Netherlands and of our Netherlands colleagues particularly sad and a great anxiety to us. Many of our friends of many years standing are thus affected, and our sympathy goes out to them.

'We meet this year', he added, 'in the midst of a devastating war. I can only hope that when next we meet, we shall do so under happier and less anxious conditions.' The fulfilment of that hope was a long time coming: it was not until 13 December 1945 that he and the other British directors were able to meet their Royal Dutch colleagues at The Hague.

In 1942, reviewing Shell's expanding range of chemical products, Walter remarked: 'We look forward to the time when the manufacture of all these materials for war purposes is no longer necessary; they will, however, each be able to play a more constructive part in peace-time.' He never expressed the slightest doubt that when peace eventually returned, it would be with an Allied victory; but there were times when that belief could seem wildly optimistic, and in 1943, deep in the weariness of war, *Shell Magazine* carried a wistful article by an employee. Set in the year 1948, it imagined a time when the war would be over, and proposed that the London head offices should be relocated to Scotland – the scene of many happy summer holidays and, with mountains, lochs and glens, almost as great a contrast with England's ruined capital as could be conceived within the United Kingdom. St Helen's Court had a stores within it where personnel could shop for oddments; the new offices, it was suggested, could have the same, as well as a post office, private telephone booths, a travel agent, a barber's, a library, accommodation, a restaurant with continuous service sixteen hours a day, a fully licensed cafe and bar on the ground floor and a garden on the roof. It was a sweet sustaining dream, and in case critics said it would be impossibly expensive, the writer had an answer. On leaving the new headquarters, he said:

> you will notice an inscription: 'Erected by the Management and Staff as a foundation for the future in memory of those who made a future possible.' I cannot tell you the cost. They did not count it, so why should we?

Captain D.W. Mason of the Ohio, awarded the George Cross for his part in Operation Pedestal

Walter was always emphatic in his praise of the tanker personnel, recognizing quite rightly that their bravery was the lynch-pin of every other action undertaken by Shell Transport, and he rejoiced when they were accorded public recognition. After Operation Pedestal, the dramatic and critical convoy to Malta, *Ohio*'s crew became one of the most decorated of all: Captain D. W. Mason, was awarded the George Cross and Chief Engineer J. Wyld thc DSO while other members of the ship's company received five DSCs and seven DCMs. But they were far from the only Shell Transport people to be publicly honoured for their deeds: by the war's end over three hundred personnel had been given medals, citations, Mentions in Dispatches and other awards. Out of all those, perhaps just three of the most outstanding may be mentioned.

At sea, there was 19-year-old Apprentice D. O. Clarke. In 1941 he had already shown great gallantry when, during the blitz of Liverpool, he saved the life of a drowning dockyard worker. In October 1942, Clarke's ship was torpedoed two days out of Trinidad while carrying a cargo of petrol. Clarke was one of only eight to escape, and one of only three capable of lifting an oar, though he was extremely badly burned. No one else knew quite how bad his injuries were, but when he died his hands had to be cut from the oars: the flesh had all melted from his palms, and he had rowed for two hours on the bare bones of his hands. He received the George Cross, posthumously.

In the air, there can be little doubt that the most distinguished Shell Transport pilot was one whose name, unlike Clarke's, became world-famous: Douglas (later Sir Douglas) Bader. Invalided out of the RAF in May 1933 following a crash in which he lost both his legs, he joined the Aviation Department of Asiatic Petroleum a month later, then in November 1939 was able to rejoin the RAF, where his legendary exploits formed the basis of the film *Reach for the Sky*. Despite his physical handicap, he gained several promotions and in 1940 earned both the DSO and the DFC before taking command of a squadron in 1941. Later that same year he was shot down in France and imprisoned, remaining a PoW for the rest of the war; but that did not seem to deter him, and while his repeated attempts to escape eventually landed him in Colditz, they also brought him bars to both his medals. The French government honoured his example too, with the Croix de Guerre and the Légion d'Honneur, and happily he was able to rejoin Shell after the war. He became manager of Group Aircraft Operations in 1952, the first managing director of Shell Aircraft in 1958, and retired in 1969.

Finally, on land, emphasizing the ubiquitous service given by Shell personnel, it is impossible to overlook Major Robert Cain. When war came he left his work as a manager of Shell Nigeria and joined the South Staffordshire regiment of the British Army. On 19 September 1944 at the battle of Arnhem, in the Netherlands, he and his men came under heavy attack from Panzer tanks. Taking an anti-tank launcher, Cain left cover alone, shot at the leading Panzer, immobilized it and then, though wounded, co-ordinated its destruction by howitzer. For this and other subsequent acts and examples of extreme courage and leadership he was awarded Britain's highest military decoration, the Victoria Cross.

Douglas Bader, DSO, DFC* (centre), seen with members of his squadron*

CHAPTER ELEVEN

The Fight to Recover: 1946–1951

The return of peace brought Shell very little chance of rest or relaxation. The effort, the outpouring of mental, physical and fiscal resources demanded by the war, still continued; all that changed was the purpose to which it was directed. With the Allied declarations of victory in Europe (8 May 1945) and the Far East (15 August), Shell was at last able to begin to assess the full extent of its enormous physical losses in the war, and to start implementing plans for reconstruction. Even if all other things had been equal, reconstruction would have been a fully sufficient trial; but it took place against a background of almost universal shortages in every industrial material, and of the greatest peacetime political turbulence in living memory. It is no exaggeration to say that despite all the difficulties it had faced in the past, these years were the most challenging which Shell had ever known as a mature commercial organization. Much of that will be outlined in this chapter; and, as will be seen, the Group succeeded in meeting those challenges, and more. For although post-war reconstruction was difficult and expensive, plans did not stop there: there was also a very considerable expansion of the business, and in the process, many changes large and small.

Shell's thinking immediately after World War II followed two closely linked forms. The first had been expressed during the war by T. S. Eliot:

There is only the fight to recover what has been lost
...and now, under conditions
That seem unpropitious.

The second – more far-sighted in nature and more constructive in approach – had found expression much earlier. In 1859 (coincidentally, the very year the modern oil industry began) the English philosopher John Stuart Mill wrote: 'When society requires to be rebuilt, there is no use in attempting to rebuild it on the old plan.'

The fifth logo and fourth pecten, introduced in 1948 – the first to include the word Shell

Among Shell's lesser signs of post-war change, aimed at consolidating its public identity, were new names for operating companies (for example, the

Asiatic Petroleum Company became The Shell Petroleum Company) and a new pecten design, for the first time incorporating the word SHELL. But a much more important indicator of new times lay in the change of Shell Transport's leadership.

In 1946, Walter Samuel (who was 64 and falling ill) told his colleagues that he would like to step down, with effect from 12 July. He had been on the board of Shell Transport and Trading since 1907 and chairman since 1921 – only its second chairman in forty-nine years. His sons, Richard and Peter, had fought in the war just passed. Richard, the elder, had been severely burned in Italy when the tank under his command suffered a

Rebuilding war losses: Walter Samuel, second Viscount Bearsted (left), with his wife at the launch of the experimental and highly successful Auricula *at Hebburn on the River Tyne, 17 April 1946*

direct hit. As one of the famous 'guinea-pig' patients in the pioneering days of plastic surgery, he made a very good recovery. Through the therapeutic programme to restore the use of his hands, he became, incidentally, a skilful embroiderer; but he had no desire to enter Shell, instead taking charge of the family's banking business, M. Samuel & Co. Peter had been awarded the MC (as had Walter himself in World War I) while in the field at El Alamein, and was already a director of Shell Transport; yet the business had grown so much that, all things considered, there seemed no good reason to insist on a family succession to the chair. Walter's recommendation, which the board accepted unanimously, was that Frederick Godber (who had been knighted in June 1942) should be the next man to take charge of Shell Transport.

Frederick Godber's career with Shell could have been cut short at a very early date. He had joined Asiatic Petroleum in 1904 as (in his words) 'stamp-licker to a very peppery Dutchman', who quickly gave him notice to quit. However, the young man had simply ignored this, continued to turn up for work, and continued to get paid – 'in cash once a week', he remembered. 'We all used to line up and were solemnly handed our pay packets as though they were bonuses for good behaviour.' He had then become assistant to George Engle (the originator, subsequently, of the Lensbury Club) and was put in the Secretarial Department. A routine task was to lay out important correspondence on the boardroom table, available for any senior member of staff to read. Godber was hardly senior, but he had recognized a first-class opportunity for self-improvement, and had taken it upon himself to read every incoming letter. Thus, he said later, 'I learnt something about the business'; and he had learned more about Shell's emerging American side by writing letters for the elder J. B. A. Kessler (who, in order to concentrate while dictating, would walk up and down eating grapes). By 1919 Godber had learned so much that he was sent to the US to gain experience with Shell's subsidiary Roxana. His rise continued: before long he was made a vice president of the company, and in 1922 became its president. His stay in the United States, originally meant to last two years, became much extended – he did not return to London until 1929.

Sir Frederick Godber (later Lord Godber), chairman of Shell Transport and Trading 1946–61

He had a very personal style of management. His own copious business correspondence shows a man of quick decision, and one character assessment described him as 'a very efficient though somewhat humourless figure, firing off brisk instructions at a considerable rate'; but he did not lack warmth. Through his years in Shell he had always taken pains to get to know his staff, and during the Depression years of the 1930s, he received an enormous number of personal letters from people seeking employment. These applications were, alas, mostly in vain; but even though he was by then a member of senior management, the writers clearly felt able to approach him directly.

He seems to have viewed personal involvement as a reciprocal matter – 'the company insurance department was expected to deal with his (numerous) road accidents, as well as losses of personal belongings'. Moreover, his two daughters married the sons of two other board members. Perhaps they were taking it rather literally, but the Royal Dutch/Shell Group had already been distinguished for many years by a definite 'family feeling' – not in the sense of the family firm which Shell Transport once had been, but in an altogether more remarkable manner.

Godber's close colleague J. B. A. Kessler junior recognized the value of this intangible asset, and knew it could not be forced; as he rather disarmingly once wrote, all Shell's advantages of organization, knowledge and experience

> would be of little use if we were not able to build upon a sound foundation, and that foundation is to be found in the strong family feeling throughout the Group. Human relationships are *so* important. The Management does not beg for loyalty, but it does try to create an atmosphere in which loyalty is able to flourish.

'The Management' undoubtedly succeeded; employees had a real, and perfectly conscious, friendly pride in working for the Group. In the years just after World War II, this was accentuated both by the sense of having survived through mutual support, and by the sense of personal loss as the list of those maimed or killed was confirmed. Whether it was the son of the wartime chairman, or a Bataafsche employee starved and tortured in the East Indies, these were more than names in a dossier: they were known colleagues and often friends.

Under Godber's leadership (which was to last 15 years), the hallmarks of the immediate post-war years were reconstruction and ambitious expansion. When damage to facilities for production, refining, storage and transport was assessed, everywhere from western Europe to the Far East, it was soon evident that even reconstruction alone – merely restoring facilities to their pre-war condition – was going to be a colossally expensive task. Nevertheless, it was essential; so was the entry into new markets. All told, a dauntingly large amount of new money would be needed to accomplish these ends; but as chairman, Godber followed the spirit of Marcus Samuel and saw no point in thinking small. He and his colleagues decided that Shell Transport's authorized capital base (£43 million in 1947) must be more than doubled, to £88 million. It was more easily said than done.

There had been no need to raise new capital since 1930. If such a share issue were now made, then Royal Dutch would either have to raise a proportionately larger amount with a simultaneous issue in the Netherlands, or else agree to a rearrangement of the 60:40 relationship. The idea that the two parent companies might alter their historic relationship to a true 50:50 partnership had been aired as early as 1942. Dutch colleagues did not object in principle, and the Bank of England had seen only one difficulty:

> It is...unthinkable that the Dutch Government (now in London) would consider or deal with such questions until it shall have again established itself in Holland – if it ever does so. The moment therefore does not seem opportune.

But when 'the moment' came in 1947, the very size of Royal Dutch/Shell was more a hindrance than a help: the Group was so big that paradoxically, it could not simply do what it wanted, but had to consult with governments. As an official at the Bank of England remarked to another (aptly named Playfair) at the Treasury, both Royal Dutch/Shell and the other great Anglo-Dutch combine, Unilever, were 'private concerns. Nevertheless, both require Governmental permission on both sides to shift their assets about or to increase their overseas investments...'

That proved the sticking point. In Britain, the new Labour government limited the amounts of new capital that private industry was allowed to raise, and in Holland (as the British Foreign Office discovered)

> The Dutch Government are clearly unhappy at the possible shifting of the sixty-forty balance hitherto prevailing, since if Shell put up new money in proportions different from the above, they will claim a correspondingly increased influence in the affairs of the Group. The [Royal] Dutch directors would not (repeat not) be equally disturbed.

Montagu Norman (first Lord Norman), chairman of the Bank of England 1920–44

This was an interesting comment on the confident relationship between the two sets of directors, British and Dutch. But whereas Shell Transport was one big company among many others in the British economy, Royal Dutch had become far and away the largest company in the much smaller Netherlands economy, to the extent that the relationship between it and the Dutch government had to be seen in a special light. Moreover, in that time of postwar national reconstruction, what was virtually a planned economy had become essential. As its staff newspaper remarked, Royal Dutch could 'scarcely move a step without coming into contact with one of the many government departments', and the Dutch government feared their country simply could not afford to fund an expansion on the scale envisaged by Shell Transport's directors; so they refused to permit it.

After further discussions with the Treasury, the Bank of England and the bankers Morgan Grenfell, Godber had disappointing news for his colleagues on Shell Transport's board: 'it had become clear that the immediate issue should be confined' to the comparatively low figure of £10 million, bringing Shell Transport's capital base to £53 million.

The Hon. Francis (later Sir Francis) Hopwood, subsequently Lord Southborough

Ironically, though, we can see now that despite official worries, the original target might have been achieved, without any change in the established 60:40 ratio. John Loudon – then a new young director of Royal Dutch, and subsequently its chairman – remarked decades later that it was only by scraping up every last guilder in Holland that the necessary amount was found. Yet both issues, the British and the Dutch, were considerably oversubscribed, with many foreign buyers. In their annual report for 1947, the directors of Royal Dutch noted that 'the capital market was invited to furnish an unprecedented volume of funds', adding proudly that their company's importance 'for the Netherlands foreign exchange economy was strikingly illustrated by our issue of shares, since it yielded 125,000,000 guilders of foreign currency.' After its battering in the war, it is not surprising if the Netherlands government had lacked a degree of confidence; but had it realized how keen non-Dutch investors were, it could probably have permitted a larger issue without upsetting the historic ratio. And if it had been willing for the ratio to be changed from 60:40 to 50:50, it could definitely have permitted much more. To a colleague in New York, Sir Frederick Godber wrote in December 1947:

> The two issues seem to be going well. When it came to making the publication the markets received them, certainly here in London,

The transportation of fractionation column to Stanlow, 22 November 1950

> with a good deal of enthusiasm. I don't think it would have been at all difficult for us to have secured the larger amount for turning the Group into a real partnership. As Bob [Sir Robert Waley Cohen] has probably told you, the whole fault in the end lay solely with the Dutch government who just dug their toes in...and blindly and, I think, somewhat irresponsibly, refused to permit any change in the relative positions.

In the context of Shell Transport's history, Godber's criticism may be taken as the official, if unpublicized, company view: after all, it came from the chairman. But in the wider context of the history of Royal Dutch/Shell, the comments are as interesting for what they omit as for what they say, and for two reasons. In placing 'the whole fault' on the Dutch government, Godber ignored the British Labour government's concurrent cap on private investment; and perhaps understandably he was writing only from his own point of view, which was not entirely shared by Royal Dutch colleagues. Indeed, as John Loudon described it later, the subject 'was not always diplomatically handled' by some Shell Transport directors, causing hard feeling between the Group's two parent boards. Loudon and an equally junior member of the Shell Transport board, Francis Hopwood (later Sir Francis, and later still Lord Southborough), viewed this with deep concern: 'Nationality-wise, it was destructive.' Being very junior, they decided to keep out of the discussions, yet they saw lessons to be learned for the future – lessons which they applied when, in the course

Drilling in the Gulf of Mexico 1950

of time, they became senior. And these (as we likewise shall see in later chapters) proved to be of fundamental importance in shaping the Group's unique character.

Returning to the postwar raising of funds, another avenue of finance came available in 1948, when through a subsidiary (Shell Caribbean) a very large dollar loan – $250 million, then equivalent to about £62 million – was negotiated with half a dozen American insurance companies; and the success of the 1947 simultaneous issues encouraged the Dutch government more than a little. In 1950 and again in 1952, twin issues of new shares were permitted. Meeting with equal success, and still keeping the ratio at 60:40, the 1950 issue raised £10 million for Shell Transport, bringing its total capital base to £63 million; and the 1952 issue raised another £25 million – which by simple arithmetic made a new total of £88 million, the original postwar target. As observed before, it was easier said than done; but the target had at last been achieved.

As it arrived, stage by stage, the new capital was put to good use. All Shell's existing refineries were repaired, enlarged and extended. In Britain, the Lancashire refinery of Heysham (which Shell had operated on behalf of the government during the war) was purchased in 1947, while in 1948 construction of two major new refineries began at Stanlow in Cheshire and on the Thames at Shell Haven (the name of which was an historic coincidence, not a new invention). New exploration was undertaken in Nigeria, Ghana, Somalia and the Venezuelan Gulf of Paria (1946), India, Pakistan, New Guinea, Burma, Sinai and Queensland, Australia (1947), Bunju island off Borneo (1949) and Tunisia (1950), as well as in the Bahamas with aerial and geophysical techniques. New productive oil fields in existing concessions were found in the United States and Venezuela, including (in 1947) *in*, rather than next to, Lake Maracaibo. Two new gas fields were found in the Netherlands (1949 and 1950) and another in Canada (1951), and with the novel method of artificial re-pressurisation by 'water-drive', older oil wells in the United States were given new leases of life. And there was much innovation, too, both on and under the sea.

A remarkable event took place in October 1947. In the Gulf of Mexico, off the shore of Louisiana, a small American oil company called Kerr-McGee drilled the world's first commercially productive offshore oil well. There had been others – tiny ones, producing perhaps two barrels a day – in California fifty years earlier, but they had been drilled literally from beach pierheads. Kerr-McGee, unable to afford a concession on land, bought one 10½ miles out at sea. In later decades the siting of rigs hundreds of miles from shore became so commonplace that Kerr-McGee's pioneering 10½ miles may not sound far, but it was quite far enough to be over the horizon, out of sight of land and surrounded by an alien environment – which was why Kerr-McGee could afford it. Because of the strange new factors involved (tides, currents, storms, supply, corrosion, the nature of the invisible seabed), many people thought that drilling so far out would be impossible, and the concession was much cheaper than an onshore one. It was good that, as in Marcus Samuels' early days, a small company could still venture successfully where the giants feared to tread. Seeing success, the giants were naturally emboldened; but that did not mean the pioneers were trampled. An old prayer of sailors still applied: 'Oh Lord, be Thou kind – my ship is so small, and Thy sea is so big.' If that emphasized the dangers, it also hinted at an equally important aspect of offshore work: like the Far East, which Marcus Samuel had long ago said was 'quite big enough for both the Royal Dutch and the "Shell" line', the sea was so big that there was plenty of room for fair competition. Highly excited at the possibilities, Shell worked quickly, and in 1949 – just two years after Kerr-McGee – drew its first sub-sea oil from the Gulf of Mexico. South of the Mississippi delta, this, the 'Main Pass' field, came under Louisiana's jurisdiction. By 1955 Main Pass had become the state's largest oil field, and, with over 125 wells there giving a daily production of over 30,000 barrels, Shell was the field's leading producer.

Meanwhile Shell ships were given a new distinctive livery, with the pecten included both on flags and funnels: a further strengthening of the

Shying away from the dangers of the deep: 'offshore' drilling began at Santa Barbara, California, in the mid-1890s, but even by 1927 – as this picture shows – rigs had barely advanced beyond the surf-line

public image. But the fleet's most pressing necessity in 1946 was the replacement of sunken tonnage. A twin programme of purchase and new construction was pursued so vigorously that within 18 months almost every lost ship had been replaced. 'You will the more easily measure the post-war recovery of this great fleet,' said Godber:

> which I remind you is one of the largest in the world sailing under one House flag, if I recall that before the War its tonnage amounted

> to 1,525,000 dwt [deadweight tons], and that during the war we lost 66 ocean-going ships of 632,000 dwt. By the end of 1946 the tonnage of the fleet was 1,487,000 dwt...

The programme continued: by the end of 1947, Group tonnage (owned and managed) had risen to 2,140,000 dwt and just one year later to about 3,750,000 dwt. During the four years 1946–9 inclusive, Anglo-Saxon's fleet alone acquired 71 new ships – an average of one every three weeks. Two of these were particularly exciting experimental vessels: *Auricula* (1946) and *Auris* (1948).

Both extended the boundaries of knowledge in a quest close to the heart of every ship-owner: the search for the most efficient and economical method of propelling his ships. That was why sail had given way to steam engines powered by coal-fired boilers. Then, at first also fuelled by coal, came steam turbine engines, which were faster; then steamships fuelled by oil, which could be handled more easily and gave greater thermal efficiency; and then diesel-engined or motor ships, which were cheaper to run than steam boilers. Diesel oil, the ordinary motor ship's fuel, was a distillate only slightly denser than kerosene. In contrast, *Auricula* was designed to burn high viscosity fuel (HVF), one of the final fractions of crude oil, so dense that hitherto it could not be used alone. But if it could be used alone, then calculations indicated that Shell could save about £500,000 a year; and *Auricula* was conspicuously successful. By her tenth birthday in 1956 all Shell's motor ships had been altered to burn HVF, and with many other shipping companies following Shell's lead, something like 500 other vessels world-wide were using HVF.

Auris was a more qualified success, at least in merchant shipping terms. In 1951, when three years old, one of her four diesel engines was replaced with the world's first marine gas turbine engine, and in 1952 she became the first ship to cross the Atlantic on gas turbine power alone. Gas turbines had several advantages over diesel engines: they were lighter, easier to maintain, simpler to operate, and could reach full power within a few minutes of starting from cold. But for a commercial company, the advantages of complete conversion to gas turbine were cancelled out by the cost; the project was therefore eventually abandoned, and the onus of developing this particular system passed to the world's major navies.

Group tonnage purchased in this period included nineteen American-built 'T2' tankers of 16,600 dwt each. Pre-war, a middle-sized British tanker would have been about 9,000 dwt. The T2s were noticeably larger than usual, marking the start of a trend which would continue for thirty years; and Shell's new construction orders in 1947 went still further than the American ships, to include three of 28,000 dwt. For their time these were so large as to warrant a new generic name: supertankers. The first,

Maiden voyage of Auris

Velutina (launched by HRH Princess Margaret in April 1950 and completed the following August), was the biggest merchant vessel to have been built in Britain.

In 1950 a further 31 big tankers were ordered from British yards at a cost of over £30 million. Even for Shell, this was a very significant investment, made possible in part because government policy for the regeneration of post-war British industry included a tax allowance (the 'Initial Allowance') of 40% on the cost of new capital assets. But shortly after the orders were placed, the allowance was suspended. Godber was deeply concerned, not only for Shell but British shipbuilders, and with foreboding observed to shareholders:

> while I will not go so far as to say that had we known this the programme would have been modified, I do submit this case as an example of the severe impact these proposals must have upon necessary building programmes for shipping in particular and industry in general.

Subsequent decades proved his warning right as, faced with mounting foreign competition benefiting from government subsidies, Britain's shipbuilding industry slowly shrivelled to a tiny, humble fraction of what once it had been. But that lay in a greyer national future. Retired Shell seafarers accurately recall the post-war years as a bright, exciting time

6 January 1947: a heavy snowfall in London was fun at lunchtime …

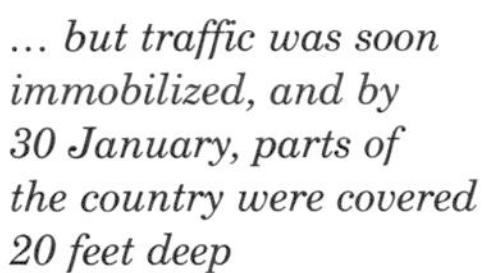

… but traffic was soon immobilized, and by 30 January, parts of the country were covered 20 feet deep

of enormous expansion as every part of Shell regained and then far exceeded its pre-war size.

Raising and investing new capital were only two of the themes of these years. Approaching the middle of the century, the figure 50 naturally loomed in many minds, and perhaps particularly in the minds of those in Shell. One important reason was that 1947 was Shell Transport's jubilee year, marked in proud but practical manner by the publication on 18 October (the company's 50th birthday) of a booklet, described by Godber himself as 'modest in format and volume owing to shortage of paper and to the national need for economy', in which its history was summarized. And '50' cropped up in other ways. Shortly after hopes were dashed for a reorganization of the Group on a 50:50 basis, 1948 brought two momentous agreements, both subsequently referred to as 50:50. The first was between Shell and the American company Gulf Oil; the second was a multilateral royalty arrangement with Venezuela, which affected every oil company and eventually changed the entire industry's basis. Although they occurred at the same time, it will be useful to describe these agreements separately.

The contract between Gulf and Royal Dutch/Shell arose from circumstances which may at first sound odd. Despite maintaining its global market share at about 11.5% and despite producing more than ever (an average of 630,000 barrels per day in 1946 and 753,000 in 1947), Shell was suffering from a shortage of oil. The return of peace had completely altered the shape of the market. The Allied air forces' gigantic wartime call for aviation fuel shrank rapidly. Simultaneously, there was an explosion of world-wide civilian demand for oil products – an 'astonishingly rapid expansion', said Godber. Geared to making the products needed in war, the oil industry could not keep pace.

In the United States in particular, the speedy end of rationing had released a huge pent-up desire for petrol. In 1945, there were 25 million cars in use there; in 1950, 40 million, an increase of 60%. By 1947, America's consumption of oil products was already approaching 2,000 million barrels a year – more than the whole world's consumption a decade earlier. The country was consequently having one of its periodic nightmares about running out of indigenous oil, and, following a policy of conserving national stocks, had become for the first time a net importer.

At the same time in Europe, the price of coal was very high. During the exceptionally bitter winter of 1946–7, this made fuel oil a comparatively cheap (as well as a less messy and more convenient) source of heating. As a Shell employee remarked, 'The Englishman began to realize that there was no value in being cold', and despite continued rationing, British demand for oil products soared to 40% more than pre-war. Expensive coal also contributed to a scarcity of steel in Europe, which could only be partly alleviated by transatlantic buying – Europe's depleted coffers did not have enough dollars. The lack of steel in turn impeded the construction of tankers (Shell got its new fleet, but would have liked it still faster), pipelines and the equipment both for extracting and refining crude oil. Everywhere, the cost of refinery equipment shot up: in the United States pre-war, an investment of about $9 was needed to produce a gallon of refined products a day. In 1947, it took $35 to achieve the same result.

Sea trials of Velutina, *1949*

Shell-fuelled aircraft in the Berlin airlift of 1948–9

All told, considerable difficulties of supply were rapidly developing. (At the time, oilmen called it a crisis.) This was not because petroleum itself was running out. In the ground, said Godber, 'there is ample crude oil, which, given the facilities, could be produced and refined economically'. But to meet its markets at a time when its Far Eastern wells were still not fully functioning, Shell needed to refine more crude oil than it could produce. Simultaneously, though, in the Middle East, Gulf Oil was producing petroleum which it could neither refine nor market.

To make a liaison was natural and sensible. Gulf would produce; Shell would transport, refine, and market, and after deducting all costs would divide profits 50:50. For those with long memories, the arrangement contained two reminders of times past. The amusing one was that Gulf's president was a Colonel J. F. Drake, namesake of but no relation to Edwin Drake, the top-hatted former railroad conductor and self-styled 'Colonel' who had started America's oil industry in 1859. The other, more worrying, reminder was that Gulf was the descendant of Guffey Oil, which in 1901–3 had agreed and then failed to supply Shell Transport with oil – a failure which played a significant part in Marcus Samuel's downfall as an independent oilman. But any foreboding was unnecessary: the Gulf-Shell 50:50 deal of 1948 prospered so well that its initial 10-year term was repeatedly extended, until eventually it ran for 23 years. Moreover, the assurance of supply meant that Shell could withdraw from one part of its extensive and expensive world-wide programme, and after nine years of fruitless endeavour, exploration in Ecuador was abandoned.

Twenty years earlier, in 1928, Shell had been part (22½%) of the 'Red Line' agreement, covering the whole of the former Ottoman Empire. The Red Line participants agreed to explore within the region only as a group. Astonishing as it seems today, geologists of the Red Line companies believed there was no oil in Saudi Arabia. Consequently, two companies outside the Red Line agreement, Texaco and Standard Oil of California (Socal, later Chevron), were able to gain concessions in Saudi Arabia and Bahrain, forming an organization which became Aramco. In 1948, after prolonged and complex legal arguments, the Red Line agreement came to an end; but had it not been for the new liaison with Gulf, Shell would have had virtually no part in the region's subsequent outstanding expansion.

Turning now to 1948's other 50:50 agreement (that with Venezuela), the first thing to be said is that it was undertaken only in part for similar reasons. Though it also helped to ensure supplies during these difficult post-war years, otherwise it could scarcely have been more different, both in cause and effect. Rather than a company-to-company arrangement, it sprang mainly from the Venezuelan government's desire to improve the welfare of their nation as a whole. Since the death of the dictator Gomez in 1935, contracts between the government of the day and the major oil companies working in Venezuela (predominantly Shell and Standard Oil of New Jersey) had edged steadily towards what was now ratified: an equal division of profits. This was quite readily accepted as reasonable, not least because it was better than expropriation. But soon the inevitable occurred: learning of the agreement, other oil-producing nations wanted the same benefits from the industry. By 1952, all the Middle Eastern producers except Bahrain had reached similar agreements, often following direct advice from Venezuela. Not that the advice was given out of charity: the Venezuelans' 50:50 had made Middle Eastern oil relatively cheaper and they were losing trade. Rather than retreat themselves, it seemed better to encourage the others to advance; and thus were sown – although very few people realized it then – the seeds of OPEC, the Organisation of Petroleum Exporting Countries, which in October 1973 would send the world's economy into tailspin.

Air Commodore Sir Frank Whittle, KBE, CB, FRS

A cut-away sketch of Whittle's W2/700 jet engine

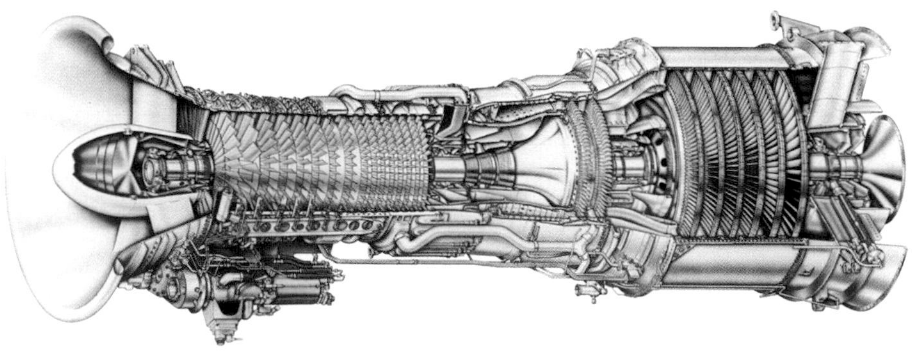

That long-term effect of the Venezuelan 50:50 agreement was something no one could credibly have foretold in 1948. All contemporary prophecies were made under the shadow of a new demon: the spread of international Communism. Immediately following the defeat of Nazism, the Cold War became the dominant theme in political thought for two generations. Its effects on Shell were swift and direct. As the grip of Communism hardened in eastern Europe, Berlin came under Soviet blockade, and for a year, from May 1948 to May 1949, its citizens – until so recently the enemy – were sustained only by supplies airlifted in from the West. Two hundred thousand flights were made by aircraft fuelled in large part by Shell. During that year Shell's organisations in Czechoslovakia and Yugoslavia were nationalized; Astra Romana, its Romanian company, was forced into dissolution; and as Godber declared angrily, in Hungary 'Familiar processes are at work...the Group's local company is being deliberately forced into bankruptcy.'

The mutual dislike and distrust between Shell and Communist regimes had never been deeper since the first expropriation in 1918. But even when surrounded by these manifold and serious difficulties, not all was utterly gloomy. 'Now', wrote Godber in 1947, 'we are moving into still further realms of scientific discovery and application, in which Shell is playing its part...' Science generally progresses by steps rather than strides, and by 1947 it was already less and less possible for any individual, any company, or even any multinational within the oil industry, to turn out some sudden advance; but whether in large ways or small, Shell remained a pioneering member. One major development in which Shell's part was limited but crucial was a pre-war invention that bloomed when the war was over: the jet engine.

The elementary principle of jet propulsion (the burning of fuel at a high temperature under pressure, to produce a gas stream of very high velocity) had been known since the Chinese first made fireworks some thousands of years ago, but its practical development did not start until the 1930s, when simultaneous competitive experiments began in Britain and Germany. In 1927, at the age of only 20, Flight Cadet Frank (later Air Commodore Sir Frank) Whittle of the RAF wrote a prophetic paper on the subject, patenting his plans in 1930 and producing a functioning jet engine in 1937; yet the first such aircraft to fly was the German Heinkel He178, in 1939. Following the Battle of Britain, *Shell Magazine* published a photograph of vapour trails in the sky (or 'visible vortices', as they were called) with, for the benefit of readers outside Britain who might never have seen them, an explanation of what they were. 'Often enough', said the accompanying article, 'these trails are the only evidence people on the ground have that a battle is going on 20,000 feet and more above their heads...' The trails were created and the battle was fought by propellered aircraft; the difficulties of development were such that jet aircraft entered the war too late to affect its outcome. However, beginning in 1952, when the first jet airliner (the de Havilland Comet) came into service, they produced an incalculable effect on peacetime life.

De Havilland's engines, wrote one of Whittle's colleagues, 'would not have been designed but for the stimulus and information provided by the early Whittle successes.' The writer continued: 'It may be said that without Whittle the jet propulsion engine and the other applications of the gas turbine would have come just the same. They would. But they would have come much later.' And probably later still, one may add, without Shell; because it was a Shell-Mex scientist, Isaac Lubbock, who in July 1940 solved a problem that had baffled Whittle's team. Their engine's combustion chamber, crucial to the whole project, was maddeningly and puzzlingly unreliable: satisfactory results in the test-bed could not be replicated elsewhere. Lubbock invited two of Whittle's men to inspect similar experiments in Shell's Fulham laboratory. They 'went, saw and were conquered', said Whittle. Borrowing the equipment, 'they brought it back and had it rigged up for me to see the next day. Thereafter we concentrated our effort on the "Shell" combustion chamber...'

Results came very rapidly. The borrowed combustion chamber was so impressively successful that, stressing his words with italics, Whittle

A de Havilland Comet, the first jet airliner, fuelled and lubricated by Shell

wrote '*...the introduction of the Shell system may be said to mark the point where combustion ceased to be an obstacle.*' When next flying away on holiday or business, we might remember his verdict.

After retiring from the RAF, Whittle worked for Shell for eight years (1953–61) as widespread jet travel literally took off. Another pre-war invention which Shell helped spread in peacetime, and which altered the oil-refining industry just as much as jets altered the travel industry, was catalytic cracking. The theory of catalytic cracking was known as early as 1915, but remained undeveloped until 1937, when a Frenchman named Houdry announced his work to the industry. It offered considerable improvements over the earlier system, thermal cracking. Thermal cracking remained useful, because it could cope with heavy residues, but it required high temperatures and pressures. Catalytic cracking did not – instead, a catalyst was introduced to do the same job more easily and cheaply. That was the idea, anyway; but Houdry demanded such an exorbitant fee for a fully paid-up licence (allegedly $50 million) that, spurred by the pressures of wartime, Shell and a number of other oil companies decided to work out a viable alternative together. They succeeded quite quickly, and 'cat cracking', as it was nicknamed, soon became an industry standard: Shell's first cat cracker (at Wilmington, California) was ready for operation in November 1943, and in 1951 its first such plant in Europe was inaugurated at Pernis in the Netherlands, using a still further improved method – in which the catalyst was *platinum*. Fortunately the metal was re-usable.

In comparison to these great post-war developments of shared science, another which was exclusively Shell's might at first seem small – its progress in the manufacture of synthetic glycerine. But in the words of a spokesman for its main manufacturer, the US-based Shell Chemical Corporation,

> glycerine is probably used more, and talked about less, than any other comparable chemical compound in existence. Along with water, it is one of the world's most useful liquids, being a component of literally thousands of everyday products...

In the war it had been a vital component of high explosive, and for that and many other purposes – the manufacture of paints and varnishes, shaving cream, toothpaste, cakes, glue, hair tonics, soft drinks, mayonnaise, cellophane and more – it was extraordinarily valuable in peacetime too. Many people outside the industry would have been astonished to realize the number of items in their larders, refrigerators and bathrooms which contained this product of crude oil. But the industry as a whole was fully aware, and in 1948 honoured Shell with the 'Award for Chemical Engineering Achievement', an industrial Oscar.

Welcome as it was, the award was in a sense only public acknowledgement of the great success Shell was enjoying in chemicals generally. Taking just Shell Chemical Corporation as an example, this was vividly demonstrated by a glance at sales receipts over a period of 15 years. In 1931 Shell Chemical recorded receipts of just $135,000 – barely 0.08% of Shell's total sales receipts in the United States for that year, which exceeded $174.5 million. By 1946, Shell's total US sales receipts had gone up more than 2.5 times to $442 million, but Shell Chemical's receipts had multiplied over 170 times, to $23 million, or 5% of the total.

Another contemporary Shell development which deserves a niche in its history was the engine lubricant Shell X-100, evolved at the Merseyside Thornton Research Centre. Britain's wartime petrol distribution system, the 'Pool Board', was wound up on 1 July 1948. The olive drab 'Pool' pumps with their just-adequate fuel remained in operation for another five years: it was not until 1 February 1953 that oil companies were allowed to sell branded petrol again. Meanwhile, though, their other products could be branded, and Shell, eager to have its name before the public once more, devised Shell X-100. Godber rightly described the new lubricant as 'superior to anything that has hitherto been produced': compared to any other existing lubricant, it gave a measurable reduction in engine wear. From the time it was first put on sale in 1949, it proved a great success wherever it was marketed, and the following year brought an excellent means of publicizing it and other products: motor racing's first Formula One World Championship, in which Shell immediately proved it would excel. Giuseppe Farina's winning Alfa Romeo was fuelled and lubricated with Shell products – a combination which, for other drivers in other cars, over the next 45 years or so brought victory in more than 150 *grand prix* races.

Shell / BP racing service pits, Silverstone, 26 August 1950

Together with raising the necessary funds, Shell's huge post-war programme of reconstruction and expansion was as grand and ambitious a task as it had ever undertaken. Its production figures reflected this. In 1938, the last full year before war broke out, Shell companies produced an average 582,000 barrels of crude oil per day (bpd). In 1946 the figure rose to about 630,000, and by 1951 had nearly doubled to 1.2 million bpd. It could fairly be said that for drive, daring and determination under its new chairman, this period in Shell Transport's history had been matched by only two others: Marcus Samuel's audacious launch of the original fleet, and Henri Deterding's metamorphosis of the ailing "Shell" into the flourishing partner of Royal Dutch. Yet great post-war expansion was by no means Shell's sole prerogative; rather, it was an industry-wide phenomenon. Among the other British-based oil companies, Anglo-Iranian (formerly Anglo-Persian, which in 1954 would be renamed BP) doubled its production in the same period from about 390,000 bpd to about 778,000; and large as Shell's own new production statistics were, they placed it not in first but in second position among the world's oil majors. In the immediate post-war years Standard Oil of New Jersey consistently produced more, with figures rising from about 940,000 bpd in 1946 to 1.45 million in 1951.

But in May 1951 a growing storm burst suddenly and violently on Anglo-Iranian. All its assets in Iran, source of most of its crude oil and products, were nationalized by the militantly nationalist prime minister Dr Mohammed Mossadegh. It might seem at first that this had little to do with Shell. In fact, it was the first great post-war oil shock. Reverberating throughout the industry, it affected every oil company, and the politics of many nations.

CHAPTER TWELVE

All the World's a Stage: 1952–1956

Though Iran had never been a colony of the West, the country felt subordinated to British interests – especially to its oil industry, exemplified by Anglo-Iranian's refinery (the world's largest) at Abadan. Iran's action against Anglo-Iranian was both a symptom and an outcome of the way post-war politics continued to remould the world.

For similar reasons Shell had experienced a recent setback of its own in China, where it had enjoyed a strong traditional market from the Samuels' earliest days. When the Communist regime came to power in 1949, the pattern of international trade there became (in Godber's words) 'very obscure' – an understatement, as two world maps published by Shell in 1949 and 1951 demonstrated. Both showed its areas of operations, be they production, marketing or any of the other activities of an integrated oil company. In the 1949 map, printed before the Communist takeover, China's colour-coding indicated a marketing presence throughout the nation. However, on 17 June 1950 the British government banned petrol exports to China; eight days later, the Korean War began with the invasion of South Korea by Chinese-backed forces from the north; and in Shell's 1951 map, China was a blank. Even so, Godber refused to be pessimistic, believing correctly that in due course Shell would be able to resume its retail operations there.

Opposite page: Venezuela – a geologist begins work

Iran was a much trickier question, in which Shell could not avoid involvement. A embargo enforced by Britain's Royal Navy effectively stifled the country's oil exports, and national production fell from a high of 666,000 barrels per day in 1950 to a trivial 20,000 bpd in 1952. In earlier years, lack of formal diplomatic representation in distant lands had often obliged oil companies to perform aspects of intergovernmental relations. Now, their task was somewhat simpler: the maintenance of supplies to consumers. The oil industry of the non-communist world pulled together to make good the shortfall.

Mohammed Mossadegh (1881–1967), Prime Minister of Iran 1951–3

For their part, national governments had long learned the need for oil, and the crisis illustrated vividly how enmeshed oil and international politics had become. While the British government (a 51% shareholder in Anglo-Iranian) wanted 'its' company's assets back and believed the embargo would eventually achieve that objective, the US government feared that if prolonged, the embargo would tip Iran into the communist world. At one point, after eighty hours of negotiation with the violently anti-British Mossadegh, it appeared the Americans might have found a solution. With emphasis on the Group's Dutch connection, they proposed that Royal Dutch/Shell should operate the Abadan refinery, with Anglo-Iranian buying the oil on a 50:50 basis. Mossadegh agreed, as long as no British technicians were involved – a condition which so insulted the British Foreign Secretary Anthony Eden that he rejected it outright.

Sir (Robert) Anthony Eden, subsequently 1st Earl of Avon (1897–1977), was Britain's Foreign Secretary three times – 1935–8, 1940–5 and (seen here) 1951–5, when he succeeded Sir Winston Churchill as Prime Minister. He resigned abruptly from that office on 9 January 1957

Through 1952 and into 1953 the situation grew increasingly delicate. On 22 October 1952 Iran severed diplomatic connections with Britain, and while Mossadegh arrogated more power to himself, the country moved economically closer to collapse and politically closer to the Soviet Union, receiving as the new Soviet ambassador Michail Alexandrovitch Silin – who had been ambassador in Prague during the Communist coup of 1948.

The political climax came in August 1953. In Operation Ajax, Tehran experienced (as Anglo-Iranian's chairman Sir William Fraser tactfully put it) a 'change of government'. More plainly, the British and American administrations had decided to tap a groundswell of rising discontent in the increasingly impoverished Iranian people, and lent support to a movement designed to topple Mossadegh. The coup came very close to failure – indeed the Shah and his family had briefly to flee the country to avoid imprisonment at the hands of Mossadegh supporters. By the end of the month, the situation was reversed; but with the Shah back in power and his former prime minister in jail, the popular mood in Iran was still adamantly against Britain in general and Anglo-Iranian in particular. The industrial aspects of the problem remained, and Rab Butler (Britain's Chancellor of the Exchequer) remarked privately to the Secretary of Fuel and Power that he was 'completely stumped' for a solution.

Richard Austen (Rab) Butler, subsequently Lord Butler (1902–82), when Chancellor of the Exchequer (1951–5)

However, the US government's State Department had a suggestion: the other oil majors should form a consortium to carry out the production and refining of Iranian oil. But the majors were not keen on involvement in a country which they viewed as unacceptably volatile; and at that very time, the US government's Justice Department launched a criminal action against them for breaking anti-trust laws in the Middle East some years earlier, by the creation of the Arabian-American Company (Aramco) and the Iraq Petroleum Company (IPC) – even though both were comparable to the State Department's current proposal and had received governmental approval.

The State Department had not known what the Justice Department was doing. Worse, the Justice Department's suit was targeted not only at American companies, but at all companies involved in Aramco or IPC –

which in the latter case included Shell. The British government protested that the Justice Department was overstepping its area of authority. The American government concurred, and made the Justice Department modify its action. In their joint national interests, the two governments also made the proposed consortium in Iran go ahead; and so it did, though with reluctance. (An official from Standard Oil of New Jersey said plaintively, 'If the US and British governments hadn't really beat us on the head, we wouldn't have gone back.') Seven of the majors, including Shell, bought 60% of Anglo-Iranian's interests in Iran for a total of $90 million down,

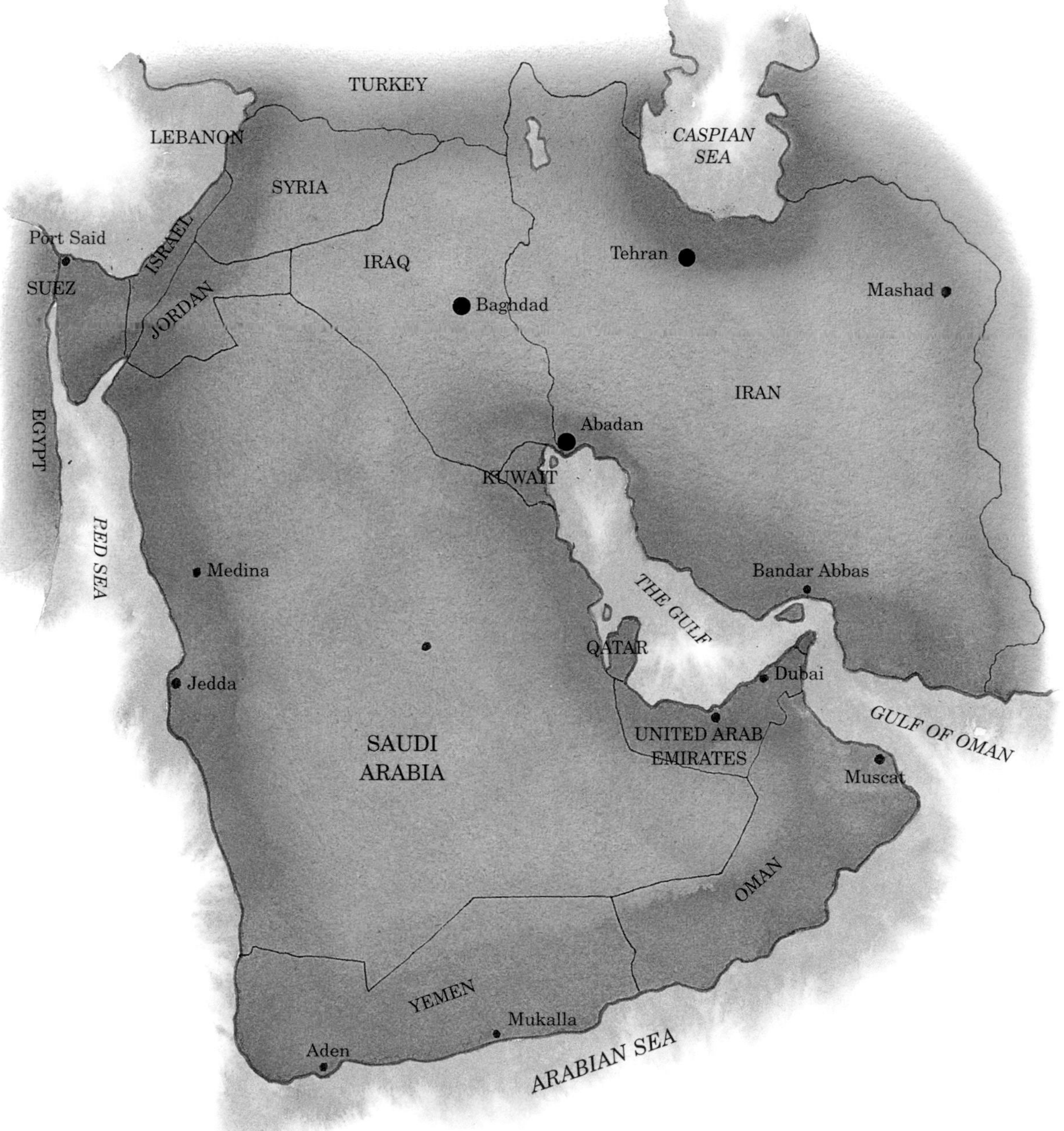

plus royalties to a further maximum $510 million. Shell bought 14% of the whole, and John Loudon, the Dutch head of its Committee of Managing Directors, remarked wryly: 'It was a wonderful deal for [Anglo-Iranian's chief, Sir William] Fraser...After all, Anglo-Iranian actually had nothing to sell. It had already been nationalized.'

In all this, there was an important lesson for everyone outside the oil industry who had ever suspected that any large oil company, sooner or later, would use its economic strength to try for political power. By traditional inclination and by learning from the mistakes of others, Shell had always been (and is) firmly apolitical. But even if oil companies dreamed

The charismatic John Loudon, left, (1905–96), head of the Group's Committee of Managing Directors 1959–65 – 'the true father of the Royal Dutch / Shell Group in the last third of the 20th century'

of political control, they could not achieve it. The old Standard Oil tried, and its efforts in that direction were one factor leading to its break-up in 1911 by American courts of law. In Iran forty years later, Anglo-Iranian was denounced as a state within the state; but it was states, not companies, which had created the crisis – and states which resolved it. The events of 1911 and 1951 emphasized that political power remained with the politicians. If their desires and those of the companies coincided, harmony prevailed; otherwise, it was not the governments but the companies which were obliged to give way – even to the extent of buying parts of the industry which they did not want to own.

The Iranian controversy was one of the dominant constraints upon Shell and its industry at the start of the 1950s. But in reviewing the history of such an organization, it would be an easy mistake to look only at the great external formative forces. To make a geological parallel, a study of the movement of tectonic plates may explain the formation of continents and mountain ranges, but it does not describe the countryside or a culture; and Shell's countryside and culture are both shaped by its people. Time, then, to move away from the turmoil of politics, and to return for a while to Shell people.

On 27 November 1952, a few weeks after his 75th birthday, Sir Robert Waley Cohen died. He had worked for Shell for 51 years. When he joined the "Shell" Transport and Trading Company in 1901 it already had assets valued at £3.5 million in the money of the day, but it was still only an offshoot of the Samuels' family businesses, with no offices of its own – and the total staff of all their businesses numbered just forty-eight. By the time of Waley Cohen's death, the Royal Dutch/Shell Group's world-wide payroll listed more than 200,000 people.

Sir Robert Waley Cohen KBE (1877–1972), seen here in 1951

A very large proportion of those spent a very large proportion of their lives travelling; and one who did left a legacy of such beauty and enduring value that both he and it deserve particular notice. In 1931, Rodney Searight (1909–91) was sent to work for Shell in the Middle East. He came to love the region and remained there for many years, journeying to all parts, and becoming increasingly fascinated by and expert in Middle Eastern history and culture. He was also a man of great artistic sensibility, and in his travels he gradually built up a wonderful collection of 18th- and 19th-century drawings and watercolours depicting the people and architecture of the lands in which he worked. In 1985, the Searight Collection became a distinguished part of the holding of the Victoria and Albert Museum in London.

Of course, it was not only men who travelled far and wide on Shell's behalf. Throughout Robert Waley Cohen's life, the great majority of positions of responsibility were indeed held by men, with female staff filling junior roles as secretaries, telephonists and so forth; social factors of education and expectation made this inevitable. A few women, but only a few, held middle-management posts, and decades more would elapse before women entered Shell's senior levels. But from the beginning, an identifiable group not included on the payroll (at least, not directly) were the ladies who, simultaneously supporting and dependent upon their menfolk, formed that nebulous, shifting yet essential part of the whole – to use the title they gave themselves, the Shell wives.

Their part in Shell's story has not hitherto been touched upon to any great degree, simply because in the early decades they left few records. One of the earliest chronicles of a 'Shell wife' participating in (she did not entirely see it as enjoying) a working journey was made in 1909, when May Abrahams joined her husband Mark – nephew of Marcus and Sam

An image detail from the Rodney Searight collection, now in the V&A Museum, London

Samuel, and the man responsible both for finding Shell Transport's first oil and for setting up its refinery in Balik Papan – on a long voyage of inspection in Africa.

'Left London on Tuesday Sep. 14th '09 with Mark', she wrote. 'We stopped in Paris for two days, then on the 16th in the evening came to Marseilles where we went on board S.S. Prinzessin (German East African Line) – Our berth is a nice one & we hope to be comfortable.' Corsica, Sardinia, Naples and Pompeii followed ('Have *not* felt seasick'), with binoculars out to view the distant 'towns spoilt by earthquakes last year – the ruins are pitiful to see...' Arabs at Port Said were 'quite amusing, but try to cheat or toss one for everything...I only bought a Maltese lace scarf & had to bargain for it for half an hour...' The Red Sea 'deserves its bad name, for it *is* the hottest place it has ever been my luck to strike...the perspiration runs off me in perfect rivers.' Then on to Aden ('Did not land, but from the sea at daylight nothing to please the eye – just a barren bit of world with a few houses scattered round') and Mombasa, where she was carried ashore through the surf by two local boys ('the sensation was not agreeable'); then further on to Zanzibar and all the way down Africa's east coast, returning home at last on 23 January 1910 after 'a miserable journey, owing to the floods in France...crossed the Channel in an old

Cargo boat – waited three hours on Dover pier (shivering with cold) before train left & arrived in London at 4 a.m....'

Mrs Abrahams' experience of travelling with her husband was rare then, but it was one which many a later Shell wife would recognize with some sympathy. May knew she had committed herself to no more than a long journey. In the event she did not like it very much, which was unfortunate, but it only lasted a few months. The typical Shell wife of later decades put up with (but often enjoyed) much greater upheaval, not just once but repeatedly. In 1993, some of their characteristic experiences were gathered together and published in a book with a very apt title: *Life on the Move*. One contributor recorded setting up home in four successive houses inside two years. Those more fortunate might find themselves moving every two or three years: a posting to Venezuela (which in its boom years, the 1950s, was the most exciting place in the world for anyone in the industry) could be followed by second to Borneo, a third to China, a fourth to some part of Africa and a fifth to London or the Hague. All great fun if you enjoyed travelling; but before you knew it, *voilà* – ten or fifteen years of one's life were gone. And where exactly was home?

In some places the expatriates' way of life could be enviable, even luxurious. In 1929–32, for example, Shell's 'up-country' manager in China was based in the International Settlement in Shanghai, and more than sixty years later his son still clearly remembered the household: to look after the family of four, there were two cooks, three houseboys and two amahs (nannies), as well as a chauffeur to drive the Essex motorcar – an often-used vehicle, not least because when Father was at home, both parents would be out almost every evening, Father in evening dress with tails. That particular Englishman and his family were in China for twenty years, which afforded good background continuity. But the average three-year stints in different places still applied to them, and in their twenty years 'Uncle Joe' (as Shell was often nicknamed) required them to move house seven times within China's sprawling regions.

Left London on Tuesday Sep: 14th 09
with Mark & Mr Comerford – We
stopped in Paris for two days, then
on the 16th in the evening came to
Marseilles where we went on board
S.S. Prinzessin (German East African Line –)
Our berth is a nice one & we hope to
be comfortable –
The Steamer leaves Marseilles on Sep: 18th
at [illegible] – It is pouring wet [illegible] – we
lunched at the Hotel de Louvre but
returned to the boat as it was too
wet to remain ashore
Sep 19th – Lovely weather, passed [illegible] &
Sardinia this [illegible] – Have not felt
sick up to the moment – The Steamer very
clean, but food rich & greasy, as usually
is on German Boat – plenty of Germans
& a funny mixture of folks altogether –
Several nice English people, quite [illegible]

The opening page of May Abrahams' 1909 diary

Shanghai then was one of the best postings one could hope for outside Europe, offering good pay, domestic help which was both plentiful and cheap, and a stimulating combination of the civilizations of east and west. Expatriate life elsewhere was much more for the young and hardy, without children. Typical accommodation was in simple wooden bungalows in the jungle. (In Sarawak, the jungle came so close that one could shoot wild

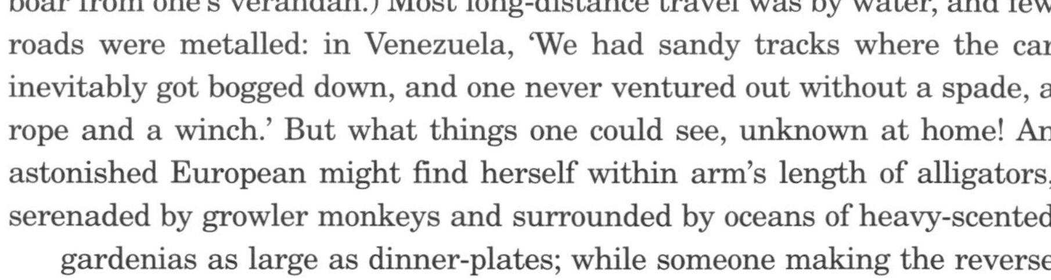

boar from one's verandah.) Most long-distance travel was by water, and few roads were metalled: in Venezuela, 'We had sandy tracks where the car inevitably got bogged down, and one never ventured out without a spade, a rope and a winch.' But what things one could see, unknown at home! An astonished European might find herself within arm's length of alligators, serenaded by growler monkeys and surrounded by oceans of heavy-scented gardenias as large as dinner-plates; while someone making the reverse journey could admire the tulips of Holland, revel in the public transport system, and even *enjoy* crossing the English Channel by ferry.

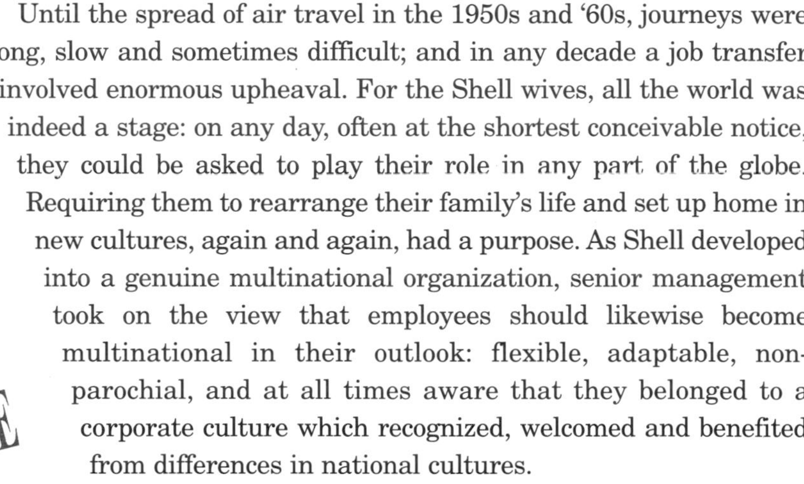

Until the spread of air travel in the 1950s and '60s, journeys were long, slow and sometimes difficult; and in any decade a job transfer involved enormous upheaval. For the Shell wives, all the world was indeed a stage: on any day, often at the shortest conceivable notice, they could be asked to play their role in any part of the globe. Requiring them to rearrange their family's life and set up home in new cultures, again and again, had a purpose. As Shell developed into a genuine multinational organization, senior management took on the view that employees should likewise become multinational in their outlook: flexible, adaptable, non-parochial, and at all times aware that they belonged to a corporate culture which recognized, welcomed and benefited from differences in national cultures.

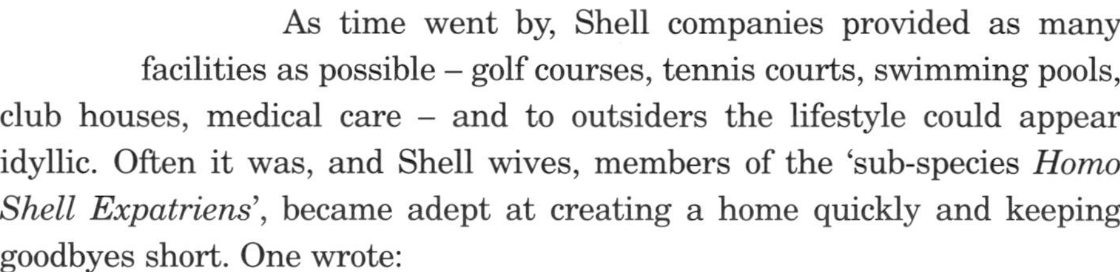

As time went by, Shell companies provided as many facilities as possible – golf courses, tennis courts, swimming pools, club houses, medical care – and to outsiders the lifestyle could appear idyllic. Often it was, and Shell wives, members of the 'sub-species *Homo Shell Expatriens*', became adept at creating a home quickly and keeping goodbyes short. One wrote:

> It has always been give and take...I believe I got as much as I gave to Shell – or gave as much as I got. Shell has meant far more to me than just being my husband's employer. Shell provided me with a training, a job and continual travel in comfort. Through Shell I met my husband and most of our international circle of friends. I have never felt recognition for my role as a Shell wife to be lacking.

But the frequent house-moving was extremely stressful too, sometimes inducing an overwhelming sense of rootlessness, isolation and superficiality – perhaps particularly in those who had given up a satisfying job at home so that their husbands could accept a career opportunity, and who now found themselves with no outlet for their talents and no established friends in a small claustrophobic expatriate community. 'Damn you, Uncle Joe,' another Shell wife burst out in the early 1990s. 'We are only a pawn on your chessboard, you control our lives. *"Such is life with Shell."* Oh, why?'

She answered her own rhetorical question with absolute frankness: 'Because of the good money we earn, which we cannot do without any more. Because of obligations: children's education, old age, loyalty to the Company.' No one could argue with her order of priorities, but it was interesting to see that even at such a low moment and at such a late date, loyalty to the company still figured. Deep-rooted and tenacious, it did not always flourish, but it did survive. 'Besides,' the lady concluded philosophically, 'nothing is perfect.'

Returning with that inescapable truth to the early 1950s, it would be idle to pretend that Shell is or ever was perfect. Its goal has always been to be the best possible, and if that is short of perfection, it is ambitious enough. An example came in Britain in 1953, when after eleven years of restriction originating from World War II, oil companies were at last allowed to sell petrol under their brand names. This was a great day psychologically – one could drive when and where one chose, limited only by personal budget, not by government restrictions – and Shell in the UK acted to reclaim its popular pre-war image. Preparing the public for 'the day', Shell's advertisers decided to take over the entire national consciousness about the end of rationing. Never mind a return to the free market; more importantly, 'Shell Day is Coming!' Similarly, echoing the popular pre-war slogan 'That's Shell – That Was!', people were now told 'That's Rationing – That Was!', and smiled at the thought that the world was returning a little closer to normal.

The campaign succeeded. In Britain's renewed free market, Shell's market share showed no decline. This was partly because consumers enjoyed the advertisements' simple lightness of heart, taking them back to more innocent days; but the success was supported by more than nostalgia. During the years of Pool products, Shell chemists had continued working to make a better petrol. The habit of trying at all times to improve the product could be dated from 1901, when Robert Waley Cohen's inquiries showed that with simple distillation, Shell's Borneo oil produced a petrol which was markedly better than anything else. (Shortly before he died, Waley Cohen liked to relate how all those years ago, a sceptical Marcus Samuel had put some in his motor lawnmower as 'an empirical test'. The machine responded so vigorously that Marcus, running behind it, could not keep up.) This time, the improvement was the addition of tricresyl phosphate (TCP), which

reduced the fouling of spark-plugs. After bench tests and road trials totalling over 70 million vehicle miles, in 1953 the premium fuel was first marketed in the United States as 'Shell with TCP'. In Britain (in order to avoid confusion with a well-known household antiseptic) it was renamed at its 1954 launch and called 'Shell with ICA', the letters standing for Ignition Control Additive. As competitors rushed out their own versions, each claiming qualities beyond the others, Sir Frederick Godber chose to emphasize in his annual review the highly competitive state of the marketplace, declaring mildly that the new additive was 'believed to be one of the most important advances in this field during the last 30 years'. No doubt it is wise to avoid showing self-satisfaction, but he could have: for whether it was Shell with TCP or with ICA, consumers queued to buy it, pushing Shell's sales in America up by 30% in a single year.

EVERYWHERE YOU GO
YOU CAN BE SURE OF SHELL

Yet it was neither easy, nor by any means cheap, to maintain the production levels demanded by the market and forced by competition. New commercially viable sources of crude oil and gas had to be found regularly; new or improved methods of refining had constantly to be developed; and new marketing devices and systems had to be invented, for one could not rely upon the sale of high-priced premium grades alone, nor on invariable customer loyalty. In this period, not only Shell's construction of new plant but also its exploration, in all parts of the world, was at a new peak. In Canada in 1952–3, concessions were acquired on large tracts of tar-bearing sands at Athabasca, oil was found in Saskatchewan and a gas field was found at Okotoks in Alberta, complementing another discovered in 1944 at Jumping Pound. There, a big desulphurization plant was brought on stream in 1952; once extracted, sulphur – an unwanted impurity in gas – could readily be sold to manufacturers of sulphuric acid. Gas was also found at Denekamp in the Netherlands, while at Rijswijk, just outside the Hague, the first oil in western Holland was discovered. Oil concessions were acquired in Qatar; two new oil fields were found in Trinidad; and literally hundreds of wells were drilled every year in the United States, their numbers rising from 391 in 1951 to 1,040 in 1956. At Weeks Island in Louisiana there were nine which were over 15,000 feet deep – a record, for there were only three other wells in the world at such depths. Similarly, there were only ten fields in the world as productive as Iraq's giant one at Kirkuk. Operated by the Iraq Petroleum Company (in which Shell had a 23.75% stake), the field had been inaugurated by the first gusher at Baba Gurgur back in 1927. By 1953 it had produced more than 700 million barrels of oil. Seven hundred million barrels in 46 years was pretty impressive; but in 1954 and '55, the field produced half as much again.

In the five years 1952–6 Shell Companies spent a total of £721 million sterling on world-wide exploration and production – on average, £395,000 every single day, year in, year out. With exploration becoming more and

Experimental work in the research laboratory

The Carrington site, 1956

more scientifically based, it was already well known that palaeontology – the study of fossils – could reveal clues to the presence of oil. In 1952 Shell augmented its existing palaeontological laboratories in Britain, Europe and the United States with three new ones in Egypt, Tunis and Nigeria. Alongside palaeontology came palynology, the study of living and fossil pollen cells. This was a branch of science still in its infancy (work in it had begun only shortly before the outbreak of World War II), and Shell pioneered its use in oil exploration. A new laboratory was set up for the purpose at Seria in Borneo in 1952; but it was the existing laboratory, in the Hague, which in that year received samples of fossilized pollen from Maracaibo, Louisiana and Nigeria. In association with BP (as Anglo-Iranian became in 1954), Shell had undertaken surveys and test drilling in Nigeria as early as 1938, but without success. However, its pollen samples were assessed as 'very promising' – and in 1953 Shell discovered onshore oil there, the first in a country which would become one of the world's major producers.

Exploration in New Zealand's North Island, which had been abandoned in 1915 after five years' fruitless effort, was restarted in 1955. At the same time, Turkey was surveyed, both geologically and geophysically, in its west and south-east; concessions were gained in Libya; and – an unusual event then – a sound-proofed drilling rig was used in Venezuela, to prevent undue 'noise nuisance'.

That was also the year when, in the Niger delta, Shell sank Nigeria's first offshore oil well. Remembering that the world's first offshore oil had been struck in commercial quantities only in 1947, when the small American company Kerr-McGee went a few miles out from Louisiana, it is astonishing how quickly and thoroughly Shell joined in this new part of the business. By the end of 1955, only eight years after Kerr-McGee's opening strike, Shell had more than 300 offshore wells world-wide. Nearly 200 of these were off the coasts of Louisiana and Texas; one, established in 1952 a mile off Borneo, was the first major submarine operation in the whole British Commonwealth. (The same one had a further distinction, hair-raising and probably uncomfortable: its sole means of transport to and from the shore, for people and equipment alike, was an aerial ropeway.) With offshore business rapidly becoming big business, 1952 also saw

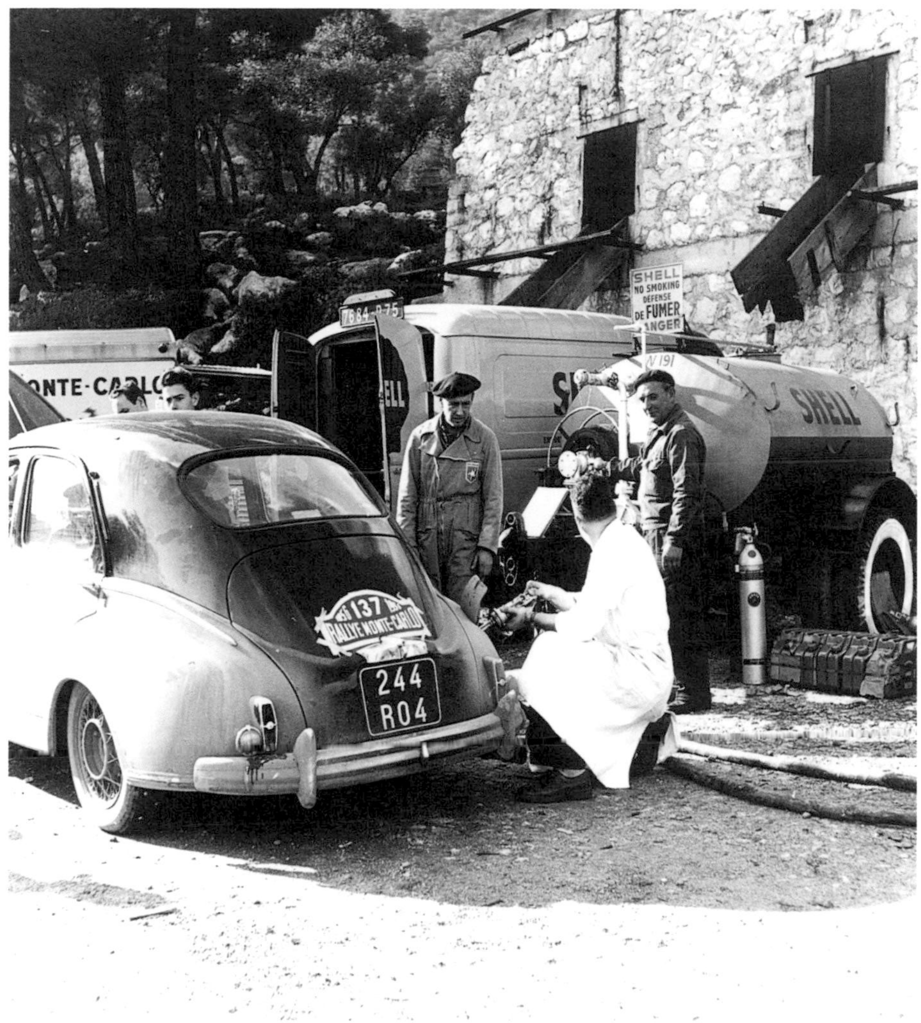

the introduction of a British-built diesel-electric drilling barge on Lake Maracaibo, while 1954 marked the appearance of a 400-ton moveable platform drilling a test well five miles off Qatar.

When that test proved unsuccessful, the platform was shifted in 1955 to another site 40 miles further out: a somewhat impractical proposition for ropeway communications. Instead, the platform was supplied by helicopters, which along with amphibian aeroplanes were becoming indispensable for the exploration and supply of inaccessible areas, such as in a geophysical survey of New Guinea in 1952. The same country provided in 1955 a notable example of the helicopters' speed and efficiency, when it was decided to drill an exploratory well deep in the jungle. The traditional approach – cutting a road to the site – would have taken 18 months before construction of the rig could even commence. With 1,100 helicopter flights, the 650 tons of necessary equipment were on site in just 22 weeks.

The colossal expenditure of £721 million on exploration and production was the largest single set of bills Shell had to face in 1952–6, but by no means the only one. Oil refineries and chemical plants swallowed another £361 million; marketing, a further £109 million; the fleet, £101 million. Even the 'Miscellaneous' category (practically speaking, the petty cash) included costs amounting to £13 million. It would have been from there, incidentally, that money came for the commissioning in January 1955 of an exhibition in the Royal Watercolour Society's galleries. Shell had invited the artists to visits refineries and other installations, giving them carte blanche to paint and draw whatever they chose. The 90 resultant works of art constituted (in the view of the Slade Professor of Fine Art at Cambridge University) 'an extremely interesting example of intelligent patronage by an industrial organization.' He need not have sounded *quite* so surprised.

In among the large bills there was also the funding for agricultural research. At Sittingbourne in Kent, Shell's Woodstock House farm estate (bought in 1945 for £45,000) was quickly becoming a world centre of excellence in this branch of science, specializing initially in the synthesis and formulation of pesticides. Its first laboratory had been in the old dairy, only eight feet square. By 1956, however, its staff had grown to nearly 100 with upwards of 40 scientists, and in May of that year Sir William Slater, FRS, secretary of the Agricultural Research Council, officially opened its impressive new buildings: 25,000 square feet of laboratory space and 4,600 square feet of glasshouses, constructed at a cost of £100,000 – more than twice the original price of the whole 332-acre estate.

Comparable progress in oil and lubricant research was being made at Thornton near Liverpool. From those laboratories, birthplace of Shell with ICA and Shell X-100, came new two-stroke oils, the improved 'Multigrade' and 'Ashless' car-engine lubricants, and in 1955 a new marine diesel lubricant with the somewhat poetic name of Alexia. (It is said that the marketing people were not wholly impressed by this product, commenting, 'Magnificent – but could you make it look less like salad cream?') At the same time Thornton's remit was extended beyond fuels and lubricants to include metallurgy, in order to look at and diagnose the reasons for engine failures, especially for Shell customers. All sorts of engines were included – 'anything', as its Head of Engineering said, 'which went up and down or rotated.' While a sister organization in The Hague investigated hydrocarbon corrosion and refinery failures, Thornton's metallurgists had a very wide clientele: airlines, the military, railways, shipping, the police. The research into engine failure soon had nearly fatal consequences. On trials in the Irish Sea, the engines of a passenger liner called *Reina del Pacifico* exploded, killing several crew members. Accepting a request to investigate the cause, Thornton engineers replicated the conditions – and, unintentionally, the explosion as well, even though their own engine,

Calder Hall in Cumbria

supplied by Shell Tankers, was fitted with relief valves. Fortunately no life was lost, and the engineers did solve the problem eventually; but by coincidence, one of Shell's visiting artists was present, and afterwards painted a watercolour of the spectacular damage – a gift of a subject, if not exactly the planned 'intelligent patronage'.

Thornton's most unusual venture at this time was into radioactivity. This took two forms: the use of radioactive isotopes to monitor wear on piston rings in car engines, and the provision of lubricants for atomic reactors. The latter came about almost by chance, when samples of oil were subjected to radiation from Cobalt-60 to see the effect on the hydrocarbons. What was discovered was that one sample proved impervious to radiation, remaining liquid throughout the experiment. As it happened, a nuclear reactor was being built at Calder Hall in Cumbria, barely 90 miles from Thornton. Lubricating the reactor's graphite moderators was a problem no one had yet satisfactorily answered, but Thornton's fortuitous find was the basis of the solution. The peaceful use of nuclear energy – in later years a highly contentious subject – was widely viewed in the early 1950s as the world's best bet for future energy resources. Even then, Frederick Godber suspected that this optimism 'suffered from some exaggeration'. But when Calder Hall was opened by the Queen on 17 October 1956, it was the world's first commercially operating nuclear reactor; in the context of the time it was a source of legitimate pride for Shell to have gained the contract for all its lubrication.

As chairman of Shell Transport and Trading, Godber began his annual statement to shareholders in 1956 with remarks which were more usually kept to the end: the praise of staff. (Since it was the staff upon which the entire operation depended, this was perhaps an overdue re-ordering.) 'As you will have seen', he said, 'the post-war years have been a period of continuous progress and almost dramatic expansion'. Recognizing that this would have been impossible with the 'immense drive and ability unsparingly applied by the staff', he continued:

EVERYWHERE YOU GO
Clive Gardiner
BOURTON-ON-THE-WATER
CLIVE GARDINER.
YOU CAN BE SURE OF SHELL

> The good relations which exist within the Group are a subject for frequent comment. In a big organization which is still expanding, bigness itself can create problems and there could be a danger of losing sight of the individual; it is therefore our prime concern to watch the progress and well-being of each individual...

The re-ordering of Godber's annual statement was not only an appropriate recognition of the staff's centrality; it was also a reflection of two new thoughts on the question of bigness. In the interests of simplification, Shell Transport and Trading was beginning to re-order itself and the Group of which it was a parent. During this process, the name of the Anglo-Saxon Petroleum Company (the British operating company established in 1907) disappeared in 1955; its business and assets were vested instead in The Shell Petroleum Company. Moreover, it had dawned upon the London-based Shell Transport that to have staff scattered throughout the capital in 30 different buildings was inefficient: there should be a focal point in which all London staff worked, a centre. On the south bank of the Thames a large acreage of slums had been cleared in preparation for the 1951 Festival of Britain. Shell Centre would be built there.

Read out at the AGM on 30 May 1956, the written version of Godber's statement for 1955 contained another small but important indicator of change: his signature. Rather than 'Frederick Godber', as it always had been, it now read simply 'Godber'. On seeing the surname *tout court*, anyone in his audience who did not already know rapidly realized that their chairman had been ennobled in the 1956 New Year's Honours, and was now Lord Godber. Shell people, shareholders and staff, liked that: it seemed eminently suitable once again to have a lord as chairman, and it certainly reflected well on their company.

Shell's survey of South Georgia, 1955–6

Thus, shaped by its people, Shell's countryside was all in all a happy one, energetic, busy and with a firm sense of purpose and direction. But the earth beneath their feet was about to give another mighty tectonic heave.

After a coup d'état in Egypt in 1952, a republic had been declared there in 1953. In 1954 President Gamal Nasser came to power, and, on 26 July 1956, nationalized the Suez Canal. On 29 October, Israeli troops invaded Egypt's Sinai peninsula. On 30 October, Britain and France announced their intention of occupying the Canal Zone. On 31 October, to strong American censure, their aircraft bombed Egyptian airfields; and on 1 November, Egypt blocked the Canal with sunken ships, broke diplomatic relations with Britain, and seized Shell's Egyptian properties. Since 1892, when Marcus Samuel despatched Shell Transport's first tanker *Murex* to inaugurate the bulk carriage of oil through the Canal, the Middle Eastern trade had grown so much that two-thirds of all oil consumed in western Europe reached the Mediterranean through either the Canal or the Iraq Petroleum Company (IPC) pipeline; but the day after the Canal was blocked, the pipeline's pumping stations were blown up. With western Europe's winter fast approaching, the world's second post-war oil crisis had arrived.

CHAPTER THIRTEEN

Suez and After: 1957–1962

Sequestration, expropriation, nationalization – legally, their definitions were different, but in practical terms they all meant the same: the seizure of assets, loss of control, loss of investment, and the prospect of only limited compensation (if any) at some far-off date (if ever). It was not an experience on which any oil company wished to become an authority, but Shell was; and because of that, as far as its history is concerned, two of the most notable aspects of the Suez Crisis of 1956–7 are how Shell analyzed the crisis when it was breaking, and the speed of its repaired relations with Egypt when the crisis was past. Both stood in marked contrast to the British government's management of events.

Firstly, Shell's analysts did not consider it likely that anyone would go to war because of the nationalization of the Suez Canal: on such an issue, they estimated, 'No civilized nation will choose military action if it can avoid it.' They were sure the problem could be solved by patient negotiation; but negotiation was far from governmental thinking in France, Britain and Israel. In all three, the nationalization raised ugly memories of Hitler's early annexations, and negotiating, patiently or not, seemed likely to give Nasser undue credibility in the Arab world. And so there was war.

Opposite page: After two months trapped in the blocked waterway, a Norwegian tanker – seen from the Port Said lighthouse – sails free of Suez

Although the conflict was brief and the Canal was reopened in April 1957, the bitter residue clouded governmental connections between Britain and Egypt for much longer: it was not until 1 December 1959 that the re-establishment of diplomatic relations was announced. Yet by then Shell had been back in Egypt for over a year. Its exploration and production there was carried out by Anglo-Egyptian Oilfields and its marketing by the Shell Company of Egypt, both of which had been sequestered. But in December 1958, Anglo-Egyptian was granted new concessions; and at the end of February 1959, all Shell's Egyptian properties were fully desequestrated with an appropriate financial settlement. Moreover, it was soon found that (largely through the efforts of a local employee, one Kamal El-Din Korra) the Shell Company of Egypt had stayed solvent throughout the sequestration. If nothing else, it showed where he felt his loyalty lay, and he was duly rewarded with promotion.

As well as Shell's lofty disbelief in the likelihood of war and its rapid post-war return to normality, the archives reveal a third intriguing aspect of its thinking on Suez: if a certain party had taken its advice, there might have been no war at all.

Three years before it occurred, Shell's analysts had forecast that the Suez Canal might be nationalized. They had also worked out a way to avoid such an event. After Egypt's coup d'état, but before Nasser came to power, one of Shell's experts (John Loudon, subsequently a Royal Dutch chairman) met the French president of the Suez Canal Company to discuss the matter. From Russia to Mexico, Shell people knew about such things: a nationalist government, especially one looking for funds, was likely to cast a predatory eye on profitable foreign-owned assets lying within its own borders. To pre-empt any risk of nationalization before the Canal Company's operating licence expired in 1968, Loudon suggested it would be wise to consider 'turning over the ownership of the Canal to the Egyptian government, and do a lease-back deal'.

> I thought that if they made a deal whereby they would satisfy the nationalistic aspirations of Egypt by giving back the Canal and then establish a long-range future for themselves, whereby they would be reimbursed through Canal dues, that would fit in very well in the picture for the future. But his reaction [that of the Canal Company's president] was very negative. I think he found it such a revolutionary idea. The Company had always been run under such conservative lines that it was quite hopeless.

Of course the Canal Company had no experience comparable to Shell's, but if its president had responded a little more imaginatively to Loudon's suggestion, Nasser could have gained the source of revenue he sought for Egypt's economic redevelopment; the Canal Company would have continued to operate the waterway; and the Suez War might never have been fought.

While it lasted, the crisis was a matter of extreme and urgent daily concern for Shell; and when it was over, although recovery was quick, it left a lasting influence, as we shall see. During the critical months the main problem was of course the maintenance of supplies to a wintry western Europe. Shell's archives contain a telling series of cables – the record from its viewpoint of the Anglo-French armed intervention, beginning on the day France and Britain announced their belligerent intentions.

> 31.10.56: We have cabled specifically to all own and time-chartered ships in our programme in or approaching Suez Canal area instructing them to endeavour to get out of Suez Canal area, Egypt and

Israeli territorial waters. Ships in Mediterranean proceeding East and due Port Said up until November 3rd have been stopped where they are and similarly loaded ships north-bound in Red Sea... Based on advice from Minister of Transport we are not diverting tankers round Cape [of Good Hope] at present in his expectation that Canal will be open to traffic within 3 or 4 days... Will keep you posted.
31.10.56: Further mine of today, latest assessment is seven-day closure after which there would be unscrambling phase taking minimum seven days. We are weighing pros and cons of re-routeing and will advise soonest...
1.11.56: Further my 31st, latest assessment is minimum seven days and maximum unprecise but at least twenty-one days closure plus unscrambling phase... Have decided divert round the Cape loaded tankers which have not yet passed Aden northbound... BP and Esso re-routeing similarly.
2.11.56: Have just heard that block ship has been sunk in Canal and prospect any quick re-opening has disappeared.
5.11.56: Latest estimate of closure is two months...
7.11.56: Would be grateful for whatever information you can let us have in regard to damage to IPC pipelines...
8.11.56: Although we along with other majors as well as cargo liner companies and in fact British shipping as a whole are pressing for this vital information, neither Ministry of Transport nor Admiralty feel able to give any official view as to when canal will be reopened or factual assessment of blockage. ...There seems no alternative but to work on the most popular Press estimate of re-opening namely 3 months from now. As regards damage to IPC pipeline system no precise information available but believe pumping installations and oil storage tanks at all three pump stations in Syria have been seriously damaged if not destroyed...

Another part of the archives holds an equally eloquent but different sort of evidence – the absence of records. In the file for the Shell Company of Egypt, routine entries on various subjects continue until 26 July 1956, then come to an abrupt halt. The next entry is dated 19 December 1958. In between lay silence: the time of sequestration.

On 17 December 1956, less than seven weeks after the Canal had been blocked, petrol rationing was re-imposed in Britain. The allowance, sufficient for an individual motorist to drive about 200 miles a month, was 'considerably more generous than that which had prevailed in any part of the ration period during and after' World War II; but it evoked terrible memories of that war. Equally unwelcome to the motorist was the sudden 34% jump in the retail price of petrol. This did not signify that oil companies were quickly cashing in: two-thirds of the increase was a new government tax, and the other one-third was entirely swallowed up by increased transport costs, because the only possible means of importing Middle Eastern oil was to send tankers around the Cape of Good Hope. For a vessel going from Abadan to London, this meant that a voyage which through Suez would have been of about 5,000 miles became more like 11,000. Extra ships could not be conjured out of the air, and existing ships could neither go any faster nor carry any more than usual.

This leads us to the lasting influence that Suez had upon Shell: namely, the firm decision that for security of supply with economy of scale, tankers must be bigger – much bigger. The thought had been stirring in Shell for at least five years, developing as the lessons of 1951 were absorbed. BP's Iranian debacle had shown yet again that refineries built in oil-producing countries could be at risk: the safety of the investment was directly related to the stability of the regime. It was therefore decided that any new refineries (or extensions to existing ones) serving the European market should not be built near the oil fields, but in Europe itself. From that decision, another evolved: there was no reason to limit the sites of new refineries to coastal areas with tanker terminals. Instead, fed by a Europe-wide network of new pipelines from the terminals, they could be located wherever consumers required them.

All told, this was to be a major shift in construction policy, but in addition to enhanced safety for refineries it had three other attractive aspects. Firstly, any interruption of supply from one source could more readily be made good by supply from another; secondly, transporting products from refinery to market-place would be much cheaper; and thirdly, tankers would no longer have to be designed to carry several different grades of oil. Instead only one grade, raw crude, would need to be shipped. This too would be more economical – and even more so if new single-cargo tankers were bigger than traditional ones.

Such was Shell's reasoning prior to the Suez crisis. However, there was then a limit to the size of new tankers: the size of the Canal. When first opened in 1869, it could accommodate vessels up to 7,000 deadweight tons. Having been enlarged on several occasions, it could now accept fully laden ships as large as 32,000 dwt. Some tankers were already approaching that size, so ideally, from Shell's point of view in 1953, the Canal should be enlarged again. John Loudon raised the topic with the Canal Company's president at the same meeting during which he warned of the Canal's possible nationalization – and received an equally dusty answer. Such a project was not on the Company's agenda: with its concession expected to end in 1968, the expense was not worthwhile.

Disappointed but not very much surprised, Shell proceeded pre-Suez to plan and order new ships, bigger than any in its fleet. Between the two world wars, a typical tanker was in the range of 10–15,000 dwt. In 1955, the first of a new class of eight ships of 33,000 dwt was delivered, while Eagle Oil (in which Shell had a minority stake) was expecting delivery in 1957 of two at 38,000 dwt. The intention was that they would transit Suez partly laden, having some cargo taken off by smaller vessels before entering the Canal and put back after leaving it. But the basic assumption behind the rationale for these vessels – that the Canal would always be open – was demolished by the war.

Serenia – *one of the first supertankers*

The Canal was the origin of Shell Transport's fortunes, but fundamental as it always had been, Shell now understood it must no longer be regarded as a dependable factor in long-term planning. Perhaps unexpectedly, the understanding was much more stimulating and liberating than painful; for if one looked at the Canal in this new and warier light, one could see that its limited size had become both a physical and mental restriction. The desire for larger, more economic tankers had been blinkered by the belief in a secure Canal. Its closure let the existing idea expand. By ceasing to consider it as the best route between Europe and the Middle East and beyond, technology became the only real constraint on the size of Shell's tankers; and suddenly, they began to grow very fast

indeed. In 1958, two further ships which Eagle had on order were redesigned to be more than twice their originally planned size. In the summer of 1959, Shell acquired Eagle's assets and business. On 1 January 1960, Shell Tankers took over the control and management of the Eagle fleet, and on 18 October 1960 the first of the two redesigned ships was launched. Built by Vickers Armstrong in Newcastle, the 71,250dwt *Serenia* was the biggest tanker in the world. She was a giant for her time, yet, within just five years of her launch, even she would be dwarfed by Shell's latest: *Naticina*, of 117,206 dwt. But it was by breaking the 70,000dwt barrier with the S class that Shell inaugurated the shipping world's new age – the age of the supertanker.

Quite apart from the food for thought provided by the Suez crisis and its ending, the year 1957 gave an opportunity for some reflection. Edwin Drake's discovery of oil in 1859 marked the start of the modern oil industry, which was still just under 100 years old; yet Shell Transport and Trading was now 60, and the Royal Dutch/Shell Group was 50. Stepping back from the daily problems of living and working, *Shell Magazine* reviewed where the companies had come from, what they had achieved, and where they were heading.

Though it was an informal exercise, the assessment of the past was extremely positive: by no means 'Where have the years gone?', but rather, 'Have we really done all that?' There had been many rough patches and several prolonged bad times, yet Shell had demonstrated it could not only endure and survive them but prosper and flourish when they were over. It had enviable records of successful pioneering in business, and of research into and development of oil technology and products. It had earned a satisfying reputation for stability and straight dealing; it could truthfully say it was a good corporate member of society, and had learned through long practice how to operate in many different countries for the mutual benefit of shareholders, staff,

Naticina

customers and the local population. But how could those records and that reputation be maintained and built upon? Its existing organization had evolved in a fairly *ad hoc* manner; would that be enough to sustain it, and better, for the coming decades? Some people were beginning to think not.

In 1957, *Shell Magazine*'s own forecast of the future was limited to world-wide generalities – a discussion of predicted global population growth, the likely increase in energy consumption, and possible developments in international relations. But at the same time, knowing full well that the answers to their questions would have far-reaching implications, Shell's senior management was addressing the subject of the Group's future in particular.

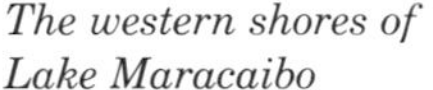
The western shores of Lake Maracaibo

At the time of Shell Transport's alliance with Royal Dutch, no formal division of departments or responsibilities had been made between the new twin head offices in London and The Hague. Some of their departments were identical but separate; for example, each had a production department. Similarly, some regional responsibilities had no particular logic. Although it made sense for Royal Dutch to run operations in the Netherlands East Indies, it also ran Argentina and Peru while Shell Transport ran Venezuela, Brazil and Mexico, with very little co-ordination between the head offices. Apart from the inherent inefficiency, this could be embarrassing too: the two parent companies did not co-ordinate their annual reports and often held their annual general meetings on different days, so it could happen that replies given to similar or even identical questions from shareholders were not exactly the same.

Looking back it seems obvious that some significant re-ordering was overdue; but thinking of change (let alone introducing it) was not entirely straightforward, for two simple and very human reasons. Firstly, though the system was disorderly, it had worked well so far. Secondly, Shell Transport's two most influential men were growing old and disinclined to change. At its head, Lord Godber had worked for Shell for 53 years and now, at the age of 69, was the only person left on the Board who had known Shell Transport in its days before the alliance. Alongside him was Sir George Legh-Jones, a brilliant negotiator and the man said by some to be the dominant force in post-war Shell Transport. This 'tough Welshman' (whose photographic portrait was taken by Karsh of Ottawa – a measure of status comparable to having one's portrait painted by Rembrandt) had joined Asiatic, the early marketing company, as manager of its Fuel Oil department in 1919 at the comparatively late age of 29, going on to join Shell Transport's board in 1944. By 1957 he was 67 years old and had worked assiduously for Shell for 38 years. No one wished or sought to disparage the enormous contributions he and Godber had made, but they were undeniably of the older generation. There is no record of any serious boardroom dispute about the need for change, and it is unlikely that any occurred. But there was lengthy discussion, which was not easy: thirty years later, John Loudon said tactfully, 'It took a little convincing of my colleagues that we should take the step'. Nevertheless in 1957, largely at Loudon's instigation, the step was taken. Enter McKinsey & Company, the distinguished American team of management advisers, with the senior managers' invitation to review and make recommendations upon Shell's entire system of internal organization.

Sir George Legh-Jones, MBE

Delivered in 1958 and implemented on 6 April 1959, the McKinsey Report was, to quote one commentator, a 'sensational step...by far the biggest investigation ever commissioned in Britain.' Its recommendations decisively shaped Shell's structure for the next 35 years.

Within the Group a new tier and terminology of companies was created. Between the parents and the operating companies outside the US came a new level: the 'service companies', of which there were initially four, two being London-based and two in The Hague. Each city had one company (respectively, Shell International Petroleum and, with a new spelling, Bataafse International Petroleum) to handle oil matters, and, recognizing the growing value of chemicals, another (Shell International Chemical and

Frederick Stephens, Chairman of Shell Transport and Trading 1961-7

Bataafse International Chemie) to handle those. 'These companies,' Loudon explained to employees, 'although separate and located in different countries, are complementary parts of an international organization whose job it is to co-ordinate the activities of Group companies (outside North America) and to provide them with advice and services'.

The top management structure was also altered, to produce the Committee of Managing Directors, or CMD, composed from the most senior personnel of the two parent companies. At its inception, this body included seven individuals (four Dutch and three British) with John Loudon as their chairman. The three Britons were John Berkin, Harold Wilkinson and F. J. Stephens: neither Godber nor Legh-Jones was included. Though it was done with the utmost respect, they had been sidelined by the rising generation. Legh-Jones never retired, but remained a member of Shell Transport's board until his sudden death on 30 April 1960. Similarly, though less active than before, Lord Godber remained in office until his retirement (at age 73) on 30 June 1961, to be succeeded the following day by F. J. Stephens as Shell Transport's fourth chairman.

The CMD was a very unusual creation. In a sense it had always been there, insofar as the Royal Dutch/Shell Group had never had one chairman. The two parent companies had always been distinct, and, though their chairmen and managing directors had always been in contact with and influential upon one another, until McKinsey that liaison was neither clarified nor formalized. When it was, the move did not come from McKinsey, but from Shell itself. McKinsey's recommendation was that the two parent companies should share one chairman and one chief executive officer, following American lines; but as Loudon said, 'The American system is different.'

A 2000-pound unexploded bomb was found on Shell Centre's site

Shell had developed almost unwittingly its own distinctive system of leadership, which could be summarized as constructive mutual tolerance. Rightly seeing this as a special strength, its directors declined the suggestion for an American-style management structure, choosing instead to refine the existing arrangements. Thus it came about formally that the Royal Dutch/Shell Group has no one chief executive; instead its nearest equivalent is the chairman of the CMD, who, though his voice may carry greater weight, remains only first among equals.

The CMD's role was to 'consider, develop and decide upon overall objectives, policies and long-term plans to be recommended to the operating companies', deliberately avoiding involvement in Shell's day-to-day activities. Each member would have specialized spheres of interest and would remain in close touch with the affairs of Group companies, but the burden of daily executive action was to be taken up by a large number of 'co-ordinators', with one branch being devoted to oil and another to chemicals. One level of co-ordinators covered marine, research, exploration and production, manufacturing, personnel, marketing and lastly finance; another level covered the different geographical regions of operation; and both reported to their Director of Co-ordination (either the DCO for oil or DCC for chemicals) who in turn reported to the CMD.

Though the new arrangements were fairly straightforward in practice, they looked daunting on paper. In a parody of the diagram which was

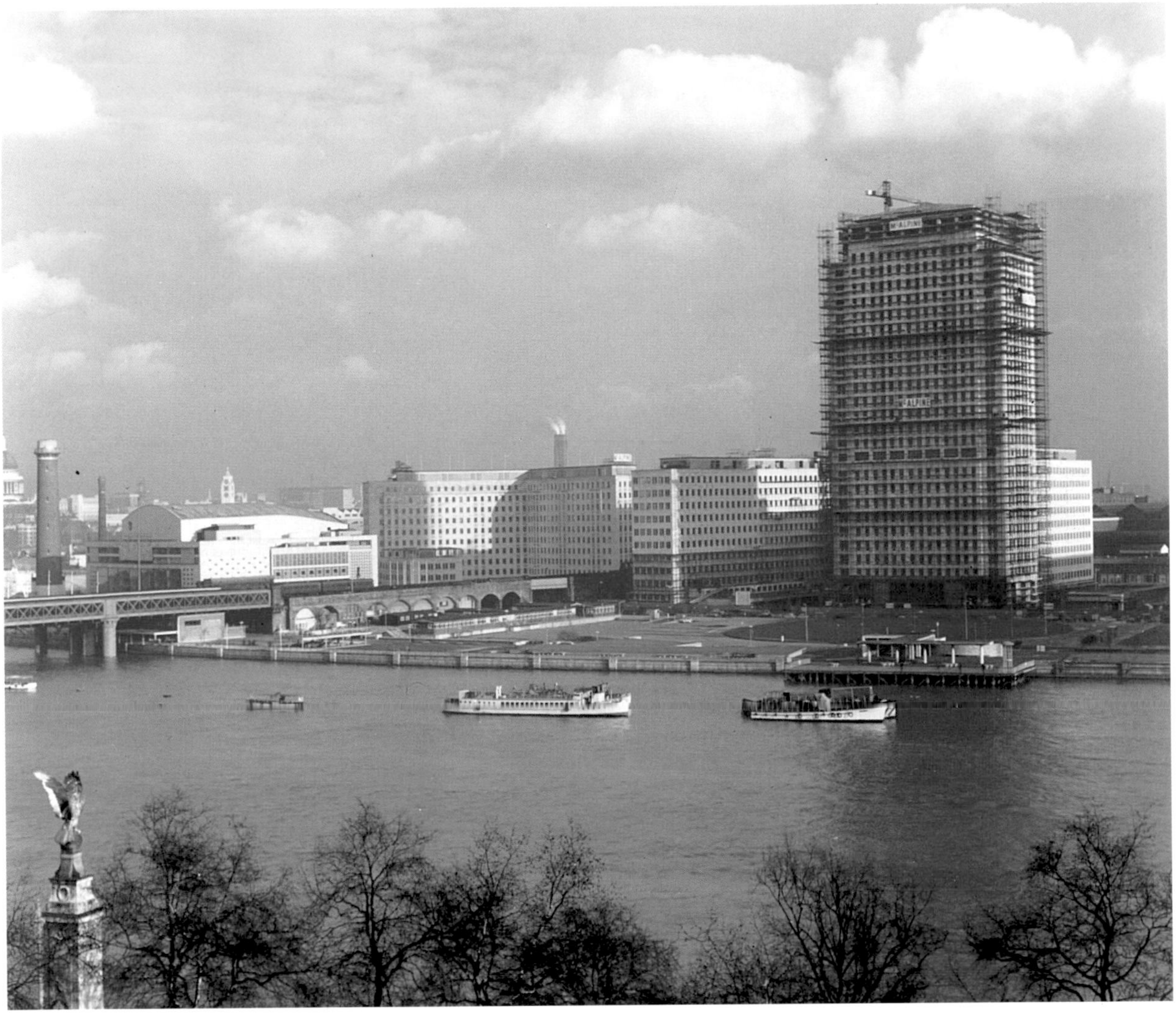

Shell Centre under construction

supposed to make everything clear, *Shell Magazine* published a recipe for a powerful cocktail called the 'McKinsey Mixture', based on large quantities of rum with 'a lot of stirring... pineapple, oranges, Maraschino cherries and whatever you like that's in season.' The editor concluded: 'Two glasses of this and you will be in a complete maze.' Yet the re-organization achieved its objectives, clarifying the relationships and both easing and accelerating communication between areas of responsibility, as well as giving the CMD the business leaders' single most valuable asset: time to think. And as ever with Shell, there was a great deal to think about.

Underpinning it all was the question of money. If the plans for new refineries, the European pipeline network and very large ships were to be realized, and if exploration were to be extended, research expanded and world-wide production increased, then formidable amounts of cash would be needed. Where to find it? Long-term borrowing was to be avoided; there were few financial institutions which even in concert could supply the necessary level of funds, and besides, the cautious policy established in the Depression years still held good. No: the answer must be to develop that policy and become as far as possible self-financing, attracting not lenders but new shareholders, and relying upon retained earnings as the primary source of investment funds.

Even before the CMD first came into formal existence, this had been extensively discussed, and acted upon, by its first members. Royal Dutch had been listed on the New York Stock Exchange in 1954. An American business magazine described the Group as the 'Two-Headed Giant of Oil with 440 Arms', the arms being the operating companies then in existence.

Going on to explain how the Group as a whole had no existence in law, the magazine said it was 'not a corporation, not a legal entity, but a condition... This condition is international business in its most highly developed sense.' Prospective shareholders homed in, and in 1957 Shell Transport joined Royal Dutch in the New York market. After authorization was

received from the Exchange authorities on 14 February, trading in its shares began on 13 March. They immediately showed their popularity: the volume bought and sold was a record for the first day of a newly listed stock. This was followed eleven months later, in February 1958, by a £45 million rights issue in London to Shell Transport's existing shareholders – the biggest such issue that had ever been seen in Britain. Again, it was a resounding success, with shareholders buying all but a fraction of 1% of the amount on offer. On the open market, the balance was bought at once. Shell Transport's authorized capital base was now £113 million.

Construction of Shell Centre began in August 1957, and by 1959 the ambitious plans were progressing well. On the South Bank of the Thames, a 7½-acre site would provide 43 acres of floor space. As one of Shell's publications explained, the towering edifice was to be built on London's characteristic soft foundations, 'river mud and clay'. The new headquarters were accordingly supported on bell-shaped concrete piles 110 feet deep and 15 feet wide at their base. Because of the clay, London has always been a comparatively low-rise city, and at 351 feet high the 26-storey structure would seem quite small in Manhattan; but for some years after its completion it was one of London's tallest, and the largest office building in Europe.

The complex was divided into two main buildings, with names which anyone unfamiliar with oil terminology would have assumed must refer only to their geographical locations. Opposite and overtopping the Houses of Parliament, the Upstream Building would have three wings surrounding a square courtyard and the central Tower. Five hundred yards away, but joined to it by a tunnel, stood the curving Downstream Building - which was literally downstream. While the taller Upstream was still taking shape, a partial move into Downstream from St Helen's Court (which was to be sold to an insurance company) took place over the weekend of 14/15 October 1961.

At the same time, other important new main offices were being built around the world, with Melbourne, Toronto and Caracas being just three sites; but large as they and Shell Centre were becoming, their construction required only a small proportion of Shell Transport's new funding. As well as the major areas of contemporary expenditure which have already been indicated – new refineries, big ships, new pipelines – some details of just a few of Shell's other current activities may be mentioned.

As world-wide exploration continued, Shell's position in the Middle East was strengthened in 1960 by the discovery of oil 55 miles off Qatar; by an exclusive agreement with Kuwait for the development of its offshore oil; and by the first discovery (in 1962) of oil in commercial quantities in Oman. This last provides a good example of the difficulties and frustrations of oil exploration, and has a niche of particular interest in Shell Transport's history as well.

By the early 1950s, geologists from several oil companies were certain that beneath the deserts of Oman (then a feudal, almost medieval society) oil lay hidden. In 1954, the Iraq Petroleum Company (IPC) began exploration there. Jointly owned by Standard Oil of New Jersey, BP, Compagnie Française des Pétroles, Mobil and Shell, the company formed for the purpose was called Petroleum Development (Oman) Ltd, or PD(O). In the whole country there was only one three-mile stretch of metalled road: travel was extremely difficult, and the heat severe. Like most oil exploration, this was work for very fit young men. In 1956, 150 miles into 'the Interior' – the high, hot, featureless and (then) largely unexplored desert plain behind the coastal mountain ranges – the first well, named No. 1, was drilled in a wide, flat, barren valley surrounded by rocky ridges. It was dry, and most of the IPC partners withdrew in exasperation.

However, Shell took over their interests in PD(O) and proceeded as 85% owner of the company. In 1962 gravimetric surveys pinpointed another area 25 miles from Fahud. Yibal No. 1 was drilled, but caved in and was abandoned. The rig was moved a short distance; Yibal No. 2 was drilled; and on 18 September the historic find was made, with 22 barrels of clean oil being recovered in little more than an hour. Since then Yibal has become Oman's most prolific field; but, emphasizing how much luck as well as judgement was involved in exploration, a return to Fahud in 1964 revealed that it too was an oil-bearing region after all. Across one of the ridges surrounding Fahud No. 1 and barely a mile away from it, Fahud No. 2 was drilled, and oil was found. Eight years and £12 million earlier, they had been so near, yet so far; and an even sharper irony followed. Other wells were drilled to determine the extent of the field; someone decided to have another go in the original stony valley; and Fahud No. 18 was obligingly productive. It was only 200 yards from Fahud No. 1.

Yibal's niche in Shell Transport's history cannot be ignored – not only because it was a Shell achievement from which the whole of Oman derived great prosperity, but also for an interesting individual reason.

Shell's support of academic research dated back to 1919. In 1958, funded by Shell, one of its new employees began a three-year postgraduate research programme at Edinburgh University. In 1961, at the age of 24, he obtained his doctorate in geology; in 1962, he was chosen to be well-site geologist at Yibal, and was there when the historic discovery was made.

The experience of helping to bring an entirely new oil country into production did the young man no disservice: in 1993, after thirty-five years in a typically varied career within the Group, Dr. John Southwood Jennings CBE FRSE became Shell Transport's tenth chairman.

After exploration, transport. Among the more spectacular developments of the time, the enlargement of Tranmere port (close to the Merseyside Stanlow refinery) began in the summer of 1958, with facilities for tankers up to 65,000 dwt fully laden being opened on 8 June 1960; and Shell's pipelines world-wide grew from 17,300 miles in 1955 to 27,350 miles in 1962. Bringing crude oil inland to new refineries far from tanker terminals, its European network (with sections from Le Havre to Petit Couronne, Marseilles to Karlsruhe, Karlsruhe to Ingoldstadt, Genoa to Rho in north Italy, Rotterdam to Kelsterbach and Trieste to Vienna) was on course for completion, at 1,372 miles, in the middle 1960s. More new refineries were going up in other parts of the world – Durban, Singapore, the Philippines, Northern Ireland, Turkey, and elsewhere; and simultaneously, under the leadership of Lord Rothschild, Shell's research programmes were becoming steadily wider. As assistant director of zoological research at Cambridge, Rothschild came to Shell in 1959 as a consultant; by 1962 he was chairman of Shell Research Ltd, and in 1965 he became the Group's Research Coordinator. He was a most extraordinarily talented man, someone who might have stepped out of a novel: apart from being a good cricketer and pianist (classical and jazz) with a liking for pungent Turkish cigarettes, in wartime he had been awarded the prestigious George Medal for his great gallantry and courage in defusing unexploded bombs. As a lieutenant-colonel he had then worked in Military Intelligence, earning the American Bronze Star and Legion of Merit; and for his outstanding work as a zoologist he had been elected a Fellow of the Royal Society (FRS).

John Jennings in Oman

Colleagues at the Sittingbourne laboratories in Kent were naturally proud of the Royal Society connection, and gave private circulation in a house magazine to

> previously unpublished reports from the Internal Auditors and the Chemical Industries Training Board indicating that there was a crisis at Sittingbourne in that they were down to their last FRS... The situation is now desperate...

Five Fellows of the Royal Society, all closely connected with Shell: (L-R) Lord Rothschild, Dr G J Popják, Dr M Sugden, Sir Robert Robinson and Dr J W Cornforth

Actually, Rothschild's was far from the only top-class brain being attracted to Shell. A photograph exists of him talking to four other men closely connected with Shell who were all Fellows of the Royal Society: Professor Sir Robert Robinson, Dr John (later Professor Sir John) Cornforth, Dr M. Sugden and Dr George Popják. Moreover, Robinson and Cornforth were both winners of the Nobel Prize for Chemistry. Robinson had won it in 1947 and, for his research work with Shell, Cornforth went on to win it in 1975.

The inventions and discoveries which flowed from Shell's laboratories were much too numerous to list in detail here, but to give a hint of their range, they included epoxy resins, insecticides (especially against mosquitoes and locusts), herbicides for above ground and under water, fungicides, plastics, synthetic elastomers, diesel lubricants and liquid detergents. Coke of extremely high purity was developed as fuel for nuclear reactors; comprehensive research was also conducted into the dangerous phenomenon of static electricity in stored oil products, with a powerful static additive being developed to prevent the accumulation of such charges. Serious business, altogether; but there could still be lighthearted moments. On 10 July 1962, the foundations were laid for Shell's new Milstead research laboratory at Sittingbourne in Kent by Dr George Popják and Dr John Cornforth; and, using a spade later engraved to record the event, these two eminent men ceremonially buried there a sealed tin of popcorn.

CHAPTER FOURTEEN

An Operation without Commercial Parallel: 1963–1967

After six years in construction, Shell Centre – the new headquarters of Shell Transport & Trading, and the London central office of the Royal Dutch/Shell Group – was fully opened for business on 12 June 1963.

In 1943 one of *Shell Magazine*'s contributors had written an article dreaming of palatial new offices set in the peaceful Highlands of Scotland. The architects of the real Shell Centre, Sir Howard Robertson and R. Maynard Smith, had designed a complex which in some ways was strangely reminiscent of that dream base. Although (unlike the dream) it would not have a complete post office, overnight accommodation or a roof garden, it would have a viewing gallery on its 24th floor, while below there would be private telephones, a travel agency, a bank, a barber's shop, a market area for small traders, a very good restaurant and several licensed bars. Moreover, like old St Helen's Court, it would have a general store and a shooting range; and under the one roof it would also have an unusual number of leisure and health facilities for staff – reception rooms large and small, exhibition areas, a theatre and cinema, a well-equipped gymnasium, squash courts, a big sports hall and a swimming pool just short of Olympic size, as well as a medical and dental wing. Deep underground, there was secure parking for 453 cars, and office accommodation was arranged for 5,000 people. It was perhaps just as well that living accommodation was not provided too; if it had been, it would have been perfectly possible to live a thoroughly healthy life there, combining work, exercise, eating, shopping, health care and cultural and social events, without ever having to step outside.

For its time it provided a remarkably good working environment, and over thirty years later was still a highly efficient building. It has always had its critics; as Nikolaus Pevsner noted in *The Buildings of England,* 'The aesthetics of the Shell Centre have been much argued.' In another guide to London's architecture, its most prominent part, the Tower, was termed 'lugubrious'. Pevsner also dismissed the underground car-park as 'obvious-

ly' inadequate, 'if we consider that the buildings are for 5,000 employees' – a comment which said much about contemporary attitudes to driving in cities. But most employees did not need to drive to their work, for Shell Centre could be entered almost directly from both the Underground railway network and the mainline station of Waterloo. In both, it had its own directional signs, and visitors coming by taxi very soon found that no address was needed other than 'Shell Centre'.

As staff congregated on the south bank of the Thames in the summer of 1963, they marvelled at their new offices' quietness, lightness and simplicity of use. Its thousands of windows were double-glazed, keeping out almost all of the bustling city's traffic noise; no desk need be more than five feet from a window; and each desk, wherever it was moved, would be within 27 inches of power points. One member of the clerical staff found an unexpected bonus in the purpose-built premises. As well as being much easier to work in than anything he had known before, all his old friends were there, able once again the share the pleasure of lunching together. In many such small but important ways, the building made a real contribution to the contentment of those who worked within it, and decades later (as visitors could not help but notice) it still had a cheerful atmosphere, with the hush of concentrated work often broken by laughter – and occasionally, even by singing.

The five years from the completion of Shell Centre – 1963 to '67 inclusive – could be summarized as business as usual. But with at least its quota of dis-

From Shell Centre's 24th floor, with the Thames and many famous landmarks clearly visible, the view conveys a powerful sense of London as a truly great capital city

tinctive events, it was far from being a dull period, because for a company like Shell Transport and a Group like Royal Dutch/Shell, 'business as usual' never meant standing still or resting on laurels. Exploration, production, research, construction, transport, marketing: in each area there was development, sometimes in detail, sometimes with broad and sweeping innovation. But that was life with Shell. Some events confirmed the wisdom of earlier strategic decisions; others made it plain that further deep thought was required; and all took place against a background of sustained intense competition, with the occasional political or economic earthquake for good measure.

Before touching on those, however, it is time to introduce a theme which would be increasingly important, and which Shell began actively to address at a much earlier date than might be supposed: not the working, office environment, but the world's environment and the effects of the oil industry upon it.

Aside from fears of manipulative cartelism, the pollution of the global environment has long rated as the biggest bogey in the general public view of the oil industry. The images are all too easy to evoke, and carry a high emotional charge. Shell Transport has not always been completely blameless in this respect: in its earliest days, for example, one of its tanker captains recalled 'an article in a European newspaper published in Japan, complaining about oil contaminating sea water and preventing people from bathing, and also stating that it was killing the fish in the Gulf of Tokyo.' Enraged to realize that this was directed at his employers, he composed a no-nonsense reply:

> I said, 'Tell the people who are putting such trash in the newspapers that they do not know what they are talking about, because at Batum, where I have been loading oil for 10 or 15 years, there were no difficulties of that kind.' It was suggested to me that the articles were inspired by our competitors...

Whether the cause was underhand trickery or ignorant prejudice, this seemed to put a stop to it: 'At any rate, what I had to say seemed effective, as nothing more was heard on the subject.' But to modern ears, the good captain condemned himself by adding

Before the Clean Air Act the legendary London 'pea-souper' fog was still a painful – and sometimes fatal – reality

> This is not a new incident, because I remember that when I first discharged bulk oil on the second trip, part at Kobe and part at Yokohama, we went out to the breakwater and cleaned our tanks before taking in general cargo.

That was in the 1890s. Then and for decades into the 20th century, tanker crews from all companies habitually and legally discharged overboard the waste from tank-cleaning, just as sailors from any period in history had disposed of any rubbish. 'Ditching gash' had never mattered very much: the quantities were small, the sea was big, and the rubbish in general would rot away unnoticed. But crude oil and its derivatives were soon seen to be different. Acceptable practices for handling them at sea have changed out of all recognition, all for the better; yet as one environmentally conscious individual wrote, it was not only the sea which was at risk – 'Smoke from burning fuels, and gases from industrial processes, foul the atmosphere; sewage and industrial effluents pollute streams and rivers; and these and oil can sully the seas and their shores.'

Though well expressed, the observations may appear unremarkable; but they were made remarkable both by the perhaps seemingly early date – 1964 – and the speaker: Shell Transport's chairman, F. J. Stephens.

'Our companies', he continued, 'have long accepted their own responsibilities in preventing pollution, and welcome the growing public awareness of clean air and water to health and to man's enjoyment of his environment.' Stephens then proceeded to describe some of the most recent steps in pollution abatement for which Shell was responsible – advanced burner designs, minimizing air pollution from oil fuels; parallel plate interceptors (which had been made available to competitors) for cleaning waste water from refineries; and 'biologically soft' detergents which were readily broken down by natural bacteria.

A corporate environmental awareness and sense of responsibility is so intimately linked with changing attitudes in society generally that it is difficult to establish a precise date for its beginnings in Shell; but it was certainly no later than the middle 1950s. It was 1954 when more than 30 nations ratified the 'International Convention for the Prevention of Pollution of the Sea by Oil', a highly important agreement which was further tightened in 1962. It was 1956 when Britain's parliament passed the Clean Air Act and began to make London's infamous 'pea-soup' fog a thing of the past. Shell was involved with and affected by both of these; and it was 1957 when a Shell chemist with the improbably apt name of R. C. Tarring began to advise the government on the nuisance caused by biologically 'hard' detergents. As well as affecting the fate of fish and other marine life, these were creating a disgustingly visible problem for Britain's waterways: holidaymakers in canal narrowboats could find themselves ploughing through detergent foam as high as deck-level.

That item alone indicates why one should not be surprised to realize that Shell's concern for environmental matters is at least as old as more public concern; after all, although it is a very large enterprise, its members are members of society as well, no less interested than anyone else in subjects which affect everyone.

Since then, Shell's active interest in maximal environmental protection and minimal environmental abuse has never waned, but rather has grown steadily. It would be naive to suggest its environmental sensitivity springs only from the goodness of its collective heart, without any commercial or legal prodding; a good environmental policy is also sound business sense. But equally, it would be ignorant and unjust to suggest that this attitude has arisen only because of governmental or partisan pressure.

Oil is a contentious commodity which has become an essential, at once loved and hated for what it can do, and Shell's environmental responsibility is a subject to which we shall return in later chapters. Now, though, it is time to go back to 'business as usual' – bearing in mind that for Shell, business as usual was likely to mean change and innovation, and always a striving for improvement.

Such things were no harder to find in the middle 1960s than at any other time in Shell's history. A new generation of British petrol stations began on

Turnbull's of Plymouth, the first Shell garage in England to be converted to self-service on 28 August 1963

28 August 1963, when, following the successful pattern it had already established in the United States, Shell chose Turnbull's Garage in Plymouth as its first self-service station in England – a small yet significant change in British social history. While some people enjoyed the novelty of self-service, others grumbled, fumbling with fuel caps and lamenting the lack of someone to clean their windscreens; but no doubt they would have grumbled equally at the alternative. As the number of cars in use increased, attended forecourts were becoming more labour-intensive and therefore more expensive to run, which inevitably would force up the pump-price of petrol. In an ideal world perhaps one could have high-quality petrol at a low price and the luxury of attended service, but not in the real world.

Another development, arguably at least as useful but much less obviously from Shell, was the launch in 1964 of 'Vapona' flykiller. Replacing swats, which needed a degree of luck and skill to have any effect, Vapona was sold both in aerosols and in the form of impregnated plastic strips steadily releasing minute quantities of insecticide into the air over a period of months. In similar vein came the insecticide Azodrin, developed especially for cotton crops; the industrial herbicide Prefix; the molluscicide Frescon, which killed the water-snails responsible for spreading bilharzia; Supona, which dealt with parasites living on the skin of sheep and cattle; and Atgard V, which disposed of parasites living in the intestines of pigs. Several other such products were originated by Shell at this time, but these examples demonstrate sufficiently that its agricultural research gave real benefit to creatures valued by mankind – and very much the opposite to pests.

This progress was reflected in the production of chemicals. Crude oil had become the world's principal raw material for the manufacture of organic chemicals, and was increasingly so for inorganic ones as well; but in either

case it involved very heavy capital expenditure. In 1959, following the McKinsey Report, the London-based Shell International Chemical Company (SICC) began providing services and guidance to a number of operating companies, which dealt exclusively in chemicals, from several different countries outside North America, where Shell Chemical Company performed a comparable function. Now, under that double umbrella, isoprene rubber ('synthetic natural rubber') was being produced at Torrance in California and Pernis in Holland. Phenol (or carbolic acid – white crystals of benzene derivation used as an antiseptic, disinfectant and as a constituent of resins and explosives) and acetone (a solvent for paints, varnishes and lacquers) came from Houston in Texas, with other solvents coming from Berre-L'Etang near Marseilles in France. Glycerine (for antifreeze and sweetener) was made at Norco in Louisiana, ethylene (for plastics) at Carrington, near Manchester in England. Soft detergents were being manufactured as far apart as Shell Haven near London and Geelong in Australia; polythene at Wesseling in Germany and Yokkaichi in Japan. Contributing further to plastics and rubbers, Yokkaichi also produced ethylene and styrene (for packaging and insulation), while Clyde in Australia became a major source of Shell's heavy-duty Epikote resins (or Epon, as they were called in the US). Combined with asphalt, Epikote had proved to be an ideal paving material for airports, being sufficiently durable to resist not only cleansing solvents and hydraulic fluids but also the heat from jet engines. The resins were also 'unsurpassed as bonding materials for unlike substances – glass to brass, plastic or concrete to steel, and so forth'; and in 1963 it was announced that by chance, an extremely valuable new use had been found for them. With the correct proportions of curing agents, Epikote could easily be applied to metal while under water, proofing the metal against hurricanes, salt corrosion, drying out, and all the myriad battering which it faced at sea.

The construction of a section of the Trans-Alpine Pipeline in Austria

To an organization like Shell – founded upon surface transport at sea, and making increasing use of submarine pipelines as well as offshore exploration and production rigs and platforms – the benefits of such a compound were self-evident, whether used underwater or at the waterline, where the combined effects of wind, weather and salt were more corrosive. For ships, quicker, simpler and cheaper hull maintenance meant that intervals between dry-docking could be extended; for fixed metal structures (not only rigs and pipelines, but docks, pier pilings, buoys and bridges), *in situ* maintenance and repair could now be done without the awkwardness of installing waterproof caissons around the unit. Instead, all that was needed was a trained diver with a pair of rubber gloves. Recognizing the product's merit, shipping companies, harbour authorities, marine engineers and other major oil companies all became eager buyers.

By 1966 Shell's global chemical production far outstripped that of any other oil company. Chemicals contributed one-eighth of its total proceeds; it was about the 10th or 12th largest chemical producer in the world – a ranking which would rise; and only a little reflection is needed to realize how greatly it was shaping the world we know. Simultaneously, Shell accelerated its construction of pipelines for crude oil, oil products and natural gas:

their 1962 length of 27,350 miles world-wide rose, by the end of 1965, to 37,750 – an increase of 38% in four years. The pace continued, and the diameter of new lines grew too. Running 300 miles from Trieste in Italy, across the mountains, through Austria and on to Ingoldstadt in southern Germany, the Transalpine pipeline was completed in 1967, with its 40-inch diameter making it the widest in Europe. In the same year, construction began on the 630-mile 'Capline' from New Orleans to Illinois. Also of 40-inch diameter, it was the widest pipeline in the United States; yet both the

RHINE-DANUBE PIPELINE
WEST GERMANY
Ingolstadt
N
Augsburg
Erding
Munich
TRANS-ALPINE PIPELINE
Rosenheim
AUSTR
HAHNENKAMM TUNNEL
Elevation 1100 meters.
Kufstein
Innsbruck
AUSTRIA
FELBERTAUERN TUNNEL
Elevation 1550 meters.
SWITZ.
Lienz
PLÖCKEN TUNNEL
Elevation 930 meters.
ITALY
Udine
LEGEND
Tank Farm
Pipeline
Marine Fac.
Tunnel
Rhine-Danube Pipeline
Pump Station
Trie
Venice

Transalpine and Capline had already been 'capped' by the line from the Agha Jari oil field in Iran to Ganaveh on the coast of the Gulf. Completed in 1965, its length of 106 miles was comparatively modest; but its diameter of 42 inches made it a very big pipeline indeed – in fact, the largest in the world.

New production did not lag behind – 1967 saw Shell's first Gabonese production, at 45,000 barrels a day, and the first exports from Oman – and in that year, net of sales taxes, revenue (whether of crude oil or natural gas) increased from the £2 billion of 1963 to £2.6 billion, helped by a particularly welcome contribution from the natural gas field of Groningen in the Netherlands. Groningen provided much statistical data of the extreme kind. Discovered on 14 August 1959, it had unexpectedly turned out to be the biggest natural gas field in the non-Communist world. In 1963, when figures were first released, it was estimated to contain the energy equivalent of at least 1.1 billion tons of coal – a calculation which was steadily revised upwards, to the point where the field was described as Europe's 20th-century equivalent to Drake's 19th-century discovery of oil in the US. By 1967, a total of £240 million had been spent on it – and it was only just beginning to show a return on the investment.

Not even Shell could provide that sort of cash alone. The exploration, discovery and much of the subsequent development had been as a joint venture with Esso (formerly Standard Oil of New Jersey, and subsequently Exxon), with further development by other companies on the periphery; but Shell's involvement still amounted to around £60 million in the money of the time before any profit was seen at all. Such joint ventures would become more and more typical, as the investments and risk became increasingly untenable for any single corporation.

A precise example of this arose in 1964, when a 50:50 venture was created with Esso. By this agreement Shell became the operator of the venture, under the name Shell U.K. Exploration & Production, or Shell Expro. Building on the Groningen experience, Shell Expro went exploring in 1966 offshore from Holland, and in the southern North Sea discovered the Leman gas field – the starting-point for the entire oil and gas industry in that sea. In similar vein, as underwater exploration gained emphasis towards the end of 1967, Shell and Esso jointly took delivery of a

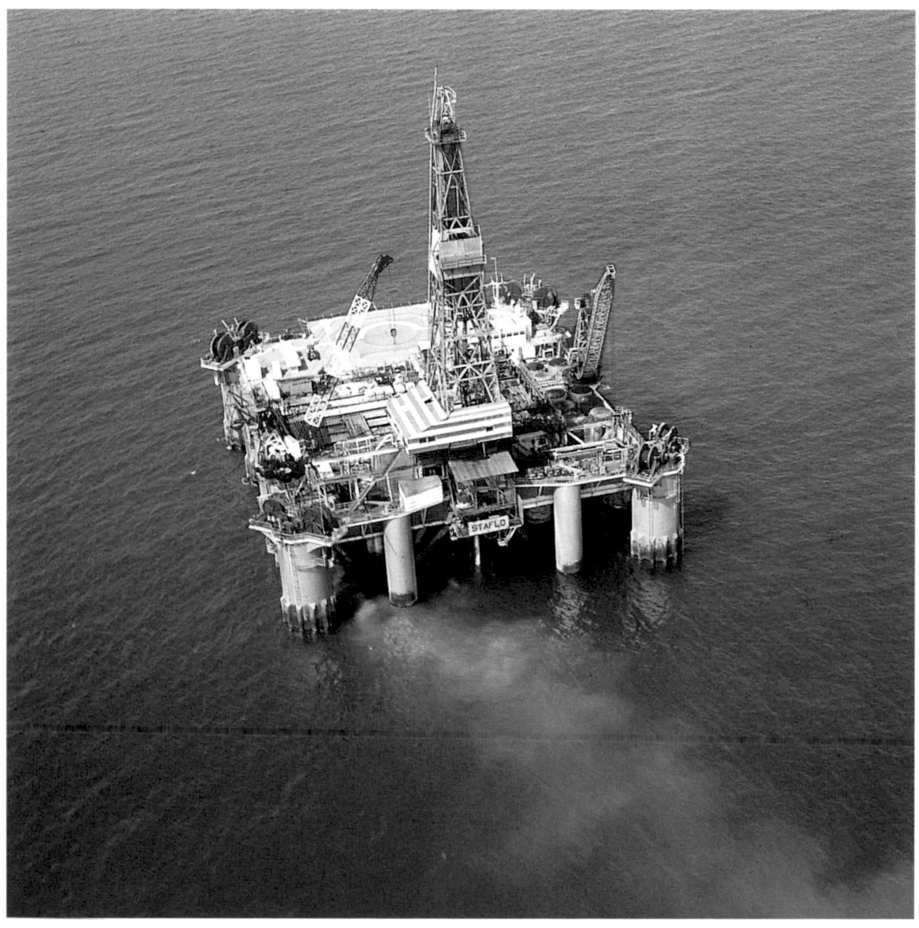

new device: Staflo, a floating drilling rig. This was a very important addition to Shell's maritime interests, but by no means the only one. Others were of such great importance that they deserve to be described more fully.

Way back in 1927, musing on the subject of gas as a fuel, J. B. A. Kessler junior had written, 'We should turn all this energy that is going to waste at present into something that we can put into packages and sell. ...The only thing we can do is to make something we can ship.' His foresight was now coming to full fruition. In 1931, the 4,700dwt *Agnita* had been delivered to Shell: the world's first ship purpose-built for carrying liquefied petroleum gas (LPG) – butane, propane and other gases from refined petroleum. Her technology, though new then, was comparatively simple, since the LPG only had to be kept under pressure in order to remain liquid. A much greater technological problem was posed by liquefied natural gas (LNG), which is mainly methane. To stay liquid, LNG must be not only pressurized, but deeply refrigerated as well, at a constant 165^{0} below zero Celsius. LNG was first made as early as 1917, but it was not until 1959 that a ship was capable of keeping it at the necessary very low temperature throughout a voyage. This first ship was *Methane Pioneer*, a converted cargo vessel belonging to Constock International Methane Ltd, which successfully carried seven trial cargoes from Lake Charles in the United States to Canvey Island in the UK. The possibilities for Shell were clear, and early in 1960 the oil giant became Constock's partner, acquiring a 40% interest in its work. Constock was then renamed Conch International Methane (a simple but imaginative way of indicating both the continuity of the original company and the nature of its new partner) and in 1964 Shell Tankers took on the management of the world's first two purpose-built LNG carriers, *Methane Progress* and *Methane Princess*.

The fulfilment of Kessler's foresight: Methane Princess and her sister Methane Progress, both managed by Shell and completed in June and May 1964 respectively, were the world's first purpose-built LNG carries

LNG cargoes are measured by volume, not weight, and these two ships carried a lot – 27,400 cubic metres each. Sailing with their icy explosive cargoes from Arzew in Algeria to Canvey Island, they produced considerable unease in Canvey Islanders: as someone said, 'The bigger the cargo, the bigger the bang', and if an LNG tanker blew up, it would certainly make a very large hole in the sea. An exhaustive four-year public enquiry confirmed that every conceivable safety precaution had been taken, and today LNG is universally recognized as one of the cleanest possible fuels, whether for domestic or industrial purposes. No less importantly, Shell ships have an immaculate record of safety in carrying it. This is all the more impressive when one considers that in addition to their numerous other owned or managed LNG tankers, a dedicated fleet of seven has been operated by Shell Tankers on a regular shuttle from Brunei to Japan since

1971. With nearly three times the capacity of *Methane Progress* or *Methane Princess*, these ships carry 75,000 cubic metres of gas apiece, and each is about the size of a 100,000dwt oil tanker. Delivering on average one cargo every 2½ days, they have made thousands of voyages covering millions of sea-miles, and have never had an accident.

At the end of its voyage, the LNG tanker's cargo was piped from storage direct to the consumer. In contrast, crude oil, after its journey by tanker and pipeline, had next to go to a refinery; and in 1965, in addition to European developments, three were commissioned in faraway places – at Abidjan on the Ivory Coast, Alese in Nigeria and Umtali in Rhodesia (Zimbabwe). Totting up the total world-wide, Shell could count 77 refineries in its operation, whether wholly owned, partly owned, planned or completed. In 1965, in his annual report to Shell Transport shareholders, F. J. Stephens said apologetically but without embarrassment,

> The enormous scale and the complex nature of the oil industry's operations lead unavoidably, in a review such as the present one, to the quotation of very large figures and the use of sometimes rather formidable technical language.

Quite so, Mr Stephens. But a lot of shareholders found the technical aspects not only interesting but exciting, and none minded the quotation of large figures, as long as they were positive; it was in order to hear such things that they had bought their shares in Shell Transport and Trading. This was business as usual, and the apology was unnecessary.

Yet (one might reasonably ask) where in all this *was* Shell Transport and Trading? Surely these were activities of the Royal Dutch/Shell Group as a whole? The answer to the latter question is, yes: no single company undertook all the activities. As to the former question, Shell Transport and Trading undertook none of the activities in operational terms; yet wherever Group work was going on, Shell Transport and its shareholders were involved as ultimate owners of 40% of the Group's part in the operation, as had been the case since its alliance with Royal Dutch in 1907. In one sense it could well be said that from then on the two parents had no separate history. But no one could buy shares in the Group as such, nor could an individual become a direct shareholder in, say, Shell International Chemical Company; the only Group companies to feature in Stock Market listings – the only two doorways into the Group – were Royal Dutch, and Shell Transport and Trading. A buyer of shares in one did not become owner of shares in the other as well, and in that important sense at least, the parents remained separate, as they are to this day.

This point, the fact that since 1907 Shell Transport and Royal Dutch had merged their interests while retaining their separate identities, has been made before in these pages; but it is so central to understanding the nature of the whole that it is well worth repeating – indeed, the annual reports of both parent companies repeat it every year.

The reports have themselves become an example of the Group's underlying concept. After the formal creation of the CMD, the Committee of Managing Directors, a number of further organizational or administrative changes occurred. One of these emphasized the parent companies' community of interests: they began to hold their Annual General Meetings on the same day, and their Annual Reports became almost completely identical, apart from the title and the first few pages of text. Since each parent drew its income from the same sources, this made good sense; but any confusion in the public mind about their separateness would have been understandable, so one paragraph which was sure to appear in each report, every year, was an explanation of the Group's nature and the relationships between its companies.

The 'Chairman's Foreword' – the first differentiated pages of the Annual Reports – gave each parent company the place to inform its own shareholders about events which related solely to that parent. But these were few, often limited to the retirement of a senior individual.

Stephens retired as Shell Transport's chairman on 30 June 1967. He remained one of its Managing Directors for the rest of the year, then, at the age of 64, retired a little further to the level of director for his final four years on the board. Described as 'the antithesis of the popular idea of the pre-World War Two oil tycoon', he had never been a domineering character, yet within the industry had become one of its best known leaders. His gradual relinquishment of responsibility was not unusual: the company continued to benefit from his experience while he gained respite from the more onerous roles. But in two ways his retirement *was* unusual.

First, his six-year chairmanship of Shell Transport was, by a long way, the briefest in the company's history to that date. Marcus Samuel had been chairman for 24 years, from its foundation in 1897 until 1921, when he was nearly 68 years old. Walter Samuel, his son and successor, was chairman for 25 years, retiring in 1946 at the age of 64 only because of ill health; and Frederick Godber, Walter's successor in office for 15 years, did not retire until he was 73. In other words, the company had had just three leaders in its first 64 years. Stephens' tenure of office brought that to four in 70 years; but in the remaining 30 years of Shell Transport's first century, it had six chairmen more, each retiring at the age of 60.

The second unusual aspect of Stephens' retirement was that in another organizational change, he became on retirement one of Shell Transport's first directors to join an entirely new level of supervision. While some companies might find it useful to have a mandatory age of retirement for chairmen, Shell placed great value in the accumulated repository of knowledge represented by its 'elder statesmen', and did not wish to lose that; so from 1961, all retiring managing directors became non-executive directors and members of 'The Conference', with which the CMD would routinely confer.

Like the Royal Dutch/Shell Group, The Conference did not (and does not) exist as an entity in law. As well as retired managing directors and non-executive directors, and supervisory directors drawn from other large corporations, its members included all current managing directors of the Group. Once a month (apart from August, the holiday month), and alternating between London and The Hague, they would meet as a body, utterly international and deeply experienced. Though advisory rather than governing or managing in character, the opinions of The Conference naturally carried great weight, and its existence made a noticeable difference to (among other things) the minutes of Shell Transport's board meetings. There, using the invariable and respectful capital initials, The Conference's first meeting is recorded as having taken place on 20 September 1961. Before that date, matters of company policy were gone through at Shell Transport board level and the outcome duly noted; afterwards, such topics were merely listed as having been discussed with The Conference.

Having said that, it is important to add that discussions with The Conference did not (and do not) compromise Shell Transport's ultimate authority over its own decisions.

Junior and middle-ranking members of staff knew only vaguely about The Conference, and in the course of time, their hazy knowledge gave it a kind of mythological status in their minds. It was somehow not very surprising that the Royal Dutch/Shell Group, one of the greatest commercial operations in the world and yet a body without a formal corporate existence, should be influenced by another such body containing one of the oil industry's greatest concentrations of expertise. Yet though it was barely visible to the majority, The Conference's supervision of policies produced some highly visible results.

Undoubtedly the most far-reaching of these was Shell's ever more deliberate internationalization. Like the establishment of The Conference, this is credited to John Loudon, described by Sir Peter Holmes (chairman of Shell Transport from 1985 to 1993) as the true father, in the last third of the 20th century, of the Royal Dutch/Shell Group.

Urbane, charming, fluent in at least five languages and generally recognized as Shell's international diplomat *par excellence*, Loudon supported the principle of internationalization untiringly and with great success. Its first application had come in the wake of the Mexican expropriation in 1938, so when Loudon became a managing director in 1947 it had operated for a decade. Even so, there had been little move towards placing local people in a given country in top positions; and at home Loudon found that between British and Dutch personnel there were still 'sometimes certain frictions... even at the top'. Setting himself to improve these conditions as far as possible, he remembered later that 'once the older generation retired there was a gradual understanding, which had to come from the top, and which was...that passports were not that important.' Thereafter, the CMD soon included an American and a Frenchman, and by the time of Loudon's retirement in 1965, Shell's internationalization was so conspicuous that it was objectively described as 'an operation without commercial parallel.' The same writer (Anthony Sampson, a commentator not always well-disposed towards the oil industry) went further:

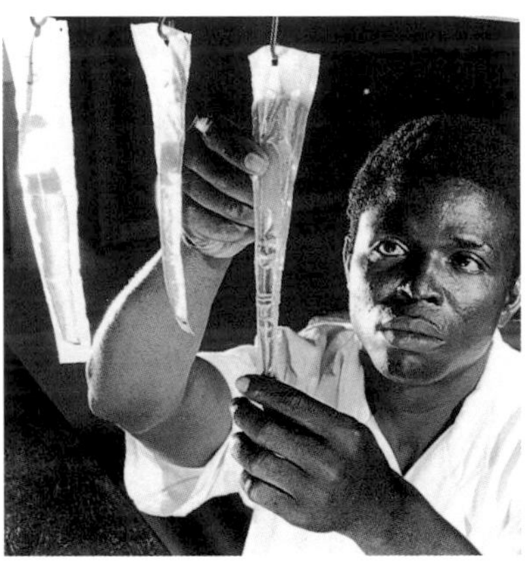

> To speak entirely in terms of English and Dutch is misleading, for Shell in its staffing is probably the most international firm in the world... They pushed through – *ahead* of politics – the quick recruitment of Asians, Africans or South Americans, giving them as much independence as they dared. They tried to avoid choosing local managers by Western or 'old boy' standards, and to accept the values of local communities. For many of the old-style administrators the change was appalling. (I remember seeing their bewilderment in East Africa in the early fifties when inexperienced Africans were promoted.) But it was carried through, helped by Shell's hard international experience...

And difficult as it was for some people to accept, it was nevertheless a change in tune with – even, as Sampson said, ahead of – the times; for the period of Loudon's high responsibility within Shell (1947–65) was precisely the period in which the old and historic British and Dutch imperial powers gave way and, with varying degrees of grace, granted independence to their former colonies. Companies of the Royal Dutch/Shell Group were similarly decolonializing, and attempting (with perhaps rather more conviction and success than their parent nations) to abandon the attitudes of Empire. Sampson continued:

> They became deeply involved in local problems... It was a painful operation, full of disappointments and mistakes, but it achieved quickly and relentlessly – and in striking contrast to BP – the obliteration of the imperial idea alongside which Shell had been built. As regionalization progressed, so Shell was able to have a hundred different nationalities on their various staffs. By 1964 they had Trinidadians in Nigeria, Indians in Germany, Venezuelans in Brunei, and a Tanganyikan [Tanzanian] in Norway.

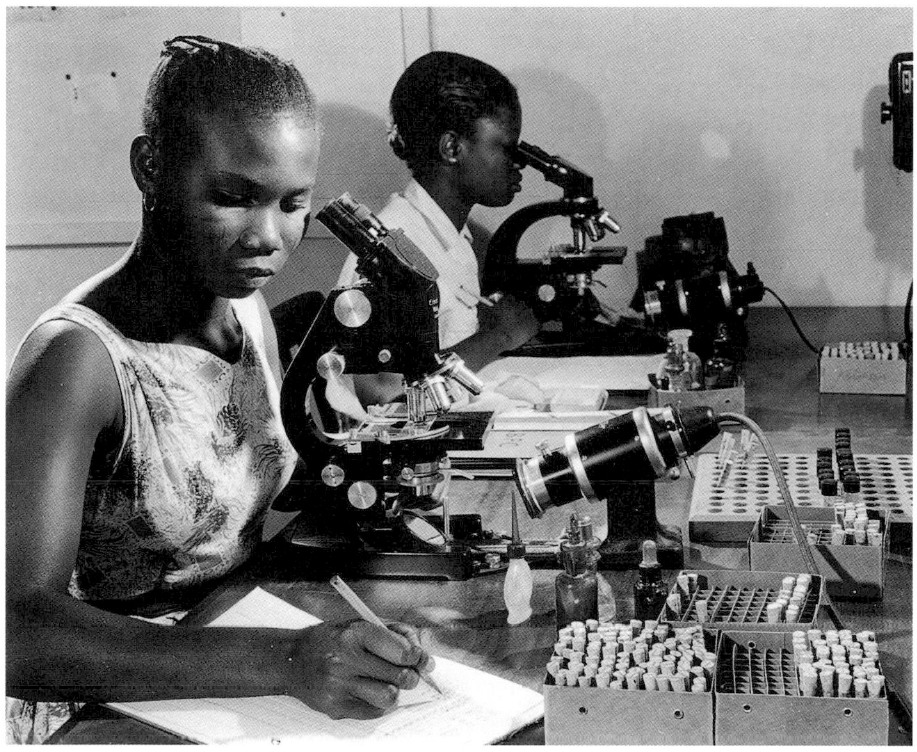

The policy continues, and Shell's continued success speaks for its validity; indeed, some would judge that in the world of the middle 1960s, becoming fully internationalized was Shell's most valuable action both in human and commercial terms, outweighing the production of many different chemicals, the laying of thousands of miles of pipelines, and the construction of increasingly large ships. Though in good years the chemical business could be very profitable, its nature was highly cyclical, and sometimes in the future the CMD would wish that Shell had never become involved in it. However, if proof were needed that the decision to girdle the world with pipelines and supertankers was correct, it came in June 1967 with the Six-Day War between Israel and Egypt. Like a recurring bad dream, the Suez Canal was closed again; and this time it remained closed not for a few months but for eight years.

KITTIWAKE L.B.
21/18
SHELL-ESSO

CHAPTER FIFTEEN

The End of the Buyers' Market: 1967–1972

The establishment of the Committee of Managing Directors (described in chapter 13) was a highly imaginative and creative step. In one way the need for the CMD, or something like it, had been obvious for some time: for despite being separate companies, Shell Transport and Trading and its sister Royal Dutch were so intimately connected, and the Group which they led had developed into such a large concern, that close regular discussions between their managing directors had become essential. But the CMD introduced something entirely new both to Shell Transport and to Royal Dutch: collegiate leadership and collective responsibility. Leadership was vested not in any one individual, but in the CMD as a whole.

This may well have been unique in the contemporary business world. Certainly it was very different to the traditional structure of a board of management – so much so that many people were deeply sceptical and believed the arrangement had little chance of success. But John Loudon, its instigator, was convinced that the range of operations shared by Shell Transport and Royal Dutch had become far too diverse and complex for any one chairman, or pair of chairmen, to grasp fully. Instead he wished to see the Group led by a group, a body of coequals.

Opposite page: Kittiwake SBM – the advantageous system of Single-Buoy Mooring, developed by Shell

> The McKinsey people wanted to have such a committee with a chairman as chief executive and a president as chief operating officer, but I felt that...as long as we had a well organised committee with a chairman who was really *primus inter pares*, you did not have to define so clearly whether one MD would be in overall charge of day-to-day operations.

The new collective structure made a colossal and lasting change in Shell's most senior levels. Through their own deliberate choice, the managing directors of the parent companies had ended the days of the powerful individual leading character – the Marcus Samuels, the Henri Deterdings. But as Loudon explained, that did not mean the CMD would be an indecisive talking-shop:

David Barran, knighted in 1971, was chairman of Shell Transport and Trading 1967–72

> I also believed that on the other hand you should not slow down the decision-making process and that if anyone was away, business had to go on. If any managing director was travelling around and something happened in his particular field and you could not get hold of him, the other managing directors would decide.

This absence of a single top decision-maker needs to be emphasized, firstly because it was such a very unusual way of governing a large organization, and secondly because from the public point of view, the change was almost unnoticeable. Shell Transport still had a chairman; Royal Dutch still had a director-general (or president, as the post was later termed); and each man, as his company's official spokesman, still led its annual general meeting. Yet – rather like a hologram, apparently solid, but requiring a special light to be seen – the distinction had become something of an illusion, only visible in the spotlight of public attention. In the privacy of their boardrooms, whether in London or The Hague, when chairman and director-general met together each week with as many as possible of their fellow managing directors, the illusion dissolved. Although there was still of course a hierarchy, and although a chairman of dominant personality could become rather more than first among equals, nevertheless the CMD in session was probably a more thoroughly collegiate governing body than anything that had ever been known in the business world.

That very important point must be remembered throughout the rest of this book, because – even though they and their counterparts in Royal Dutch had purposely renounced the power to take decisions on their individual responsibility – the chairmen of Shell Transport and Trading will continue to figure prominently in these pages. If that sounds paradoxical, it is; but then it is only one more paradox in the history of Shell Transport. The Group is one of the world's largest business enterprises, yet as an entity it has no legal existence; its governing bodies, the CMD and The Conference, likewise neither exist in law nor take decisions. With paradoxes of that magnitude, it should not be hard to absorb a far smaller one.

On 30 June 1967, David (later Sir David) Barran succeeded Frederick Stephens as Shell Transport's fifth chairman. It scarcely seemed an auspicious time to take on such a post: four important streams of European oil supply from the Middle East had just been interrupted, and a fifth, from Africa, was about to follow.

The first four interruptions, all consequences of the Six-Day Arab-Israeli War (5–10 June 1967), were the closure of Suez; the cessation of supplies to the east Mediterranean loading ports of Banias in Syria and Tripoli in Lebanon; the switching off of Tapline, the 1,040-mile Trans-Arabian Pipeline from the Persian-Arabian Gulf to the Mediterranean; and (in the mistaken belief that British and American warplanes had assisted Israel) an embargo imposed by Saudi Arabia, Libya, Kuwait, Iraq and Algeria on all supplies to Britain and the United States. The fifth interruption – an unhappy coincidence, but no less important for that – took place when tribal tensions in Nigeria boiled over into civil war and brought exports from that country to a standstill. All told, these removed some 6.5 million barrels a day (bpd) from the supply stream, or about one-sixth of the whole world's production. There was every reason for newspapers to call it a time of crisis, as they did; yet reflecting on it many years later, Barran remarked that 'the closure of the Canal, although it was tiresome, wasn't in any way crippling.'

By then, of course, he was able to see events of 1967 in a longer perspective, and to recognize that compared to events which soon followed, the closure was for Shell a fairly easily managed crisis. It did not seem so then; but at the very outset it introduced two themes – shipping, and the Middle East – which dominated Barran's chairmanship.

Shipping, the foundation of all Shell Transport's fortunes, was now the answer to its sudden and manifold supply problems. Shell's analysts had foretold the Suez closure, and in doing so, they made a flat statement that would have amazed Marcus Samuel: 'the closing of the Suez Canal does not affect oil.' The main difficulties, they said, would be 'tanker tonnage problems', and so it proved to be. Globally there was no *production* problem: Shell made good its shortfalls initially by drawing on reserve stocks and subsequently by increased production in Venezuela and the United States, as well as by bringing forward production in Gabon, western Australia and Oman – 1967 saw that country's first oil exports. Of course there were limits to what could be done at no notice: in June 1967, the amount of crude oil arriving in Europe as a whole dropped by 24%, and in the UK (the worst-hit nation) by 33%. But, even though there was much more to contend with than the closure alone, nonetheless there was plenty of oil available world-wide – if the hulls could be found to carry it.

The entire industry's marine sector immediately became the scene of intense activity. Middle Eastern oil now had to come via the Cape of Good Hope; but big ships that could make the long haul economically could not be pulled out of a hat, and within ten days of the closure, freight rates tripled. BP, with its production focus still firmly in the Gulf, reacted especially quickly to the changed situation. Inside a week, in addition to its large existing fleet, it had chartered about 1.1 million deadweight tons (dwt), or 1% of the whole world's tanker resources.

'They have so far been first in the field', a Shell Transport report admitted with rueful admiration, 'and have been chartering skilfully in a rapidly rising market.' Having entered the emergency charter market a little late, Shell found itself paying at the high end. However, following advice from former managing directors who had been through the first Suez closure, some of its collective experience now stood it in good stead. A variety of plans were prepared in case consultation with other majors or governments should become necessary, and all chartering was kept strictly short-term. In 1956, the shocking novelty of closure had generated a degree of panic within the industry. Independent tanker owners 'made untold millions at that particular time', one Shell executive noted drily, with many companies ('and unfortunately Shell was at the greater end') making long-term charters 'at what turned out to be exorbitant rates. This mistake has not been repeated today...'

There were other reasons too for Shell's more phlegmatic modern reaction. Once bitten and now shy of the Canal, it had already become accustomed to carrying far more than anyone else by the Cape route: 160,000 bpd in 1966, at a time when BP and Esso, the nearest competitors, were bringing only 65,000 bpd each, and all other majors a mere 10–20,000 bpd each. Moreover, big ships were on their way. Slated for late

Mytilus, *1969*

October, the launch of the 100,000dwt *Narica* was brought forward by six weeks at a cost of £60,000 ('money well spent'). By mid-October the new vessel was loading in the Gulf, joining two other new ones of the same size; and others twice as big were nearing completion.

Every oil major had the new-concept VLCCs (very large crude carriers) on order: Gulf Oil, for example, had six 300,000-tonners building. Shell had chosen 200,000 dwt – a figure widely agreed as a new standard – for its first VLCCs. The economies of scale were impressive: an 80,000dwt tanker taking crude from the Gulf to Rotterdam via the Canal broke even at 27 shillings and sixpence (£1.375) a ton, but a 200,000dwt vessel using the much longer Cape route was 35% cheaper per ton, breaking even at only 18 shillings and one penny (90.4p). Shell's first such ship, *Myrina*, began service as intended in 1968. Three sisters followed in that year, and the 'M' class rapidly grew as planned to 31 ships, aggregating 6.2 million dwt, by the end of 1970.

Had these ships existed in 1967, there would have been very little supply problem. As the *Financial Times* expressed it then, 'the latest closure of the Canal has come too soon'. The problem was indeed largely a matter of timing, but while it existed it was perfectly real, and posed a very serious challenge. Describing the successful counter-measures – increased production outside the Middle East, and the immense programme of additional chartering – an understandably satisfied note crept into Shell Transport's report for the year. 'With the world-wide spread of their operations and with their operational flexibility,' it said, 'Group companies were able to ensure that their customers received all the oil they required.' Had the report chosen to go further, it could have justifiably echoed the words of an independent commentator at the time:

> In defending their industry against its many critics, the major oil companies were wont to emphasise the enormous flexibility arising from world-wide deployment, which enabled it to maintain supplies whatever the emergency. In 1967, their performance certainly matched their promises.

With scarcely any governmental involvement, without rationing (which had occurred in 1956–7), and for an average extra charge to the consumer of only about twopence a gallon (0.175p a litre), European summer demand was fully met and reserves rebuilt, with ample stock for the winter. To sound a moderate note of satisfaction seemed fair enough.

But if the Arab nations' use of 'the oil weapon' for political ends had failed, that was at least in part because Venezuela had been willing to permit additional production; and no one gave much thought to what might have happened if Venezuela had joined the embargo – nor to what might happen if such a thing occurred in the future. Yet the mechanism for Venezuela to join in already existed, in the shape of the Organization of Petroleum Exporting Countries (OPEC); and OPEC, founded in 1960, was the brainchild of Juan Pablo Pérez Alfonso, the Venezuelan Minister for Mines.

David Barran had not expected to be selected for the role of chairman of Shell Transport and Trading, but later he recognized how he had been groomed for it – 'or at any rate,' in his words, 'given the opportunity to learn.' Like Sir Robert Waley Cohen, who had had the reputation of being the brainiest man in the City of London, both Barran and his predecessor Stephens had been educated at Cambridge, and joined Shell on graduating – Stephens in 1926, and Barran (with a BA in History from Trinity College) in 1934. After a few months in the London offices, Barran had spent twelve years in north Africa and the Middle East, getting to know the region very well. He then had a year in India, followed by eleven based in London. During that time he was twice given charge of delicate negotiations, including at governmental level with Venezuela: 'a deliberate opportunity', he subsequently decided, 'to try out my abilities in the negotiating field and particularly to educate me – allow me to educate myself – in the handling of international trade in oil.' The experience was extended when in 1958 he was sent to New York as chairman of the Asiatic Petroleum Corporation, Shell's 'ears and eyes on the international scene'.

For three years, 'within the permitted bounds of anti-trust, I had a lot of contact with all our main competitors and kept very much abreast of what was going on the world at large, not just in the North American scene'; and in 1961 he returned to London as a director of Shell Transport and Trading. He was still only 49 years old.

The directorship brought with it numerous additional responsibilities, for simultaneously he became a director or managing director of seven other major Shell companies – 'a daunting prospect, but with the enormously pleasant and confidence-inspiring feature that one was joining a group of friends, all people I'd worked with very closely... They were enormously helpful to me coming in as a new boy, showing me the way to go.' Their guidance was clearly good and his ability to educate himself unimpaired: the chairmanship of Shell Transport came six years later.

Barran had just turned 55. For him personally, for Shell collectively, and indeed for the industry as a whole, his five years as chairman were overall a period of great optimism, development and prosperity. However, as we shall see, they were punctuated by tragedies, and by a deeply ominous turn of events in Libya.

Before those less happy events transpired, the optimistic atmosphere within Shell Transport was well founded. Though the Canal remained shut, the embargoes had been lifted and Tapline re-opened after a couple of months. However, for the British Exchequer, the Canal's continued closure was the last straw in a series of great economic difficulties, and on 18 November sterling was devalued by 14.3%. The effect of devaluation on Shell Transport the following year was very marked – not in operational terms, but in accounting terms, because oil had always been priced in US dollars. All of a sudden, when translated into sterling, earnings jumped smartly: the company's share of Group net income rose from £106.4 million in 1967 to £144 million in 1968, an increase of more than 35%.

From the careful explanation given in the annual report, shareholders understood that although much of this was an artificial consequence of the altered exchange rate, nonetheless a healthy proportion of it stemmed from real growth. Exploration in sea areas was on a larger scale than ever; sales of oil products had increased by nearly 10%; natural gas revenues had more than doubled; and as for chemicals, 'sales were limited only by manufacturing capacity.'

Barran was justified in calling the results excellent, and in viewing the future cheerfully. Compared with any other energy source, oil was cheap. Demand was high and still rising. To cope with it, existing Shell refineries in nine other countries had been enlarged, and new ones had been commissioned in Norway and the UK, with the latter having a larger initial capacity (120,000 bpd) than any yet built there. Oil had been discovered in northern Alaska, not by Shell but encouragingly close to Shell acreage; and the new 200,000-ton M-class VLCCs were already proving their worth. One of them was the third *Murex*, capable of carrying 41 times more than Marcus Samuel's first tanker; and the carriage of every M-class cargo cost less per ton than any in history. In short, it seemed that bigger was assuredly better, in whichever sphere.

The impetus continued through 1969. Production and sales of oil, oil products, chemicals and natural gas all rose again – indeed natural gas was the highlight of the year. In Europe, its sales increased by more than 50%, following the virtual completion of the North European trunkline network (in which Shell had an interest), covering the Netherlands, Belgium and parts of France and West Germany; and in the Far East a deal of major importance was concluded with Mitsubishi and the state of Brunei. With effect from 1972, the £650 million agreement (as Barran said, 'the largest of its kind yet undertaken') would run for twenty years, in which period Shell ships designed especially for the purpose would transport over 100 million tons of liquefied natural gas from Brunei to customers in Japan.

Opposite page: Mactra, *badly damaged by a gas explosion during tank cleaning in 1969*

Very occasionally (and this was one of the occasions) one can sense, between the lines of Shell Transport's annual reports, a hint of astonishment at the successes they are reporting. But tragedy was close at hand. On 14 December 1969, the VLCC *Marpessa* exploded on her maiden round voyage. Two men died, and the vessel gained the ghastly distinction of being the biggest ever to sink. A mere fifteen days later, her sister *Mactra* blew up. She did not sink, but again two men died, and a hole some 400 feet long and 50 feet wide was ripped in her deck. And the day after that, a non-Shell ship of very similar design met the same fate, with a huge spontaneous explosion followed by raging fire.

These catastrophes threw the entire VLCC strategy into question. Yet there could be no question of abandoning it: not only because of the magnitude of the investment, effort and ingenuity involved in the design and construction of the ships, but also because of the imperative thrust of political and economic history.

Two years of intense experiment and investigation established both the cause and the cure. It became apparent that each explosion had taken place during tank-cleaning operations. The cleansing water spray had created an electrically-charged mist which, inside the vast spaces of a VLCC hold, could form a cloud large enough to generate sparks – a miniature thunderstorm, fatal in the tank's gaseous atmosphere. The cure was to flood the tanks, prior to cleansing, with an inert gas which could not explode. Put so simply, it might seem that both cause and cure should have been obvious from the start; but these were new technologies, and it was only through the trial of reality, and the hardest of error, that their faults were found.

However, everyone learned invaluable lessons; and at the same time, Shell's marine sector brought two further developments to the world of oil tankers – single-buoy moorings (SBMs), and 'lightening ship'. Neither was a new idea; an SBM is essentially the system used by the skipper of any small yacht, while lightening ship (also known as ship to ship transfer) is the process of taking some of a cargo off while the vessel is at sea, so that she may enter a shallow port. But what was new, in both cases, was the technology and the sheer scale. SBMs were highly complex and imaginative pieces of engineering which enabled a vessel to load direct from offshore wells while remaining free to swing with wind and weather. Shell first used one at Miri in Sarawak, and later in Qatar, at the world's first entirely off-shore oil field. The facility gave an immense technical advantage over competitors, and proved that the transfer of oil to or from very large ships at sea could be conducted in complete safety. This in turn led to lightening operations – which were not merely economically desirable, but essential, because many ports around the world could not yet accommodate a fully-laden VLCC. (Indeed, such vessels sat so deep in the water that they could not transit the Dover Strait.) It would have seemed incredible just a few years earlier, but (from 1968) the 70,000dwt *Drupa* and (from 1972) the 117,000dwt *Naticina* were used primarily not as long-distance tankers, but as lightening ships serving the VLCCs.

Naturally there were numerous developments in other fields: for example, the invention in 1969 of a material that significantly improved upon the qualities of titanium. Titanium, a metal light in weight but very strong and highly resistant to corrosion, is used in the manufacture of such things as aircraft parts. The new material, a composite of Shell epoxy resins with carbon fibres, was as stiff as titanium, but 40% stronger and 66% lighter. Yet even events as remarkable as that were subdominants of the period. As mentioned earlier, shipping and the Middle East were the dominant themes of Barran's time as chairman of Shell Transport; and on 1 September 1969, just fifteen weeks before *Marpessa*'s sinking, a coup took place in Libya, propelling a young army officer, Muammar al-Qaddafi, into power.

Libya by then was the source of a quarter of all crude oil consumed in western Europe. The oil was highly valued, partly because it had a low sulphur content (making it comparatively easy to refine and clean to use), and partly because Libya was on the 'right side' of the still-closed Suez Canal. The country frequently featured in the monthly meetings of The Conference. In April 1969 its members (the CMD, retired Group managing directors, and external directors) had talked about the possibility of an increased offtake of Libyan oil; in May, about discussions with the Libyan government concerning the acquisition of acreage with the national oil company, Lipetco; and in

Anglesey terminal SBM and Northia

July, about a proposed joint venture with Lipetco involving exploration, refinery construction and marketing arrangements. There was no Conference in August, but when its members met again on 10 September they were eager to hear Barran's assessment of the effects of the coup.

Barran observed that very little was known about the background of the leaders of the coup – the Revolutionary Command Council or RCC, as they styled themselves. Even their political views remained unclear. However, the first stage of Shell's recent agreement with the former government, which had been signed before the revolution, was being carried out; a statement from the new government said they would honour all oil agreements; and operations had not so far been hampered. As yet, there was no published policy regarding the oil industry's future in Libya, and although pressure from the new regime might arise in the long run, it was hard to tell. Overall, he felt, there was 'not much cause for concern.'

Any other well-informed person within the industry would probably have said the same, or something very similar; there was little reason to think otherwise, and no one would have described Barran's cautiously optimistic forecast as unreasonable. But looking back, it could scarcely have been more mistaken.

Libya's Revolutionary Command Council was composed of very young men; on average, they were about 26 years old. They had not really expected

The third Murex, *capable of carrying 200,000 tons of oil, seen 'lightening ship' – transferring part of her cargo into Drupa*

their coup to succeed, but as the reality of their success began to sink in, so came the understanding that genuine power now lay within their hands – that they could change the way their country was run. Being devout Muslims and (as soon became apparent) devoted followers of the political doctrines of Egypt's President Nasser, they had three targets for immediate achievement. Resenting Libya's past as an Italian colony, they gave all Italian residents – third-generation settlers – a month to leave. Resenting

British and American presence in the country, they closed the air bases outside Tripoli and Tobruk. And, respectful of the Koran's prohibition of alcohol, they tore up every grapevine in Libya, 'which was very sad', Sir Peter Holmes recalled in retirement, 'because Libyan wine had been very good.'

Holmes had spent four years (1959–63) in Libya as one of Shell's district managers, and found it a lovely place in which to live and work; but with the revolution 'the country *totally* changed direction.' The RCC 'were quite ruthless in their approach, with young men's impetus, and therefore they got quite quick responses. During 1970 they turned their attention to the oil industry.' And when they did, said Barran, 'the avalanche had begun.' As Holmes remembered,

> The price of Libyan Zelten oil, the marker for south Mediterranean crudes, had been fixed at something like $2.26 a barrel since it first began to flow in the early '60s. That was of course with hindsight quite a low price...and the RCC determined to get the price higher. Their initial aspirations were quite modest. The whole of the summer of 1970 was spent between the industry and government, led by [Libyan Prime Minister] Jalloud, debating five or ten cents onto the posted price of Zelten.

Libya's new ruler, Colonel Muammar Qaddafi

'Modest' was a retrospective judgement. Hitherto, any net price alteration had been in fractions of a cent per barrel, over which oil executives would spend hundreds of worried man-hours. To speak of five or ten additional cents was practically incomprehensible, seemingly tantamount to economic suicide.

As negotiations continued in May 1970, a bulldozer was 'accidentally' driven into the Syrian section of Tapline, once again closing off the vital communication between the Arabian peninsula and the Mediterranean. 'By this time,' said Barran later, 'I suppose you could say we were a bit punch-drunk – "Oh God, not another source dried up. What do we do about this now?"' With repairs to the pipeline taking an unnaturally long time, Libya's negotiators, initially tentative and uncertain, gained rapidly in confidence. Their first approach, to the industry as a whole, was turned down by the industry as a whole. Next, therefore, they made a move as revolutionary as anything they had done in politics, and cut production overall.

'For years, governments had begged, beseeched and cajoled companies into increasing production', wrote Brian Carlisle (Shell's Regional Co-ordinator, Middle East). To do the reverse and impose a cut was 'an entirely new tactic, which seemed to take the industry by surprise'. The Libyans then targeted individual companies, commencing with Occidental, the newest and (because practically all its production came from Libya) the weakest. Already suffering from the unprecedented unilateral cut, and fearful that its entire production might be shut in, Occidental eventually accepted the Libyan demands: an immediate thirty-cent increase in the posted price of oil, and – crucially – an enhancement of Libya's share of profit from 50% to 55%.

After Occidental, Oasis was next in line: a consortium of three independent oil companies (Continental, Amerada and Marathon) and, with a one-sixth interest, Shell. As Holmes described events,

> The Libyans used the same threats: Agree or you'll be shut in. The three independent companies agreed, after some hesitation; Shell said No. Shell was then put on one side – by now we're in late August, early September of 1970 – and the other companies were called in one by one. Next was Amoseas, a 50:50 combination of Chevron and Texaco. They buckled, and once they did then everyone else buckled too, so that just left Shell saying No.

This of course was an impossible situation. Noting that Shell's stance hitherto had gained it respect within industry circles, Barran pointed out that continued refusal could be 'only a noble gesture'. Libya's terms must now be accepted. By coincidence, Holmes, already a veteran of Libya and fond of it, was about to return there as Shell's general manager; so somewhat ironically, 'when I arrived in early October 1970, my first job was to go and say, OK, we agree after all.'

For the oil-producing nations, the industry's collapse, as they saw it, in the face of a small group of determined young soldiers, was utterly sensational. 'Every Middle Easterner always saw the *might* of the oil companies, and Qaddafi had shown that it wasn't quite what people had thought,' Holmes explained. 'Having worked inside an oil company all my life virtually, I never saw much might at all; there was certainly no political might.' Nevertheless, perceptions had suddenly and radically altered, and whatever simile was used – whether an avalanche, as Barran said, or a bandwagon, or the breaching of a dam, as others said – the effect was the same: every producing nation set out to gain at least as much as Libya.

A wild game of leapfrog began. First Iran and subsequently all the other Middle Eastern producers opened negotiations and caught up with Libya; and when they had, Libya renounced its earlier agreements, which had been set to last for five years, and started a new round. Further revision of the posted price, with retroactivity; compensation for the declining value of the dollar; premiums for low-sulphur, short-haul crude; and eventually 'participation', or the partial ownership of the oil within their national boundaries – all these and other bargaining counters were introduced by one producing nation or another over the next few years, and when any one was successful then naturally enough the others followed eagerly.

Sir Alec Douglas-Home, Britain's Foreign Secretary and later Prime Minister

Little by little the price edged up, until by September 1973 it had reached $2.90. Looking back, it is hard to imagine it was ever at such a low level, but compared to the original $2.26, $2.90 did seem high, destabilizing an industry which placed a very high value on stability – which, indeed, traditionally felt it could only operate efficiently and economically in a stable environment. Attempts were made to gain the support of consumer governments: in particular, Barran and his counterpart at BP, Maurice Bridgeman, asked the British Foreign Secretary, Sir Alec Douglas-Home, to persuade other European countries to accept a boycott of Libyan crude. With the experience of Suez 1967 behind them, the two oil chiefs were confident that their companies and others in the industry could make good the difference; but Douglas-Home's efforts failed, for the main consumers in continental Europe were more interested in securing their own supplies. However, there was slightly more success in the United States, where, despite the anti-trust laws, the majors were permitted to sit in committee together under the chairmanship of an eminent lawyer, John McCloy. 'He was there', said Barran,

> to see we didn't tread on the holy ground of prices in any way, but we were allowed to discuss quite freely what we could do to replace Libyan oil by increasing production in other places...and that committee showed particularly to the American government that it was possible for the oil industry to operate in collaboration without any improper breach of the anti-trust laws.

This at least was a useful result, he felt; and it did enable the different oil companies to put together delegations which could legitimately conduct negotiations from a shared viewpoint. But it made little difference in the end. As Middle East Co-ordinator in 1970–3, Brian Carlisle (a former naval officer who had won the DSC in World War II), was head of Shell's delegation in the thousands of hours of discussions. 'It was more or less a continuous negotiation,' said Holmes, adding, 'I must say it was quite a fascinating time.' Barran agreed: 'intensely interesting and fascinating' was his verdict. Carlisle's view, when he retired exhausted in 1974, was less detached: he felt the producing nations had been 'dragooning the industry' with their new-found power. But he added, more philosophically, that 'they would have to be very far-sighted and altruistic to do otherwise.'

1971

Everyone involved was aware that the nature of the market-place was changing: as Barran remarked in 1971, 'The buyer's market for oil is over.' Whether the change would be permanent or not remained to be seen. But before his chairmanship of Shell Transport ended in 1972, and with the difficult and disturbing Middle Eastern negotiations as a constant background, a number of other important events occurred.

In the international political arena, perhaps the most significant development was the removal in 1971 of all British armed forces from the Gulf, after many years of active guardianship and peacekeeping within the region. Several of the smaller Gulf states were so alarmed by this as to offer to pay for the maintenance of a British presence, a suggestion which the British Prime Minister declined, saying Her Majesty's troops were not mercenaries. Quite right; but a distinction could have been perceived, and as it was, a potentially dangerous military vacuum was created, into which any ambitious adventurer could try to step.

In Shell's private commercial arena, a number of steps were taken – some small, some very large. Although its interest and activity in the oil world was undiminished, the year 1970 brought the purchase of a Dutch-based metals mining company called Billiton – a radical move into a new area of business, intended as a cushion against any downturn in the unremitting negotiations with the oil-producing nations. We shall hear more of Billiton in the next chapter.

In 1971, the Shell logo was redesigned – an expensive exercise, but an up-to-date public appearance was understood to be an important element in the maintenance and growth of market share. This was Shell's eighth logo, and its seventh to make use of the pecten, the shape of the scallop shell. (The first logo, in use from 1900 to 1904, represented a mussel.) But while remaining faithful to the pecten concept, the new logo was for two reasons a fairly fundamental departure. Firstly, the word SHELL was removed. It had been introduced into the design in 1948, as part of the massive postwar effort to reconstruct the business and re-establish its identity in the public mind. Removing the word was a silent but clear expression of confidence: everyone knew Shell. And the other novel element was just as clearly visible. The first pecten, introduced in 1904, had contained so much detail that it could have been used in a book on marine biology. The second, third and even the fourth pectens had been nearly as detailed; it was only with the fifth and sixth, introduced in 1955 and 1961 respectively, that something more symbolic was attempted. But the new pecten was by far the simplest and boldest that Shell had ever had. The shell became a pure symbol, and instantly made every other pecten – even the most recent – look dated. Indeed, it was so modern in appearance that in the United States, Shell Oil Company felt it went too far. There, it was adopted in a modified form, and ever since, the American pecten has been slightly different to that used everywhere else in the world.

The same year, 1971, saw in the marine sector the invention and application of the concept of floating storage units, or FSUs - large ships permanently moored in position to receive and hold stocks of crude oil from

Aerial view of the Karratha onshore gas treatment plant

CORMORANT-A

an offshore platform, prior to those stocks being sucked out by smaller shuttle tankers that ferried the oil ashore; and in world-wide exploration there were amazing discoveries – four large gas fields were found off the Australian North-West Shelf, and in the North Sea, the Auk and Brent oil fields were found, followed by the Cormorant field in 1972.

Barran, who had become chairman of the CMD in 1970, was knighted in 1971 and retired on 30 June 1972, being succeeded by Frank McFadzean. In his valedictory address to shareholders at the AGM in 1972, Barran spoke of the uncertainties of forecasting. He then promptly made a forecast. Noting that energy demand in 1971 had been lower than anticipated, he added:

> this is no reason now to underestimate the longer-term potential. For instance, there can be no doubt, if the world does not fall into a severe depression, that energy demand – and therefore oil demand – will resume its growth. Because of the heavy investments they have already made, Group companies will be well placed to take advantage of this situation. That is not to say that no problems lie ahead...

Frank McFadzean, Chairman of Shell Transport and Trading 1972–6

If the world does not fall into a severe depression... Although many countries had been going through a period of 'stagflation', real economic growth in the industrialized countries was about 5.5% in 1972, as compared with 3.3% in 1971, and 'a severe depression' seemed on the whole to be an unlikely prospect. But no one – including the instigators of the event – forecast what would happen when OPEC, the Organisation of Petroleum-Exporting Countries, finally decided to take concerted action. And in the autumn of 1973, the 'First Price Shock' burst onto the industry and the entire world economy.

SHELL/ESSO
BRENT C

CHAPTER SIXTEEN

Cataclysm: 1973–1977

All the negotiations of the early 1970s abruptly halted. All the efforts had been in vain, and all the worries over a few cents here or there seemed suddenly laughable. In September 1973 the price of a barrel of oil was typically $2.90. By December it had nearly doubled to $5.10; on 1 January 1974 it more than doubled again, to $11.65; and as panic-buying hit the markets, spot prices of over $22 a barrel were bid.

There were two main reasons for this extraordinary and frightening state of affairs. To begin with, early in 1973 the United States had relaxed its restrictions on oil importation. The consequent increased demand proved temporarily difficult to fulfil, and some people began to speak of an impending energy shortage. Shell Transport's new chairman, Frank McFadzean (later Sir Frank, and subsequently Lord McFadzean of Kelvinside), was more sanguine. Under the headline 'Problem but no Crisis', one of Shell's in-house publications reported his comments at the AGM: 'Our task is to prevent a crisis arising in future. Action is required to maintain the balance of supply and demand in the years ahead.' But, while Shell and the other majors worked to re-balance supply against demand, they received for a short time unexpectedly high profits, exciting the jealousy of the producing nations with whom they were still negotiating. Then on 3 October, the fourth Arab-Israeli war began: the Yom Kippur war, launched against Israel on Judaism's 'day of atonement' by combined Egyptian-Syrian forces. As western nations aligned themselves on one side or the other, Arab states registered their disapproval of those (particularly the USA and the Netherlands) which supported Israel, by cutting supplies by 25%; and at the same time, refusing further negotiation, they raised prices to the minimum they felt acceptable.

The immediate effect, in every oil-consuming nation around the world, was an abrupt collision with the unwelcome truth that for the time being at least, the age of cheap energy had come to an end. The longer-term effect depended on how each nation reacted, but overall it was to a greater or lesser degree the same: a long period of severe economic depression and financial inflation.

Before the price shock occurred, Shell had recognized that the industrialized world had allowed itself to become unhealthily dependent upon the Middle East as a prime and easy source of oil. In a confidential warning to heads of government it pointed out that any serious instability in the region could result in a dangerous and difficult scramble for oil supplies. It also began to campaign for an international agreement for shared supplies in time of crisis, and began planning just such a system. But now the crisis had arrived; there was no agreed method of sharing; and in Shell's view, which the other majors supported, 'the only defensible course if governments were not to agree on any preferred collective system' was 'equal misery'. Until another system was established, there would have to be proportionate cuts all round, including to their own refineries.

However, every consuming nation was eager to see that its own supplies were undiminished, whatever might happen to its neighbour, and several brought direct pressure to bear. In Britain in particular, pressure on Shell Transport came first from both the Ministry of Defence and the Department of Trade and Industry. When that proved fruitless, McFadzean and his counterpart at BP, Sir Eric Drake, were called to a stormy interview with the prime minister, Edward Heath. By unfortunate coincidence a nationwide coalminers' strike was about to take place, and with winter close, Heath demanded a private promise from the two oil chiefs that normal supplies would continue. They would not give it: the situation was not of their making but arose from governmental failure to heed warnings and prepare, and, if they now favoured one country over another, their subsidiary companies in other countries could be destroyed by nationalization. They could only comply, they said, if the demand was made into an official government order, which might enable them to plead *force majeure* with other customers; but Heath (who had just succeeded in negotiating Britain's entry into the European Common Market) could not bring it to an official level, for fear of the effect that such an order would have on Britain's foreign relations.

So Britain suffered equally; but the attempted intervention was far from unusual, not only because of the natural desire to ensure supplies, but also for another less palatable reason. If a government could get its way in private, it could leave a company to carry the can in public. Such efforts had been made before and would be made again; and as we shall shortly see, it was not always possible to withstand them.

On the occasion of the interview with Heath, McFadzean viewed the prime ministerial demand for undisclosed special treatment as indefensible; and because Heath was unwilling to give his demand the force of law, McFadzean was able (with the powerful support of Drake) to resist it. A man of less rigorous belief might well have done as he was asked. But had the prime minister been better briefed on Frank McFadzean's background, he might have approached the interview in a different manner, knowing that this was not a man to be coerced or suborned.

Frank McFadzean was Scottish by birth, an economist by training, and, before his Shell career began, a member of the British Civil Service. In that capacity he had worked firstly for the Board of Trade and subsequently for the Treasury, fully absorbing the Civil Service's best tradition of impartiality. These factors altogether formed a character of great integrity – knowledgeable, experienced and insusceptible to corruption. Such a character fitted well with Shell's established methods of behaviour. Unusually, though, McFadzean became a part of Shell comparatively late in his life, joining only in 1952, when he was already 37 years old; and after his retirement from the chairmanship of Shell Transport in 1975, he became a director (and later chairman) of British Airways. (Incidentally, his life peerage, created in 1980, came after his active career with Shell had ended.) In short, he was not a cradle-to-grave Shell man, as many had been. But all the upbringing and training of his first 37 years made him a natural within Shell.

Edward (later Sir Edward) Heath, MP, British Prime Minister 1970–4

Even so, in the unstable atmosphere created by the first oil price shock, he found it hard in 1973 to forecast the industry's future. When asked what level he thought prices might reach, he said simply, 'I think it's anybody's guess.' All that seemed certain was that prices would continue to rise, and that host governments would probably continue to pitch their ancillary demands more and more highly as well.

Shell's 1970 purchase of the Billiton metals-mining company was only part of a new survival plan for the decade and beyond: a strategy of diversification. While remaining firmly in the oil market, Shell began actively to seek new areas of business. In one sense, its traditional work could be viewed as mining for natural resources – hence the Billiton acquisition. In another sense, Shell's traditional business could be viewed as energy, so, soon afterwards, twin moves were made into other forms of energy, old and new: coal and nuclear power. These were absolutely mainstream, compared to some other ideas which were discussed at the time – for example, looking at the growth in cheap air transport, senior management seriously wondered if Shell should go into tourism. Nothing came of that, nor of the equally off-beat thought that they might invest in hotels in Iran. (Not that Shell had a monopoly in unusual ideas; around the same time and for similar reasons, Gulf Oil tried to buy Barnum and Bailey's circus.) Though metals, nuclear energy and coal were adventurous enterprises, they did not seem rash; yet, underlining the ever greater difficulties of forecasting, none of them prospered quite as much as had been hoped. The outline story of each outstrips the chronological bounds of this chapter, but is worth giving at once.

To take Billiton first: Sir David Barran in later years described it as 'very much a Dutch-run venture.' Indeed, out of all the myriad companies in which Shell Transport and Trading has or has had an interest as one of the parents of the Royal Dutch/Shell Group, Billiton's background was probably the most purely Dutch. It took its name from the Indonesian island of Belitung in the former Netherlands East Indies colonies. Tin ore had been found there in 1851, by a small pioneering expedition which coincidentally included another John Loudon – John Francis Loudon, great-uncle of the John Hugo Loudon who did so much to reshape Shell in the decades after World War II. By the time of its purchase by Shell, Billiton was already a very old company. After its formal foundation in 1860, it concentrated on tin-mining until the 1930s, when the mining of bauxite and production of alumina were added to its operations. Such was the situation when, in 1969, Dutch members of the CMD proposed its purchase.

'They made a very good case for doing it,' Barran recalled, 'and we said "Yes, fine", and more or less told them to go ahead and get on with it.' So in 1970, they did: the Shell touch was applied, and an important but obscure mining company, focused in a few far-distant islands, grew into a large international group in its own right, with more than 80 operating companies whose world-wide activity encompassed exploration, mining, manufacture, non-ferrous metals and the recycling and marketing of minerals. A brief history of the company, published in 1985 (its 125th anniversary), declared that its name had changed 'from a geographical location into an industrial concept', which was acceptable, if somewhat pretentious; yet Billiton never seemed to sit entirely easily under the Shell umbrella. Like chemicals, metals-mining (and especially tin-mining) is a highly cyclical business, subject to great fluctuations of profitability, and in the early 1980s slack demand and global overcapacity led to serious financial losses throughout that industry, with many companies being forced to close. Billiton survived, but in 1989, pondering the company his great-uncle had helped create, John Loudon remarked:

> I don't know whether in the long run it is going to be that important to us. Many companies have to get rid of certain parts of their

> business which they've had for years, which are not making enough money. I can see a group like ours at a certain stage saying 'Well, we might as well sell our Billiton interest...'

That stage was reached in 1994. After a brief upturn in the late 1980s, the beginning of the '90s saw a further slide in the metals market. In May 1993 a South African mining company, Gencor, made an unsolicited bid for Billiton, and after prolonged negotiations its sale was agreed and completed in November 1994.

Looking back over Shell's history, it is noticeable that wherever it has worked and whatever it has done, its commitment has always been either complete or non-existent, all or nothing; and after the sale of Billiton, its first and major interest in metals, all its later metals interests were swiftly disposed of as well, with the process being completed in 1995. An experiment lasting a quarter of a century could not be called a passing fancy; but an experiment it had been, and though it had not failed, it simply had not succeeded well enough. In 1994, remembering the original Billiton purchase, Barran described it as 'a sensible move; we certainly haven't made enormous sums of money out of it, yet I think over the period we haven't suffered. But it never really quite achieved the synergy we'd hoped for. We were looking around for what was to be the next thing, and it didn't go quite as much hand-in-hand as we had hoped.'

The venture into nuclear energy was even less successful – indeed, considerably less so. In the 1950s, when nuclear power began to generate electricity for civilian use, Shell was delighted (as we saw in chapter 13) to gain the contract to supply all the lubricants used in Calder Hall, Britain's first commercial nuclear power station, and proceeded additionally to produce coke of extremely high purity for use in reactors. At the time there was a good deal of concern among shareholders that nuclear power could become a competitor to oil. Lord Godber (Shell Transport's then chairman) dismissed these fears as exaggerated, but a watchful eye was kept on the nuclear industry's development. On 2 April 1958, Shell Transport's Minute Book recorded that 'A paper on Atomic Power was placed before the Board and was the subject of a general discussion.' Less than a year later, on 16 March 1959, John Berkin – one of Shell Transport's directors – reviewed for the benefit of his colleagues on the board a 'Memo on Atomic Power...with particular reference to its cost compared with that of power from conventional fuels.' The wisdom then was the same: there was no foreseeable likelihood of nuclear power even coming close to overtaking oil as a cheap and convenient source of energy. But by the early 1970s that view had changed. A toe was put in the nuclear water with the purchase of a 10% interest in

In appearance oddly reminiscent of Shell's pecten logo, the Doublet III experimental nuclear fusion device was developed by General Atomic

David Barran (left) and John Berkin

a Dutch company called Ultra-Centrifuge Nederland, part of a British-Dutch-German arrangement for developing the centrifuge method of uranium enrichment; and in 1973 Shell announced its 'first big step into nuclear energy'. In a 50:50 partnership with Gulf Oil, two businesses – General Atomic Company in the United States, and General Atomic International elsewhere – were established to develop, manufacture and market the second-generation High Temperature Gas-cooled Reactors (HTGRs) and their fuels.

The initial cost to Shell was $200 million, with all subsequent costs to be shared equally with Gulf. For its money Shell acquired interests in a small 40-megawatt experimental plant in Peach Bottom, Pennsylvania; a commercial-scale 330-megawatt plant in Colorado; six other larger HTGRs which were on order; and two more on which options had been taken. Nor was that all. HTGR technology was set to be introduced into France and West Germany, and possibly into the UK and Japan; and (as Shell Transport's annual report for 1973 recorded) General Atomic was already working on several other developments, including *inter alia* an HTGR closed-cycle gas-turbine power plant, a gas-cooled fast breeder reactor, the use of HTGR heat in industry, nuclear fusion research and 'the construction of the largest industrial light-water reactor fuel reprocessing plant in the United States.'

In the annual report there was, with all this, a blissful lack of technical explanation, even in the simplest terms. Probably few shareholders had any clear idea of the differences between types of reactors, or between nuclear fusion and nuclear fission as sources of power; but an annual report is hardly the place to attempt such explanations, and anyway – O brave new world! – they may not have wished for elucidation. Especially when set against the worrying and unfamiliar background of high-cost oil, it was enough to feel that their company was, as always, in the vanguard of modern energy supply.

At any time in our lives, we all (or most of us) do the best we can with the knowledge and tools currently available, and to many specialists and non-specialists alike, Shell's entry into the nuclear field seemed a sensible idea. Proponents of nuclear power saw it as the clean, simple, eternally

Three Mile Island, near Harrisburg, Pennsylvania

renewable fuel of the future, and nuclear fusion (the process at work in the sun) may yet prove to be just that. But the existing method of nuclear power generation (nuclear fission, the principle of the atomic bomb) was already a publicly contentious issue, soon exemplified – long before the much greater disaster at the Russian plant of Chernobyl in 1986 – by the episode at Three Mile Island near Harrisburg, Pennsylvania, when, on 28 March 1979, the cooling system of the plant's No. 2 reactor failed and led to a leak and partial melt-down of the uranium core, with radiation detectable over twenty miles away.

Three Mile Island was a great leap backwards for the nascent nuclear industry, hardening feelings that having a nuclear reactor on one's doorstep might not be an unmitigated good. It was followed, moreover, by a series of five smaller but similar accidents in the US, which led the Nuclear Regulatory Commission temporarily to cease licensing the construction of new reactors. Although General Atomic was not involved in any of these, Shell read them as a clear warning and decided there was not enough to be gained from remaining in an industry which was so expensive, so politically vulnerable, and so much the target of public protest. Those factors were quite sufficiently present in the oil industry anyway; one would have to be a glutton for punishment to seek them elsewhere as well. So, resolving to remove itself from active participation in the nuclear industry, Shell sold its interests in both General Atomic companies to Gulf Oil in 1980. Lasting a mere seven years, nuclear energy had been a short and costly byway – one which Shell would not follow again for a very long time, if ever.

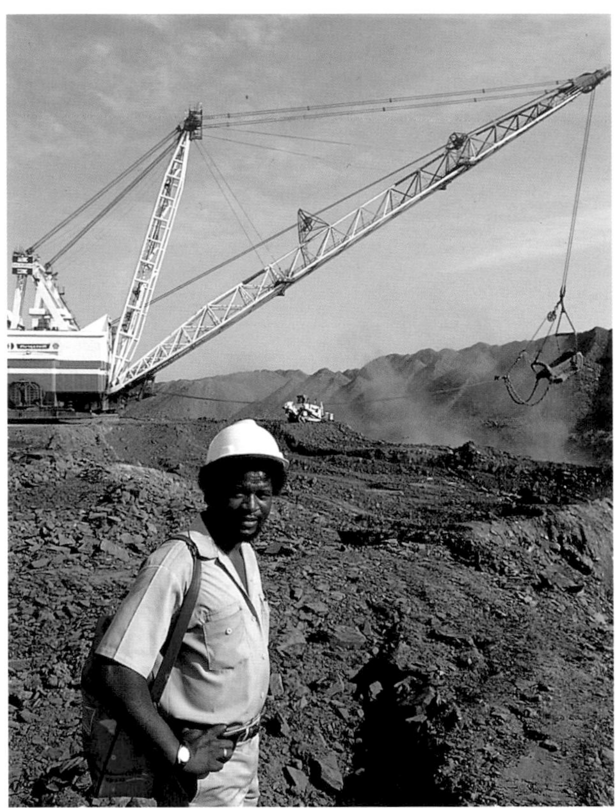

Turning briefly to coal, the third element of Shell's diversification in the early 1970s: Shell Coal International was established in 1974 as a bridge to the future – the future being a place in which, underpinned by metals, the ancient and modern energy sources of coal and nuclear power would push expensive oil into second or third place. This move was better judged than the others, for though metals and nuclear have fallen away, it is coal – oil's oldest rival – which has lasted the longest as part of Shell's post-shock portfolio. Nevertheless, coal has never yet become as dominant as, in the 1970s, it was thought it might be. Shell first made a profit out of its

coal business in 1981. Since then coal has been overall a reliable trade, sufficiently profitable, and sufficiently close to Shell's traditional role as an energy provider, to be maintained *pro tem.*; but not the imagined vital bridge to a transmogrified future. Instead, the future turned out in a quite different manner.

After the first price shock of 1973, the second (in 1979–80) brought another huge hike, with barrels of oil changing hands at $45 apiece on the spot market, and with OPEC prophesying that a single barrel would soon command at least $60. But then in 1986 came the third price shock – a dramatic reversal and collapse, which, strange as it may sound, was just as hard to accommodate as had been the earlier swingeing increases. However, these are stories of the future, to be told in later chapters. Here we must revert to the events and the people which influenced Shell Transport in the mid-1970s.

Despite forays into other businesses, there was never any question that Shell should alter its fundamentals. Its business was crude oil, petroleum products, natural gas and chemicals, and being dragooned by the producing nations was not going to change that. What *did* change was the focus – and to an extent the nature – of its never-ending hunt for crude.

In the decades since World War II, Shell Transport and its associated companies in the Group had acquired 'the reputation of being crude-hungry'. This stemmed initially from the wartime destruction of Group sources in the Netherlands East Indies, and later, when those were restored, from its lack of large, easily worked reserves in the Middle East. That had been partly alleviated by agreements with other companies, especially the 1948 50:50 agreement with Gulf Oil, whereby Gulf produced and Shell transported, refined and marketed. However, one result of access by such means was that by the 1970s more than 80% of the oil which Shell lifted in the Middle East came from sources over which it had no direct control. On the other hand, Shell's very lack of its own Middle Eastern resources had already proved a driving force in other parts of the world. Whereas BP, for example, was able (and preferred) to work with large, simple geological structures, Shell was more or less obliged to take greater chances, and to seek its own oil in more complex structures and more risky environments. Its willing acceptance of the obligation was not just from necessity; it also reflected Shell's long-standing enthusiasm for interesting technical challenges. That attitude meant Shell had long been a leader – often *the* leader – in the techniques of secondary and tertiary recovery (to extract more crude from existing fields) and in deep-water technology; and only a few hundred miles from Shell Centre, the fields of the North Sea beckoned.

Michael Pocock, CBE, chairman of Shell Transport and Trading 1976–9

In the summer of 1976, Frank McFadzean retired from the chairmanship of Shell Transport, being succeeded by Carmichael ('Michael') Pocock. By that time, as operator for a 50:50 partnership with Esso, Shell was concentrating 80% of its world-wide production expenditures for oil and gas outside the United States in the North Sea. The region was one of the most difficult it had ever tackled. As part of Europe's continental shelf, the North Sea is comparatively shallow, rarely exceeding 200 metres in depth; Yet *depth* of water was not the real difficulty. In some respects, searching for oil under the sea is very like the parallel operation on land – for example, in the techniques of magnetic and seismic surveying. Moreover, once the sea-bed is reached, the principles of drilling offshore are exactly the same as onshore. But the big obvious difference, and the real difficulty of offshore exploration and production, is the sea itself.

Offshore exploration is somewhat like surveying the land from an aircraft flying above a permanent bank of storm-clouds. The surface is always shifting, and subject to violent changes, and far underneath, the ground (or sea-bed) is simply invisible. The same adverse conditions apply to offshore production; and in addition the producer must contend with the long-term effects of erosion, corrosion, tides, currents and storms. Inevitably, the operational cost of work at sea is far greater than on shore. For example, Shell pointed out in 1973 that North Sea platforms had to be able to withstand waves 100 feet high and winds of 160 miles an hour, while to bring oil to the shore, a 96-mile submarine pipeline would have to be laid at a cost of £800,000 a mile. Overall, the exploration and production investment in the region was estimated at £1,200–500 per barrel per day; yet in the Middle East, the comparable figure was a mere £100.

Nevertheless, the new cost of oil was so astronomical that the North Sea was becoming an economically viable prospect – something that might not have happened for many years if OPEC had stayed its hand. Moreover, the region had two distinct advantages: it was surrounded by politically stable states, and it lessened international dependence upon the Middle East. Those factors, and the increasing number and size of fields discovered (Ekofisk in 1972; Dunlin 1973; Statfjord and Tern 1975; Eider 1976), justified the enormous investment necessary. First production came from the Auk field at the end of 1975 – nearly five years after its discovery – and from the much larger Brent field in November 1976.

Shell / Esso Eider platform

Shell / Esso Tern production platform

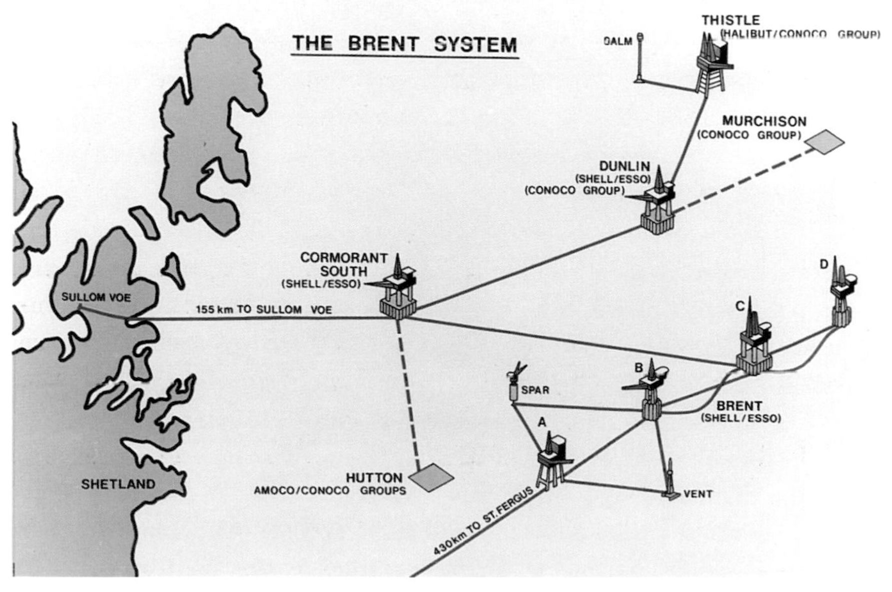

Shell / Esso Eider production platform

The Sullom Voe terminal in Shetland

Two systems were to be used to transport oil and gas ashore. One was the pipeline, noted above, which would connect the field's network to Sullom Voe in Shetland (the closest part of the British isles) where, after long consultation with the local council, Europe's largest oil terminal was being built. The second transport system, designed to act in conjunction with the pipeline, began at the Brent Spar.

As with all offshore structures, there was much more of the Spar below the surface of the sea than above. It was 463 feet high, with about 100 feet protruding above the surface at any given time. However, it was not (as many people supposed) a production platform, nor was it standing on the sea-bed; rather, it was a floating storage tank, permanently anchored by massive chains to six 1,000-ton concrete blocks. From this came its unusual name. Its concept emulated that of the traditional aid to navigation called a 'spar buoy' – a long piece of timber, painted in distinctive identifying colours and anchored at one end so as to float upright. The major part of the Brent Spar floated underwater, vertical but invisible: a giant cylinder, 300 feet high and 96 feet in diameter, capable of holding 300,000 barrels of oil.

Built in Rotterdam and completed in Erfjord, Norway, the Spar was three years in construction. Its installation took place in June 1976; its operational life began the following December; and it remained in commission for nearly 15 years, until August 1991. As its decommissioning approached, the question of how best to dispose of it was addressed. In the summer of 1995, Shell's solution – to sink it in the deepest part of the Atlantic ocean – provoked a considerable international furore. That will be considered in the final chapter of this book. Here, though, it is chronologically

The giant storage unit Brent Spar, being towed out on 31 January 1975

Brent Spar being ballasted into its upright position

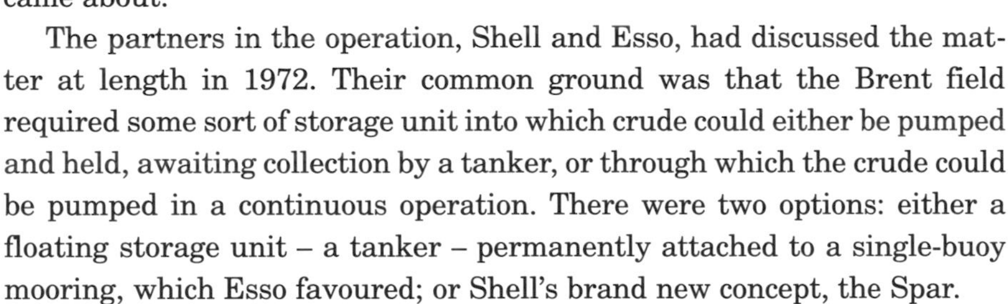

appropriate to record a fact which in 1995 was unknown to protestors: namely, that the decision to build the Spar in the first place was crucially influenced not so much by cost – it was not the cheapest option available – as by Shell's concern for its environmental implications. This is how it came about.

The partners in the operation, Shell and Esso, had discussed the matter at length in 1972. Their common ground was that the Brent field required some sort of storage unit into which crude could either be pumped and held, awaiting collection by a tanker, or through which the crude could be pumped in a continuous operation. There were two options: either a floating storage unit – a tanker – permanently attached to a single-buoy mooring, which Esso favoured; or Shell's brand new concept, the Spar.

Brent system pipeline

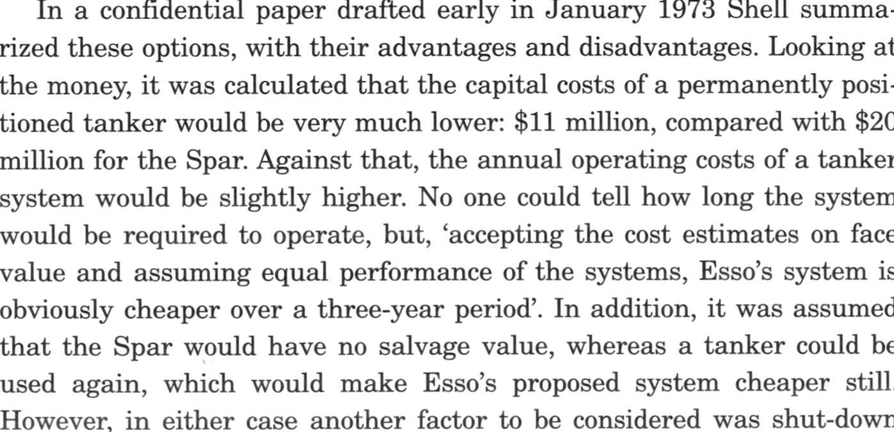

In a confidential paper drafted early in January 1973 Shell summarized these options, with their advantages and disadvantages. Looking at the money, it was calculated that the capital costs of a permanently positioned tanker would be very much lower: $11 million, compared with $20 million for the Spar. Against that, the annual operating costs of a tanker system would be slightly higher. No one could tell how long the system would be required to operate, but, 'accepting the cost estimates on face value and assuming equal performance of the systems, Esso's system is obviously cheaper over a three-year period'. In addition, it was assumed that the Spar would have no salvage value, whereas a tanker could be used again, which would make Esso's proposed system cheaper still. However, in either case another factor to be considered was shut-down time – periods when the weather would be so severe that mobile tankers could not take crude from the storage unit. On this aspect, Shell's summary said that the tanker was sure to be shut down more often, and if limited to three years' operation, 'the apparent cost advantages of the tanker system would be completely offset by the probable penalty'. In other words,

Inside the Spar

over the first three years neither system would have any significant cost advantage over the other.

Looking further ahead, the summary recorded that if the system were to operate for 'a total of five to eleven years', then even taking shut-down into account, the tanker arrangement would be cheaper than the Spar. Overall the financial arguments were thus very much in favour of Esso's fixed tanker idea. However, under the heading 'General Considerations', which included all the above, Shell listed first and foremost the risks of pollution. Existing British governmental regulations on the subject were 'rather indeterminate'. That is not surprising: the prospective business was so new, and the government was so preoccupied elsewhere (as governments tend to be), that very little if any official thought had been given to the matter. The oil partners had to make up their own minds, and, with reasons derived from sound historical evidence, beyond capital costs, running

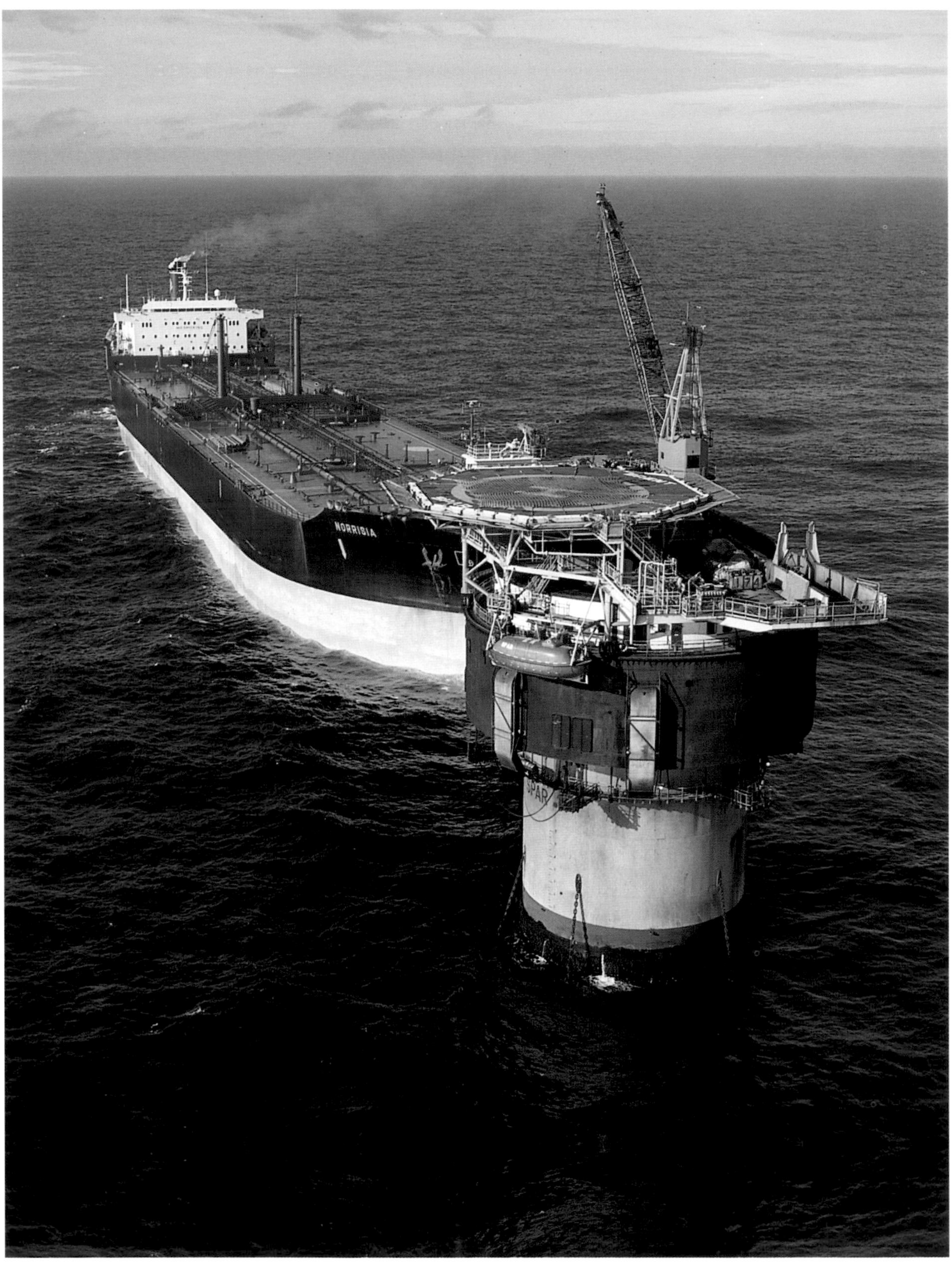
NORRISIA
SPAR

costs or fuzzy political thought, Shell decided that the single most important factor was this:

> The risk of pollution of the sea...will be significantly higher in the SBM/Storage Ship case than in the Spar case.

As equal financial partners in the North Sea, Shell and Esso had agreed to divide both costs and profits equally. But with its British-European base, Shell was to be the operator, so in these early vital discussions its voice carried more than equal weight. On the best estimates of the North Sea's yield and the Spar's effective lifetime, the Spar was going to be an expensive system. But in Shell's view it was not merely a matter of money. The oil could be extracted, but the sea should be kept as clean as possible, even if doing so cost a lot more than was strictly necessary in commercial terms. Without the slightest political pressure being exerted upon it, that was Shell's advice and preferred choice; and it was accepted.

Strange, decades later, to consider these complicated sums and simple conscientiousness, and to note how little consideration they were given by others when the time came to dispose of the Spar. Such prejudice has been a fairly constant part of the industry's history: oil is good, oil companies bad. And at this very time, when the North Sea fields were coming on stream and when Frank McFadzean was still Shell Transport's chairman, Shell and BP were having to do their best to explain some alleged very bad behaviour: corruption in Italy and sanctions-breaking in Zimbabwe, or, as it was then called, Rhodesia.

The 'Italian corruption' turned out to be nothing more than a seven-week wonder. A story published in a British newspaper alleged that over a five-year period, 1969–73, both Shell and BP had made corrupt payments to Italian politicians; and since many people viewed bribery and corruption as being as much a part of the Italian way of life as pasta, the allegation was at first widely assumed to be true. Going back into deep history for a moment, Marcus Samuel made a similar allegation against Standard Oil as early as 1902. During an investigation into Standard by the US Department of Commerce and Labor,

> Sir Marcus asserted that the Shell Company had been kept from doing business in that country [Italy] for several years through the machinations of the S.O., and he charged that the S.O. had bribed the officials high and low who were concerned with that trade.

Considering Standard's corporate culture at the time, let alone Italy's relaxed attitude towards irregular payments, the charge was very probably true. For its own part, Shell Transport in 1902 took a policy decision to try and set up a kerosene trade in continental Europe similar to its trade in the Far East. This was a complete failure: in 1906, as the company's losses approached critical mass, it withdrew completely from the continent. However, the trade was resumed after the alliance with Royal Dutch, and in 1911 the two companies together built their first installation in Italy, at the northwestern port of La Spezia. Gradually the involvement grew to include refining, marketing, chemicals and exploration; and the more recent history was as follows.

In 1969, at a time of stringent price controls and steeply rising costs, Shell Italiana (the refining and marketing arm) began making serious and continuing losses. At the end of 1973 its three refineries and its marketing chain were sold to the state company ENI. That might well have been that – except that, when going through the Italian books in 1974, Shell accountants found sums which they could not identify with confidence: items invoiced as advertising and publicity services where none had occurred. Outside auditors, when called in to check, came to the same conclusion. In 1975 a 'special comprehensive investigation' was established, which revealed that in the five years prior to the ENI purchase, influential politicians of several different parties had extorted political contributions totalling £2.5 million from Shell's local management. Similarly, BP had paid out a total of £800,000 in the same period, and many other foreign companies had suffered in proportionate degree.

Since Shell's general manager in Italy had agreed to make the payments (albeit under duress), had not sought advice from London or the Hague, and had misrecorded the payments, he was sacked. Shell's managing directors decided the time had come to formulate and promulgate the standards they expected Shell employees to meet in business; and again, that might have been that, except that an eager but suspicious journalist heard of the matter, misinterpreted it (whether wilfully or not) and scented a possible scandal.

McFadzean was furious: not because the story had come out, but because it alleged bribery, when the extorted payments were effectively protection money. He was also angry because it was suggested that only the oil multinationals were involved, whereas many other foreign businesses in Italy had been so affected; and mostly he was angry because as far as Shell was concerned the story was a non-story: it had already been dealt with.

As the truth became publicly accepted, so the story evaporated. But the extent of its initial acceptance should have reminded Shell's disgruntled directors of one of the industry's seemingly eternal truths: that some newspaper writers liked nothing better than to search for oil scandals, and a large section of the public would always immediately believe the worst. However, that lesson was overlooked. Just eight weeks after the originating newspaper had reluctantly laid the Italian story to rest, it came up with another. One of Shell Transport's managing directors remarked confidently that since there was nothing to it, the new story would also blow over in a few weeks. But Rhodesia was not Italy, and this affair ran for three years.

Part of Shell's refinery at La Spezia, Italy

At least two books, each longer than this one, have been written on the subject; but despite the risk of oversimplification, it can scarcely be overlooked in the general history of Shell Transport – not least because it is a prime example of how governments in conflict may request and require a commercial organization to conform with the contrary demands of both, and may seek to offload the burden of responsibility onto it. The saga of Rhodesian sanctions is therefore the starting point for our next chapter.

CORMORANT-A

CHAPTER SEVENTEEN

Slings and Arrows: 1978–1982

The political background was that in 1965, under Prime Minister Ian Smith, the white minority government of Rhodesia (then a self-governing colony of Britain) unilaterally declared its independence. The British government denounced this as a rebellion, stated the Smith regime was illegal and imposed trade sanctions, intending to prevent the import of (among other goods) oil and oil products into Rhodesia.

The Rhodesian civil war ensued – a long, bloody struggle in which many lives, black and white, were lost. Legal rule was not restored until 1979. On 21 June 1976, however, a pamphlet entitled *The Oil Conspiracy* was published in America, alleging that Mobil had consistently broken the sanctions, and hinting that the other majors, including Shell and BP, had done likewise. On 30 June that year, the last day of his chairmanship of Shell Transport, McFadzean responded to this with a long letter to the Foreign and Commonwealth Office (FCO). In part, he said that

> while we are still satisfied that we are adhering to the law by every means in our power – i.e. no company in which we have an interest is supplying to Rhodesia – it is self-evidently true that oil is still reaching Rhodesia from South Africa and we cannot guarantee that none of it stems from customers to whom our South African affiliates sell within the Republic. Nor are we in any better position now than we were in 1968 to impose conditions of sale on such customers.

It fell to McFadzean's successor, Michael Pocock CBE, to field the issue further. In America, where it had first been aired, there was little immediate interest in the subject. In Britain, however, there was public concern that the continuing Rhodesian civil war might be being fuelled in part by British oil companies. This grew to the point of debates in parliament; a report commissioned by the FCO (the 1978 Bingham Report); and a vote in Shell Transport's AGM of 1979 on a resolution tabled by four angry shareholders in condemnation of their company's suspected activities.

In the view of Pocock and his colleagues, the resolution was being 'promoted for what are in essence political aims'. They recommended shareholders to vote against it, and it was turned down. Whether many, or indeed any, of the voters had read the original 505-page Bingham Report is unknown. However, its main author was Thomas Bingham, QC (later to become Lord Bingham, Lord Chief Justice of England and Wales). Bingham was a man of unquestioned integrity and eminently suitable qualifications: he was a scholar not only of law but also of modern history, and had graduated with first-class honours from Oxford. His scrupulous Report was correspondingly valuable, although, being unindexed, it was not an easy document to study. Nevertheless, on a subject which could provoke violent passion, it provided dispassionately a very wide range of evidence from all involved. Its main findings may be outlined as follows.

Thomas Bingham, QC, later Sir Thomas and subsequently Lord Bingham, Lord Chief Justice

The illegal Rhodesian regime of 1965–79 may now be seen as a classic model of one of the most difficult dilemmas a multinational organization can face. With operating companies in Britain and in Rhodesia's neighbours South Africa and Mozambique (which was then a Portuguese colony), Shell and BP were obliged to obey the laws of each country; but those laws were in direct opposition. British law required oil companies to cease supplying Rhodesia; the laws of the two African nations effectively demanded that supplies should be maintained. In particular, the rigid South African regime forbade oil companies to impose conditions of sale upon customers, so it was impossible to prevent a suspect buyer from selling his oil on to whomsoever he chose; and virtually any discussion concerning oil was a punishable offence under South Africa's Official Secrets Act.

From March 1966 until June 1975, Britain's Royal Navy maintained a coastal blockade of Mozambique. At a probable cost of over £100 million, this was very expensive, yet little more than symbolic, since it could not affect communications across shared land frontiers, such as that between Rhodesia and South Africa. From the start, sanctions could only have succeeded if the British government had been willing either to blockade both South Africa and Mozambique, or to legislate against any sale of oil to them. But these were moves it would not make, for fear of provoking economic confrontation with South Africa, which, it was assessed, could result in the nationalization of all British-owned businesses operating there. Privately, the British government therefore became less concerned with successful oil sanctions, and more with being able to say truthfully in public that no British companies were involved in Rhodesian supply. To that end it took, as will shortly be seen, a peculiar step.

As commercial, non-political organizations, Shell and BP would under the circumstances have welcomed a British blockade of South Africa or an absolute ban on sales there, for then, without fear of reprisals, they could have quite properly refused supplies on the basis of *force majeure*. Life would have been relatively simple. As it was, under South African law they could not refuse supplies to anyone who could pay for them. To withdraw from the trade altogether was not a realistic option: their properties would probably be expropriated, another supplier would certainly step in, and no good would be done for anyone.

They could think of only one method that would legally resolve the incompatible demands with which they were faced: a swap. Any orders they received in South Africa could be supplied by the French oil company Total, with Shell and BP making good the appropriate quantities. If any oil sold by Total in South Africa was subsequently forwarded to Rhodesia, technically it would be from a French source.

But while this was indeed a legal solution, it was scarcely a satisfactory one. Although swaps are part of ordinary oil-trading, they hardly ever take place against such a sensitive political background. The Total swap would satisfy Britain's sanctions in the letter, but it would clearly be against their spirit, for supplies to Rhodesia would be little impeded; yet without considerably firmer support from the British government for the carrying out of its own demands, there was no alternative. All the same, a sense of unease prevailed. Before he became chairman of Shell Transport, McFadzean explained the dilemma to members of the British government and informed them of the swap method, remarking (as the Bingham Report noted) 'that while the position was legally sound, the argument still seemed pretty thin to him'.

Yet it was solid enough for the government ministers present, even though it meant oil would continue to flow to Rhodesia. They had realized that sanctions were impossible to enforce; but if the oil was not overtly British, that was good enough for them. They approved the swap method, and indicated that the government would not make it public. Bingham noted: 'It was undoubtedly felt to be desirable that the Total arrangement should not be publicly disclosed...because of the use which could be made of the information by critics of British sincerity.' By privately approving the swap, the British government contradicted its own public policy.

That was far from the end of the matter. In 1971 the Total arrangement lapsed – and Shell's general manager in South Africa, himself a citizen of the country, maintained supplies without informing Shell Centre. 'It has not become clear to us', said the Bingham Report,

> why he did not do so. He may have wished to avoid embarrassing the Shell management in London. He may also have wished to safeguard the business he was running in South Africa. Awareness of the very stringent South African official secrets legislation may have played a part. We do not think these considerations justify his conduct even if they explain it.

From 1971 until March 1976, supplies from South Africa to Rhodesia continued via Shell Mozambique. Though run by Portuguese nationals, this company was registered in London, which meant the oil was once again 'British'. Mozambique then achieved independence and closed its border with Rhodesia. Supplies still continued, now direct from South Africa – but still without Shell's most senior management knowing of the series of changed circumstances or being able to inform the British government accordingly. If the South African general manager had 'wished to avoid embarrassing the Shell management in London', as Bingham charitably suggested, then he failed absolutely. Safeguarding his business and obeying his own country's laws ring more truly as reasons, as does one which Bingham did not propose: that in common with much of the white population of South Africa at the time, the general manager feared the possibility of black majority rule in any African nation, and took the opportunity actively to support the Smith regime in Rhodesia. All that was needed was to tell no one. 'The criticisms which we have made', said the Bingham Report in conclusion,

> have related in the main to failures to disclose, either within the [Shell and BP] Groups or by the Groups to HMG [the British government]. We do not regard these failures as in any way unimportant. ... In the event both HMG and the top management of the Groups...were ignorant of facts which should have been the subject certainly of consideration and possibly of action. This ignorance led HMG and the top management of the Groups unwittingly to make

> statements and give assurances which they would not have done with full knowledge of the facts.

McFadzean's letter of 30 June 1976, in which he said 'no company in which we have an interest is supplying to Rhodesia', became demonstrably untrue; but as the Report stated, he had signed it in good faith, and there is a difference between lying with full information and telling the truth as far as you know it.

As to the British government's side of the matter, the same could be said – except that it was HMG which approved the Total swap arrangement, in order to be able to declare that no *British* oil companies were involved in supplies to Rhodesia. As soon as that was done, and whatever happened next, the official maintenance of sanctions became a pretence. The Bingham Report was delicately scathing:

> This...raises certain questions involving the evaluation of government policy upon which we do not feel entitled under our Terms of Reference to comment.

Perhaps, then, it is not surprising that while there were calls in some quarters for the oil companies' prosecution, HMG chose not to go to law.

Because it was Britain that had imposed sanctions, most of the uproar about Rhodesia occurred in Britain; and because of that, and the widespread lack of public understanding about the Royal Dutch/Shell Group's structure and relationships, much of the media attention was focused not on the Group or the CMD but on Shell Transport and its chairmen of the time, Frank McFadzean and Michael Pocock. In a sense, either man could have quite fairly pointed out that if there was any culpability, no more than 40% could attach to Shell Transport; but of course they did not, any more than their Royal Dutch colleagues would have done had the situation been reversed.

Nevertheless, the late 1970s and early '80s were very difficult years for Shell. In trading terms, far and away the most serious problems were caused by OPEC's pricing policy. We shall return to these shortly, and to the various strategies Shell adopted to manage and overcome them. But in the same period, with the Rhodesian sanctions scandal still in the headlines, Shell had to cope with a second blow to its public image: the loss of the *Amoco Cadiz*.

On 16 March 1978, the 240,000dwt VLCC suffered a steering gear failure off the coast of France. With 223,000 tons of crude on board, the ship was blown onto the rocks. On 17 March she broke her back and spilled a quarter of her cargo, covering 100 miles of coastline. Eleven days later her forepeak broke off and the rest of the cargo flooded out – a total of nearly 1.6 million barrels, the worst accidental spill of crude oil there had ever been. (At the time of writing, it is still by far the worst, and we may devoutly hope it remains so.) The direct commercial consequences of the disaster – severe loss of sales in France, and years of litigation – fell upon Amoco as owners of the ship and employers of her crew. But the *cargo* belonged to Shell, and in the mood of extreme national outrage in France, that fact brought them harsh criticism as well, and made them the victim of hostile and sometimes violent acts.

To place *Amoco Cadiz* in some perspective, there are three points which are well worth making.

Firstly, few people ever reflect that on any given day, more than half the seaborne trade around the world is oil, and that annually, many millions of tons – tens of millions of barrels – are transported across the seas in perfect safety. *Amoco Cadiz* was only one VLCC, and as Captain Alec Dickson (then head of Shell International's Marine Operational Services) remarked, 'The fact that on any day of the year there may be five VLCCs arriving at Rotterdam is, unfortunately, no news at all.' Shell attempted to make this point at the time, but understandably it cut little ice with a temporarily unreceptive public, and so was not pressed. Nevertheless, in this respect the whole industry's safety record is extremely good (and Shell's, one may add, is unequalled). Secondly, it is important to recognize that such spills are *accidents*. No sane businessperson – indeed, one might suppose no sane person – would do such a thing on purpose. But it can be done, and the world's worst oil spill was quite intentional. In the Gulf War of 1992, on the orders of Iraq's president Saddam Hussein, the oil wells of Kuwait were either set ablaze or opened to flow without control into the waters of the Gulf. The long-term environmental effect of the burning is not calculable. However, approximately a million tons of oil (at least four times that from *Amoco Cadiz*) poured into the Gulf. News footage was made all the more distressing by the knowledge that the catastrophe was deliberate; and yet (and this is the third point) within only a few years, the Gulf ecology had almost completely re-established itself.

The same was true of the *Amoco Cadiz* catastrophe. Because crude oil looks so unpleasant, it is easy to forget that it is itself a product of nature, not of an oil company. It is not conjured from some poisonous crucible, but formed in and by the earth; and that which the earth has composed, it disposes of as well – usually a great deal more safely and effectively, although it will take time, than by any short-term human intervention. Of course, that is not to suggest that in the event of an oil-spill, one should sit back and wait for nature to take care of itself. Apart from anything else, although crude oil is relatively biodegradable and of comparatively low toxicity, the seriousness of a spill is related not only to its size but to the sensitivity of the local environment. When such terrible accidents occur, the urge is to do something that will clean the shorelines and save sealife with immediate effect; but it is important that remedial action should be correct and make things better, not worse than they already are.

Before the particular scientific and the broad public understanding of these things improved, much of the public activity around oil wrecks served only to make things worse – for example, by using solvents and dispersants which were actually more toxic than the oil they removed. This was something which Shell took extremely seriously. As noted in earlier chapters, its active interest in environmental protection stems from at least the middle 1950s; and under the heading 'The environment', Shell's annual report for 1978 indicated its response to the *Amoco Cadiz* calamity:

> While notable progress has been made in such fields as environmental conservation, safety and health, more needs to be done to attain the performance and standards desired. Progress comes from teamwork and a series of initiatives painstakingly implemented, rather than from spectacular breakthroughs. Shell companies' efforts have been considerable, both in money and in specialists' time. .... These efforts are linked to national and international agen-

> cies through industry associations in which Shell people are very active. During the year, efforts ranged from collaboration with international regulatory and advisory bodies on improvement in marine safety to the development of detailed standards for plant further to reduce air and water pollution.

Since then, there has been enormous growth in the range of physical, chemical and biological techniques for cleaning contaminated sea, soil and ground water; and much of that growth stems from the work of Shell scientists at the Sittingbourne laboratory. In 1978, moreover, while rethinking its former willingness to use third-party carriers for its cargoes, Shell took an industry lead again by promoting marine accident prevention measures, in particular calling for international agreement on uniform standards of competence for all seafarers. This was a call which all the responsible oil-shipping companies heeded: both for conscience and commerce, none of them wished to have an *Amoco Cadiz* on their hands.

To have been embroiled in these episodes was very painful for people in Shell, not least because they felt that criticism, especially blanket criticism, had been earned much more by association than by culpable action. In each case events had moved out or been taken out of Shell's control, yet in public opinion it was left with responsibility; and in the pithy words of one executive, 'Today's opinions become tomorrow's laws.'

A good reputation could not be overvalued. Shakespeare had called it 'the immortal part of myself'; Shell too saw it as 'a precious resource', going so far in later years as to call it 'a licence to operate'. For Shell in the late 1970s, the questions thus became: how could its reputation be restored, and public opinion brought back to an attitude of broad support?

As time went by, public common sense came to recognize that Shell's involvement in the *Amoco Cadiz* wreck, as owner of the cargo, was sheer bad luck: the cargo could have belonged to any oil company. Inevitably the Italian and Rhodesian entanglements had a more lasting adverse effect. Though the former was baseless, the latter was such a slow and complicated matter to unravel that it left a widespread assumption that to some extent at least, Shell and BP must have been up to no good. But few people realized the factual elements which the Italian and Rhodesian episodes had in common.

Both had been brewing at the same time, and, on the authority of the general managers involved, both had featured mis-reporting or non-reporting of crucial events. In each case this had been possible because of Shell's decentralization. Established by the CMD in the early 1960s, the principle of decentralization had granted – to a large extent on trust – considerable autonomy and authority to general managers in a given country. But other

All change in Iran: the Shah departs, the Ayatollah returns

countries, other customs; and with Rhodesia/South Africa and Italy as stinging examples, Shell's managing directors drew an important moral consequence from the mid-1970s: namely that if decentralization were to be maintained, then explicit guidelines of acceptable business behaviour must be applied to it. Enter the *Statement of General Business Principles*.

The Statement was first drafted in 1976 as a direct result of the Italian bribery allegations and the emerging suspicions about Rhodesia. It was not very long – just ten main paragraphs. These have sometimes been described as 'Shell's Ten Commandments', and it is true that not only in their number but in their tone and simplicity they had something in common with the biblical precedent. Many other corporations have followed the example, but even people who joked about 'the Commandments' acknowledged that, coming from one of the world's most high-profile international business enterprises, the Statement was a very unusual document for the time, and its ten paragraphs were read with attention and respect. They were so carefully thought through, and so simply and clearly expressed, that they were both comprehensive and unambiguous.

The Statement was initially intended only for circulation to staff, but since 1981 it has been freely available to the public. From time to time it has been revised, but without material alteration, and anyone who reads it will be struck by its themes – integrity, honesty, fairness, responsibility, free and ethical competition. These were values which had been part and parcel of Shell Transport from its very earliest days, but they had never been written down before, and doing so was a worthwhile exercise. Giving definition in print to the core of its culture ensured that every member of its world-wide Group shared and aspired to the same values. By going further and communicating those values to the public, Shell clearly accepted a risk: it would be expected ever after to live up to the Statement, or by its own judgement be found wanting. But that self-imposed challenge was a further spur, if one were needed, in the quest for high standards. The Statement would become a corporate credo, a welcome point of definition in very uncertain times.

It was no great consolation that uncertainty then was the common lot of every part of the industrialized world. On 17 May 1979, Michael Pocock chaired Shell Transport's 81st annual general meeting. As well as announcing plans, the purpose of an AGM is of course to review events affecting the company in the preceding calendar year, in this case 1978; but Pocock began his address to shareholders with a reference to even more recent events.

On 1 January 1979 the second oil price shock had begun. Following the shock of 1973, panic buying had pushed the price to $22. It had then stabilized at about $13, or roughly four times its pre-shock level, and the world economy was still reeling. Now its new 5% increase came with the promise of further rises every quarter. Simultaneously in Iran, great civil disturbances against the Shah's increasingly arbitrary regime were coming to a head. On 16 January, in the midst of demonstrations and strikes, the Shah left Tehran airport, officially on vacation, but never in fact to see Iran again.

Within a month, the Iranian revolution was complete – at least 'in one sense', as Ian Skeet observed. Skeet (one of Shell's Middle East experts) lived through, analyzed and wrote on OPEC's heyday. The rise to power of the Ayatollah Khomeini was the end of the political side of Iran's revolution. The effects on its oil industry were only just beginning.

In September 1978 the country's wells had produced somewhat over 6 million barrels per day. By January 1979 that had plummeted to just 0.5 million bpd. February brought a small restoration, to 0.7 million, and March a further and more hopeful increase, to 2.4 million bpd. But that was still so far from its normal level that industry stocks had been drastically depleted, to 'the order of 500 million barrels less than might normally have been expected', in Skeet's words. 'The uncertainty was such that Iranian oil suddenly became an unknown and, possibly, non-existent factor in international oil supply. Nobody could be sure of anything.' At the AGM in May 1979, a sombre Michael Pocock told shareholders:

> The consequences of the disruption of Iranian supplies are now potentially serious. There has been a marked reduction in industry

inventories and consuming countries are already taking first steps to scale down demand...moreover, there has been a sharp effect on prices in the spot markets for crude oil and products.

And there was worse to come. In September 1979 the Iran-Iraq war began, removing at a stroke over 4 million barrels from the world's daily supply. The combined effect of OPEC's edict and the war was to double oil's base price, over the calendar year of 1979, from $13 to $26 a barrel. This was followed in 1980 by a further OPEC-instigated rise of $7, and another $4 a barrel in January 1981. By then the base price of crude oil, which just eight years earlier had been less than $3, had reached $37 a barrel; and every single day, millions of barrels had to be bought.

Sir Peter Baxendell, Chairman of Shell Transport and Trading 1979–85

Incomprehensibly large fortunes flowed into OPEC coffers and out of the customary routes of international circulation; but Michael Pocock did not live to see it happening. On 12 October 1979, entirely unexpectedly, he died of a heart attack. He was only 59 years old.

His colleagues were stunned by the loss – 'an abrupt and tragic end to a remarkable career', as his obituary in *Shell World* said. Pocock had survived one earlier attack, in his mid-thirties, but no one suspected the

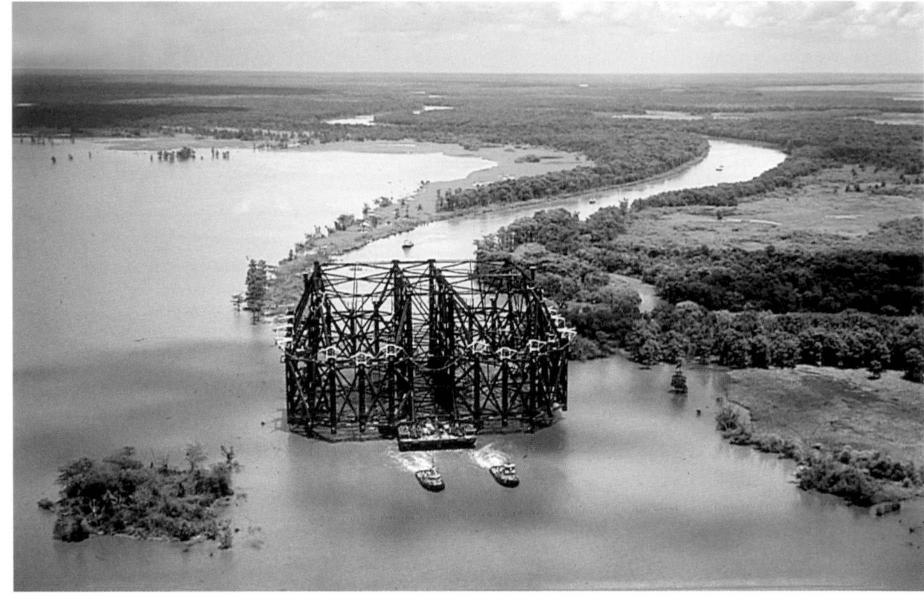

The Cognac platform being towed out to the Gulf of Mexico...

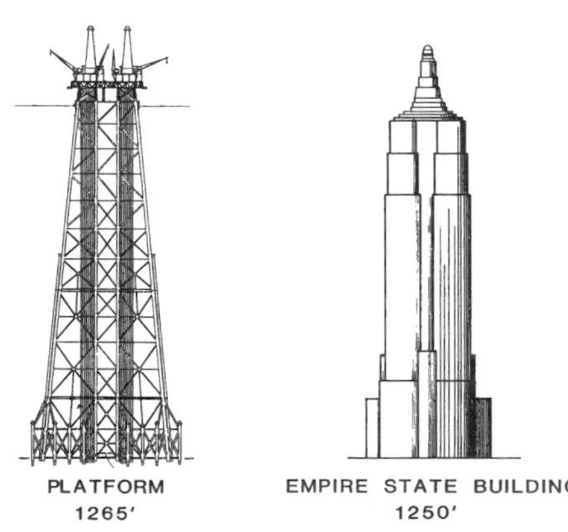

...and in comparision to the Empire State Building

imminence of another. He had been an immensely energetic and very outdoor man: his hobbies were sailing, water-skiing and mountain walking. He had also been an enthusiastic proponent of business education: he was chairman of the Council of Industry for Management Education for the seven years before his death, and chairman of the London Graduate School of Business Studies for three years. Just the week before he died, he had delivered a speech in Paris on the need 'to tread the most sensible path between continued growth and restricted resources', and the very day before his death, 'in great form and good spirits', he had been at work. No other Shell Transport chairman had died so young, and none while still in office.

The chairman is dead: long live the chairman. In any monarchy such immediate succession is customary. Shell Transport had never had to face the experience before, but was similarly prepared, and Pocock's vice-chairman Peter Baxendell at once took over the top responsibility.

Pocock, during his chairmanship, had supervised the various strategies the Group adopted to manage and overcome the serious problems caused by OPEC's pricing policy. Though he was now dead, nothing else had changed, and Baxendell's first task became the maintenance and development of those strategies. They fell into three broad categories: cost-saving, the search for non-OPEC sources, and further diversification.

Cost-saving was being achieved in the first resort by the method that any business will use when faced with emergency – that is, the unwelcome loss of staff and the equally unwelcome increase in the burden upon those who remained. But other avenues were used too, such as the decommissioning of surplus plant, as rising prices reduced demand; and as an organization created for and devoted to the supply of energy, Shell was perhaps especially aware of the expense of energy. In a move which was certainly environmentally sound, if somewhat paradoxical for an energy supplier, it tried to encourage its customers to be more energy-conscious. This made progress, but more slowly than Shell would have liked: 'There is a large potential for energy saving on a community basis,' one annual report grumbled, 'but there are difficulties in persuading individuals to act.' However, where Shell companies could take a useful initiative, they did (as, for instance, in France and Sweden, where waste heat from Group refineries was used for neighbourhood domestic heating), and Shell overall became far more energy-efficient, reducing its costs in that field by 16% in one year.

As to the two other broad strategies (the search for non-OPEC sources of energy, and further diversification), a certain amount of overlap existed, because in part they had a common goal – the maintenance of the cheapest and most efficient energy supplies. But their ways towards that goal were quite different. Let us look first at some of the more 'traditional' ones.

In his last statement as chairman, Michael Pocock had reported on many important Shell accomplishments of 1978. Three of the most notable were the acquisition of a 15% interest in Abu Dhabi Gas Industries Ltd; an

A BIG / GT project in Sweden

agreement with the Saudi Arabian national oil company Petromin to design its new refinery at Al Jubail ('Shell expertise continues to be in demand', Pocock commented); and the Cognac drilling and production platform. This was, in every sense, a towering technological achievement. Installed in the Gulf of Mexico, the Cognac platform was 1,100 feet high – a record-breaking height, not only for Shell but for the whole industry. The feat was duly honoured with the American Society of Civil Engineers' 'Outstanding Achievement' award.

Peter Baxendell and his colleagues in the CMD built soundly upon the foundations left by Pocock. 'The supply shortages', said Baxendell, 'reinforced the need to seek new reserves and more diversified sources of hydrocarbon', and in his first 18 months as chairman of Shell Transport, new exploration interests were acquired in 15 countries. These included the purchase (with effect from 10 December 1979) of the American-based Belridge Oil Company, for $3.6 billion – at that time, the biggest corporate acquisition the world had seen. Shell's enhanced oil recovery techniques were applied to the Belridge fields and brought a 26% increase in production within a year, to 53,000 bpd, with further improvement anticipated.

Other 'conventional' achievements of 1980 included the shipping of the thousandth LNG cargo from Brunei to Japan; the commencement of oil production in south Oman, Zaire and the Ivory Coast; and the expansion of the North Sea oil and gas fields, at huge expense, as new sub-sea technology was introduced, including (in the Cormorant field) a remote-controlled gathering station 490 feet underwater. These are just a sample of

the many established methods which Shell developed in its hunt for new resources; and there were several less conventional ones as well.

It will be recalled that at this stage, Shell had three major new business areas (coal, metals and nuclear power) which were still very much part of its portfolio, if not exactly flourishing. Concurrently, other new ventures were started, grouped - since at first they were all small – under the heading of Non-Traditional Business, or NTB. Among these, one was solar heating, in the shape of a 50% interest in the Australian company Solarhart, the continent's largest manufacturer of domestic solar heating systems, with sales in 40 other countries; and another non-traditional business was (surprisingly) forestry, which developed successfully into a business in its own right. This was not rain-forest hardwood logging, which was viewed with as much dislike inside Shell as outside; rather, it was the efficient production of softwoods, especially pine and eucalyptus, for paper, construction and fuel. As Sir Peter Holmes said during his chairmanship, 'We regard ourselves as farmers, not fellers'; and we shall return to this subject, for in the 1990s Shell's forestry began to offer a viable new source of replaceable energy - BIG, or Biomass Integrated Gasification.

At this stage, though, in the late 1970s and early '80s, it was in the extension of conventional areas that Shell shone. And in doing so, it helped to shape the next extraordinary scene: the downfall of OPEC.

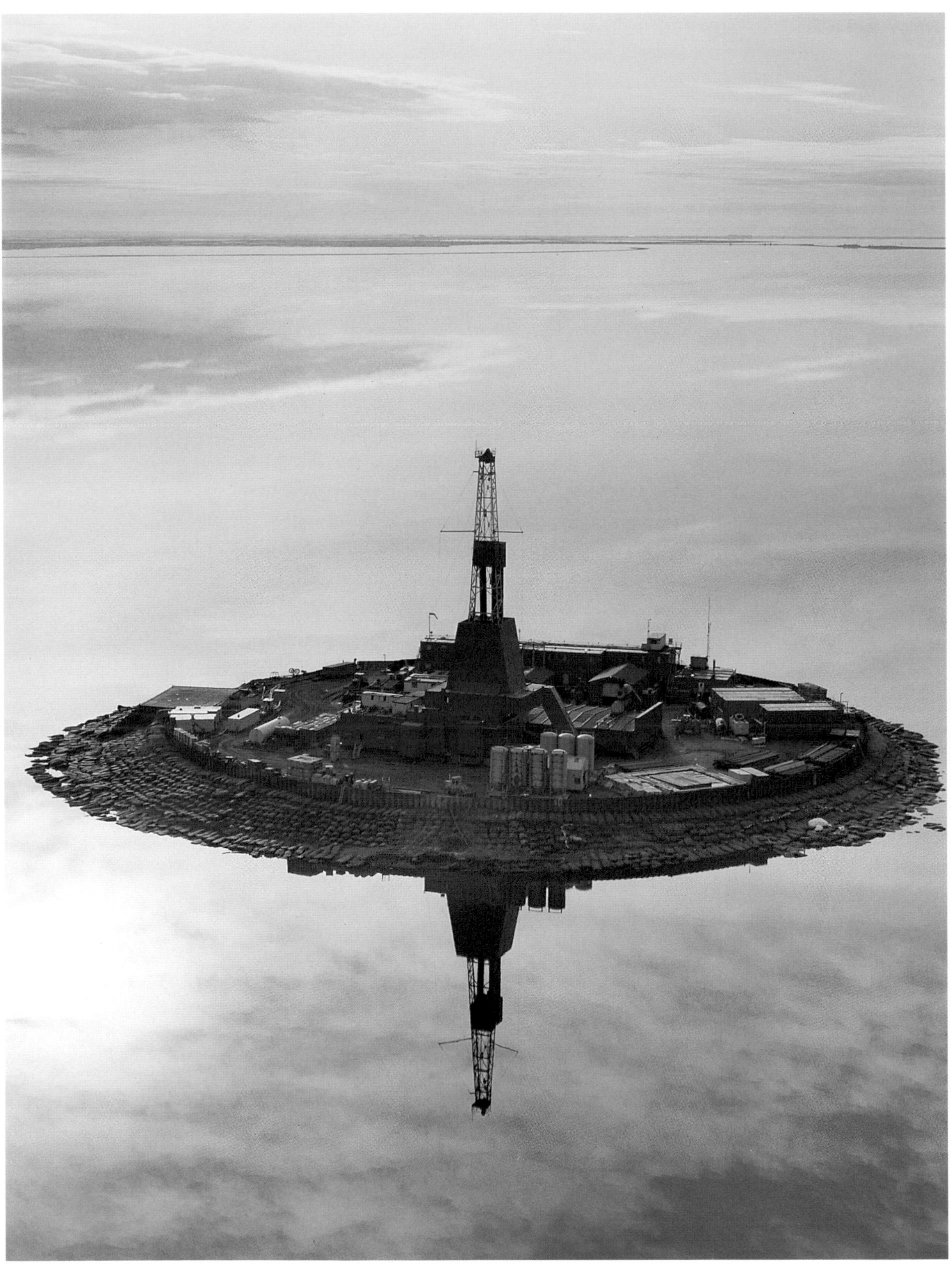

CHAPTER EIGHTEEN

Exit OPEC: 1983–1987

Before the gigantic price hikes of 1973 and '79, Shell's reputation within the industry had become that of an organization which was extremely good at marketing crude oil and oil products, hungry for more crude, yet somewhat averse to going out and finding the crude. It was not a very fair assessment: Shell's aversion was not towards exploration *per se*, but to spending large amounts of money which did not need to be spent and which seemed unlikely to yield a profit.

When oil was cheap, before 1973, there had been good reasons for taking a cautious view of the expensive upstream operations of exploration and production. People in Shell had indeed discussed whether or not they should reduce recruitment in the upstream sector, move gradually away from it altogether, and simply buy their supplies. This, interestingly, would have been a reversion – at least in part – to the business which Marcus and Sam Samuel had in mind when they established the "Shell" Transport and Trading Company in 1897; but the price shocks so altered every part of the world economy and the oil business that exploration in difficult and costly areas became not only financially feasible, but strategically essential. Sir Peter Baxendell (he was knighted in 1981) remarked later, 'The whole thing swung in the opposite direction. No longer were we just searching for very large reserves, but even small accumulations. It became a question not so much of finding giant fields as of finding the right sort of oil in the right places.'

Shell's 1979 purchase of the Belridge Oil Company marked the start of a line of similar acquisitions by other majors: it was cheaper to buy existing fields than to explore for new ones. But the new exorbitant oil prices demanded new fields as well, and Shell's successes in finding them – especially in the very hostile environments of the North Sea and Alaska – demolished the myth of its aversion to exploration. In the year 1982 in particular, a record number of North Sea wells were drilled, while an artificial island was created as a drilling base in the Alaskan Beaufort Sea. Baxendell viewed these developments with great satisfaction,

Opposite page: The artificial Seal Island in unusually calm weather

not only because of his role as chairman but for a more personal reason: ever since joining Shell in 1946 as a mining and petroleum technologist, he had always been an upstream man, and he was glad to see Shell being prominent in such operations.

Baxendell (who became chairman of the CMD in 1982) also closely monitored the overall consequences of the price shocks. Not only Shell, but every oil company which could afford it was exploring for new fields outside OPEC areas; not only Shell, but every nation, every business and every individual with any common sense was now seeking to reduce consumption of energy in general and oil in particular. The United States improved its energy efficiency by 25% in the years 1978–85, and its oil efficiency by 32%; Japan, another major consumer, made even greater improvements, with 31% and 51% respectively. In a slightly narrower timeframe, nations in WOCA (the 'World Outside Communist Areas') consumed a total of 51.6 million barrels of oil per day in 1979; by 1983 that had fallen by nearly 6 million bpd; and in the same period non-OPEC production had risen by 4 million bpd. Heavy investment, reduced consumption, greater efficiency, greater external production: all told, a squeeze was taking shape – a squeeze on expensive oil.

Baxendell had seen a lot of interesting times already. He was one of those fortunate individuals whose career seems to move through the most stimulating areas and periods of their chosen profession. Certainly he thought so: when approaching retirement in 1985, he said,

> I suppose every generation feels that the next generation will have missed the best years. But I genuinely feel I have been through the most exciting times of the industry. I was in Venezuela during the boom years of the Fifties. I was in Nigeria when we started developing oil production there, and I was managing director of Shell U.K. just as we started to develop the Brent Field complex.

And now, in the first half of the 1980s, he was in an excellent position both to witness and indirectly to influence OPEC's collapse.

Among several interlocking reasons for OPEC's ultimate failure, the most important single one was that although it was a cartel, it was not a monopoly. The same had happened in the 1920s, when Shell's autocratic financial genius Sir Henri Deterding had gathered the bosses of the other major oil companies together and attempted to set up a global oil cartel – the Achnacarry Agreement, or 'As-Is'. The theory upon which Deterding worked was not entirely selfish. His main intention was to bring stability to the chronically unstable oil markets of the world. This, he believed correctly, would benefit consumers; but the potential benefit to the producing companies was so much greater, and 'As-Is' was so manifestly contrary to free competition, that all its participants kept the agreement as secret as they could for as long as they could – which was not very long. Nor could the agreement survive intact for long. As for OPEC, that cartel differed from the attempted 'As-Is' cartel in two significant ways: it was totally and painfully public, and it made no pretence of being anything other than self-interested. But in another way, which in the long run was more significant, it was just the same as 'As-Is'. Unless it controlled the entire world's production, OPEC could not possibly maintain the new status quo for ever. Sooner or later, other producers would challenge it, and the price mechanism would come into play once more.

That was what had happened in the years 1979–83. In 1973–9, immediately after the first price shock, international demand slackened because of the new high prices; and it remained slack as other energy sources (coal, electricity, nuclear power and natural gas) became attractive. Such was the cost of oil then that it could take only a year or so for a business to recover the expense of converting to an alternative energy supply. In short, if the world energy-market had ever been solely an oil market, it certainly was not any longer: rather, oil had become just one competitor amongst several – and anyone who still *had* to use oil was doing so much more efficiently than before. OPEC's market share in the non-communist world slipped dramatically, from 60% in 1979 to 50% in 1981, and in 1982, in an effort to keep the price of oil and their revenues up, the cartel members agreed to set quota limits on their production. But they could not limit other people's production, and in that year, for the first time, non-OPEC sources produced more oil than OPEC did. Officially, the price (based on that of Saudi light crude, the international marker) was still $34 a barrel, but in what was rapidly returning to a buyer's market, discounts were easy to find; and in February 1983, the British National Oil Company knocked North Sea crude down to $30. Within weeks OPEC responded with its first-ever price cut: 15%, no less, bringing the official price to $29. Simultaneously, its members agreed to curtail production further to a total of 17.5 million bpd – 44% less than they had produced just four years earlier in 1979. As Baxendell observed, 'OPEC was striving to maintain cohesion'; and it was failing.

Much of this was due to the North Sea and Shell's operations there. Britain, in 1981, had achieved the signal success of producing more oil than it consumed. Over the next two years it became one of the world's largest producers and a leading oil exporter. In 1983 overall, more oil was produced from the British sector of the North Sea *alone* than from

Algeria, Libya and Nigeria combined. How the British government would use this astounding windfall remained to be seen, but as the beneficial effects of the developing price and production war fed through to national economies, the world began hesitantly to move out of the most bitter and hurtful recession that had been known for fifty years. But it was not over yet.

How much is one billion? In British English, it is (strictly speaking) a million million; in American English it is a thousand million. Given that oil is priced in US dollars, it is not surprising that the American usage has become an industry standard, applied even to sums (such as development and production costs) which are expressed in sterling; and it is probably because of oil's predominance in the world economy that the American usage is becoming the more widely used. By the middle 1980s most of the industry's serious sums were calculated in billions. A few stand out, both for their size and their strategic importance.

One occurred in 1984, when Shell made a bid to purchase those shares it did not already own (just over 30%) of its main American subsidiary, Shell Oil Company. The rationale was straightforward: Shell Oil was successful, and owning it completely would expand and strengthen the US interests of the Group. The purchase, however, was not straightforward. Minority shareholders contested it in court, and it was not until June 1985 that the matter was settled in Shell's favour. The price-tag was $5.7 billion, or to put it in figures, $5,700,000,000. Not a telephone number, but the amount of money paid – and cheap at the price.

'That was one of the best deals we ever made, if not the best', said the legendary John Loudon. He was then well into his 84th year and long retired, but still mentally as sharp as ever: 'If we'd waited longer, we would have paid much more.' This was undoubtedly true: not only was Shell Oil far too well managed to go broke, but the lesson of history was there too in the shape of the Belridge Oil Company. Shell had first tried, and failed, to buy that company in the early 1920s, for about $8 million. Its successful second attempt in 1979 entailed a price of $3.6 *billion* – a very large increase, even allowing for the rampant inflation induced by OPEC.

Shell was not the only oil company seeking to grow through purchase. ' There was a common feeling in the industry that future health depended upon the extension and strengthening of one's production base. Thus, in the space of a very few years, oil companies large and small were bought by others, or even by companies outside the industry. Conoco (one of the descendants of the original Standard Oil Company) was bought by DuPont, for a massive $7.8 billion; Sohio, another descendant, was bought by BP; Marathon Oil, which had been a Standard producing company, was bought by US Steel; Texaco bought Getty Oil; General American was bought by the Phillips corporation; and on 5 March 1985 Gulf Oil was bought by Chevron, another Standard descendant, for $13.2 billion – far and away the biggest merger in history, and one which triggered a degree of shock in the oil world.

Gulf Oil was one of the 'seven sisters', the world's major oil companies, and the direct descendant of Guffey Oil. People with long memories knew that more than 80 years earlier, Colonel Guffey had made a contract with Marcus Samuel, the founder of Shell Transport and Trading, for large supplies of oil from the Texan Spindletop field – a critically important contract for Shell Transport which, through subsequent circumstances, Guffey was compelled to break. Such people also knew that Marcus had generously (and unwisely, some thought at the time) agreed not to press the matter in court: a decision which had contributed in 1905–06 to the downfall of Shell Transport as an independent oil concern. Guffey, or Gulf, had gone on to great things; yet now it had ceased to exist in its own right.

With its intimations of mortality, this death of a sister sent shivers down the industry's spine; and giant corporate acquisitions could go too far. The legality of Texaco's purchase of Getty Oil was successfully challenged by another competitor, and in November 1985 Texaco was ordered to pay the astounding and crippling figure of $10.3 billion in damages.

By comparison, Shell moved smoothly through these years. Though it too expended very large sums, it did so with more thoughtful caution: it was not the spirit of Marcus Samuel but his more prudent brother Sam which prevailed. Indeed, in the summer of 1985 an event occurred in Shell which was quite uncannily similar to Sam Samuel's view of events in 1899.

That was the year when Sam, visiting Borneo, had been away from Britain for nine months. During his absence Marcus had been faced with the threat of take-over by Standard Oil, the Nobels and the Rothschilds in association. To counter the threat Marcus had made a great show of strength, buying extremely large quantities of kerosene on a rising market. Sam, on his return, was horrified to learn of the risky outlay and the huge stocks that had accumulated; and although Marcus's gamble appeared to work in the short term, before 1900 was over, Sam's gloomy prognosis was proved correct: the price of kerosene began suddenly and sharply to fall, leaving the infant Shell Transport with many hundreds of thousands of barrels of oil which could only be sold at a painful loss.

The situation in 1985 paralleled that of 1899. Once again, albeit for different reasons, Shell had large stocks of expensive oil. This time, however, Shell Transport's chairman Sir Peter Baxendell (who was approaching retirement), and his successor-designate Peter Holmes MC, sensed the inherent risk. Baxendell had just been named as the industry's best international chief executive by the *Wall Street Transcript* and Holmes

Sir Peter Holmes, MC, Chairman of Shell Transport and Trading 1985–93

had trained as an historian at Cambridge, so it is tempting to see their action as one of businessmen carrying out of a lesson learned directly from history. As Holmes described it, they and their CMD colleagues felt that

> barring some political catastrophe, the price of oil could not go up. It was either going to stay where it was, or it would decline. We didn't know, of course. We asked our planners to have a look at a possible decline scenario. They were very good at working out scenarios – the whole Shell system works on that basis.

The planners' projections included the possibility of a price drop to $20, 'which seemed...catastrophic. We had to study then what we should do to position ourselves for this. There were quite a few things we could do, but primarily we could run down some of the stocks we had built up in previous years' – precisely as Sam Samuel had wished to do in 1899. 'And that was done.' Not a moment too soon, as it turned out, because the price crash of 1900 was repeated in 1986.

OPEC had finally succumbed to the combination of external price pressures, and internal rivalries and quota-breaking. At the end of November 1985 the organization announced it would no longer try to protect the price: instead, it would seek to rebuild its market share. Quotas were forgotten; now production was all. And of course, with the taps turned on and the cartel's artificial buttress removed, the only way for anyone to retain a market share, let alone rebuild one, was to sell cheap. Just before the announcement, with the onset of the northern hemisphere's winter, oil had been $31.75 a barrel. By the time spring returned in 1986, it was only $10, with spot market prices going as low as $6.

'It was quite extraordinary how precipitous the fall was', Holmes said. But because Shell's timely planning exercise had not only been undertaken seriously but also implemented, 'we were not totally astonished. I mean, we didn't *expect* it; but we had made provision for it. So in the downstream end of the business we were not particularly hurt.' Whether it was from sensible foresight, or the application of hindsight from experience, Shell's situation after the price collapse was very unusual. In the words of Daniel Yergin, a distinguished historian of the industry:

> when the collapse struck, in contrast to the shock observed in many other oil companies, there was an eerie calm and orderliness at Shell Centre... Managers there, as well as in the field, went about their jobs as though carrying out a civil defence emergency operation for which they had already practiced.

Sam Samuel would have nodded in approval. Yet problems may occur even in the best regulated system, and it would be a mistake to suppose that Shell sailed unscathed through the Third Price Shock. Guiding patterns of thought and action based on a world of expensive oil, patterns created with much difficulty after the First Shock in 1973, now needed to be almost completely recast. Oil had been expensive for 13 years. That was a long time in the career of any individual, even the most senior: it was approaching half the 30 years that Peter Holmes had worked for Shell. Throughout 1986, the price of a barrel of oil was intensely volatile; and although it eventually stabilized at about $15–18, nevertheless that was still, for most people, very much cheaper than anything they were used to. This meant that many operations which had been viable in a world of expensive oil were now difficult to afford, and some were impossible; and in order to succeed, one had to accept that at least for the foreseeable future, harder times were here to stay. 'The industry could not quite believe the low price was going to last', Holmes recalled.

> A lot of companies said, 'Okay, it's not $28 any more, but it's certainly not going to be nine or even fifteen; let's plan on 20 or 25.' Now, if you had invested at $15 real – in other words escalating for inflation year by year – which seemed a very, very pessimistic forecast, you'd have lost your shirt. You really would have. Because the price of oil has been $15 *simple*, it hasn't been $15 real at all; it's been $15 money of the day on average. There's been the odd aberration, but nothing like the huge escalation of 1979; and the industry has had to learn to live with low oil prices. I think it's done that extraordinarily well.

But almost before anyone had been able to assess, let alone absorb, the new reality of low-price oil, Shell endured one more deeply unsettling effect of the days of high-price oil. For three years, from the summer of 1986 to 1988, its British tanker fleet – the origin of Shell Transport and Trading – tottered on the brink of financial extinction. Its ships were too big, they and its staff were too numerous, and their very high and expensive safety standards were being undercut by less scrupulous carriers. Perhaps, at some time in the future, people might regard this last point as

an historical curiosity and wonder why an oil-shipping company should have been quite so devoted to maintaining such standards when they were becoming manifestly uncommercial. As it was, the refusal to compromise meant that low-cost competition was simply pricing Shell out of shipping. Since each company in the decentralized Group had responsibility for its own affairs, there was no scope for a Group rescue package, and by the end of 1985 drastic action had become essential.

Just a few years earlier, facing a similar situation, BP had responded by reducing its owned tanker fleet from 45 to 29 vessels and sacking 1,185 officers and ratings. Shell now began seriously to consider an even more dramatic option – 'complete withdrawal from owned fleets'. Perhaps it would be best to sell all its ships, fire the sailors and shore staff, and rely on the short-term chartering of ready-crewed vessels. Instead, though, the CMD supported the preservation of the fleet. But the only way that could be done was to re-register the ships in the Isle of Man, while their crews were made redundant and offered immediate re-employment through an

Isle of Man agency created for the purpose. (A widespread misconception was that going to the Isle of Man must indicate some kind of company tax advantage. Such was not the case, but moving to the island would bring important cost-reductions in administration.) Fifty per cent of shore staff were made redundant, and redeployed where possible into other Shell companies; but there were to be no more sea-going Shell employees.

To the seafarers, the fact that they would no longer be Shell sailors was a cause of great pain. Thirty-five per cent of them, including a large proportion of highly experienced senior officers, chose to leave rather than carry on without the familiar pecten. However, in 1988 Shell's marine sector returned to profitability, and in July 1990 the British fleet was able to restore the name of Shell to its sailors. 'Success in this change was quickly obvious', Shell Tankers' annual assessment recorded.

> Resignations diminished, recruitment of higher calibre personnel became easier even to the extent that previous employees were returning to the fold, and, most importantly, morale was boosted virtually overnight. Belonging to "Shell" was of major impact.

By the end of 1987, the industry's cumulative investment in the North Sea had reached £56 billion. Shell's own cumulative expenditure there was close to £8 billion; in the peak investment year of 1977, Shell alone had poured about £680 million into the region. But (as Baxendell reflected shortly before his retirement on 30 June 1985) if massive financial commitments had been forced by OPEC's domination of the world's oil market, massive technical progress had also resulted. Though he personally had always preferred working abroad and was very much a 'reluctant repatriate' when he returned to the UK, as a technical man he had found the challenges and possibilities of the North Sea immensely exciting; and for people like him, he said,

> One of the attractions of the Group is that it has never undervalued its technical people. Its salary levels reflect that. So does its management promotion. And in the end, we won't survive unless we have and retain the technical edge...

A glance at just a sample of Shell's achievements around this time shows there seemed little likelihood of that edge being lost. In 1984 FLAGS, the North Sea's largest gas-gathering and processing system, was completed with the inauguration of a fractionating plant at Mossmorran, near Dunfermline in Scotland, and of the adjacent Braefoot Bay shipping terminal. In 1985 the Al-Jubail refinery and two fully automated lubricant blending plants in Saudi Arabia were brought on stream, as was another blending plant in Malaysia and a fourth in Oman. At the same time Mossmorran grew to include an ethylene plant capable of processing half

Above top:
Construction of the 135-mile St Fergus to Mossmorran pipeline

Above:
Braefoot Bay: refining – pipeline – loading

a million tonnes a year, while at Deer Park in Texas, construction began of a demonstration plant for coal gasification, a possible energy source for the future; and at five separate locations around the world (Yokkaichi in Japan, Wood River in Illinois, Stanlow in the UK, Montreal in Canada and Pernis in the Netherlands), Shell showed two of the vital lessons learned under OPEC. In a world of raw competition, firstly, customers had to be offered products with added value; and secondly, every last possible bit of the barrel had to be used. Accordingly, Wood River's control systems were renewed and its cat cracker modified, 'to improve product yield and energy conservation'; Pernis's high vacuum arrangements were replaced with new ones capable of extracting more high-value products; Yokkaichi's capacity was enhanced with new fluidized-bed crackers; and at Stanlow and Montreal the decision was taken to build new cat crackers, 'to boost manufacture of higher-value products.'

Such products were typified by the launches of the motor oil Helix (1985) and the petrol Formula Shell (1986). Helix (later renamed Shell Helix) proved a lasting success, but Formula Shell produced some unexpected and annoying local problems. It had been assiduously tested over

Al–Jubail

enormous distances, both in the laboratory and on the road, and the Honda Marlboro McLaren team used it to great effect; but once launched on the market, it was found that in a very small proportion of engines, it could cause the valves to stick. There seemed to be no pattern to this. The only apparent common denominator was that coincidentally, several of the vehicles which suffered were from marques beginning with the same initial letter. But an additive was hardly likely to react against a letter, and in any case only a very few (albeit highly publicized) instances of damage occurred in any given marque.

This was extremely puzzling, because every indicator in the pre-launch tests had shown that both in performance and environmental benefit, the fuel's additive package was the biggest technical breakthrough in years. Moreover, when the problem came to light, it was extremely hard to replicate on the test-bed. Well over a year passed, a tense and tiresome time for all concerned, before the damage was successfully replicated in laboratory conditions – and this even when scientists knew what they were looking for. When at last they succeeded, it emerged that the cause was neither the fuel's additive package alone nor any other single item, but a complex combination of factors. First was the metallurgy of the given engine; second was the additive package; third was the chemical composition of the petrol (for petrol is not a constant, but contains varying amounts of elements in trace percentages, just as water does); and fourth were the driving conditions. These almost infinitely variable factors were the key. Only when they were all present could the rogue chemical reaction be created. A given

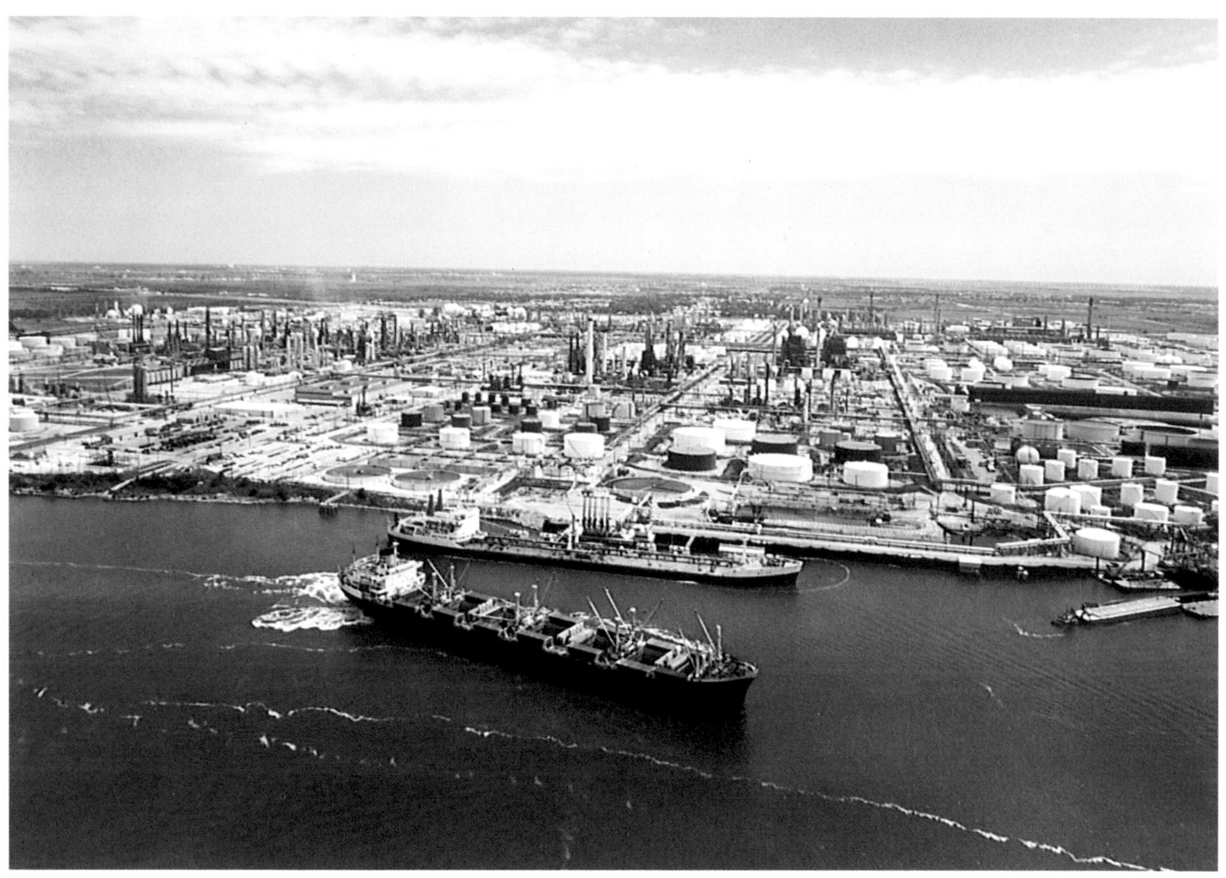

Deer Park Manufacturing Complex

vehicle could be driven thousands of miles without a hitch; another identical vehicle could rapidly develop very damaging faults. With the source of the difficulty at last identified and the unfortunate car owners compensated, Formula Shell was reformulated and in some territories rebranded as well. In certain regions it was then relaunched.

Meanwhile, following the purchase in 1985 of nearly 400 additional service stations in North America (as well as over 100 in Puerto Rico), Shell in 1986 became the largest retailer of petrol in the United States. But perhaps the most striking step taken in 1986 was Shell's signature of agreement on development plans for a new giant gas field called Troll, 50 miles north-west of Bergen, in the Norwegian sector of the North Sea.

Troll was the tenth largest gas field in the world. Covering about 480 square miles, it was located in the deepest part of the North Sea, the Norwegian Trench. From the surface of the sea to the bottom of the Trench, the depth of water was anywhere from 975 to 1,100 feet – that is, twice as deep as most of the North Sea. To make matters still more difficult, the seabed there was unusually soft, yet was going to have to support a platform weighing over a million tonnes; and the platform itself would have to withstand very low temperatures, extremely high seas and storm-force winds. Troll was, in short, the ultimate challenge the region could present.

Troll also exemplified some perennial aspects of the modern world's energy business: it was difficult and expensive to develop, and it was a very long-term prospect.

Regarding the expense of development, no single body in the world could afford to make the necessary investment or take the whole risk on its own. In the Troll consortium, which grew to include ten partners, Norway's national oil company Statoil put up and accepted the lion's share of both investment and risk; but one may gain an idea of the magnitude of the figures involved from the facts that the division of financial responsibility between the partners was worked out to the third decimal place, and that as the major private partner, Shell's share was just 8.288%. Similarly, the long term of the operation is made clear by these facts: Shell (which as exploration operator then had a 35% interest) had discovered Troll in 1979. The subsequent assessment of the field's commercial viability took another four years to complete. With complex negotiations thereafter, Statoil increased its own stake in the venture from 50% to nearly 75%, accordingly decreasing the stakes of the other co-venturers, so it was not until 1986 that Shell was able to proceed as development operator – and the field was still not expected to come on stream until 1996. Even then, it was calculated a further three years would elapse before any profit was earned at all.

The Bullwinkle under construction

Seventeen years from discovery to the opening of production, with great expenditure in between, and without any anticipated return until at least three years after production began: overall, twenty years. There can be very few private individuals or corporations which by their nature are obliged to think in such long terms. Development of the Troll project alone

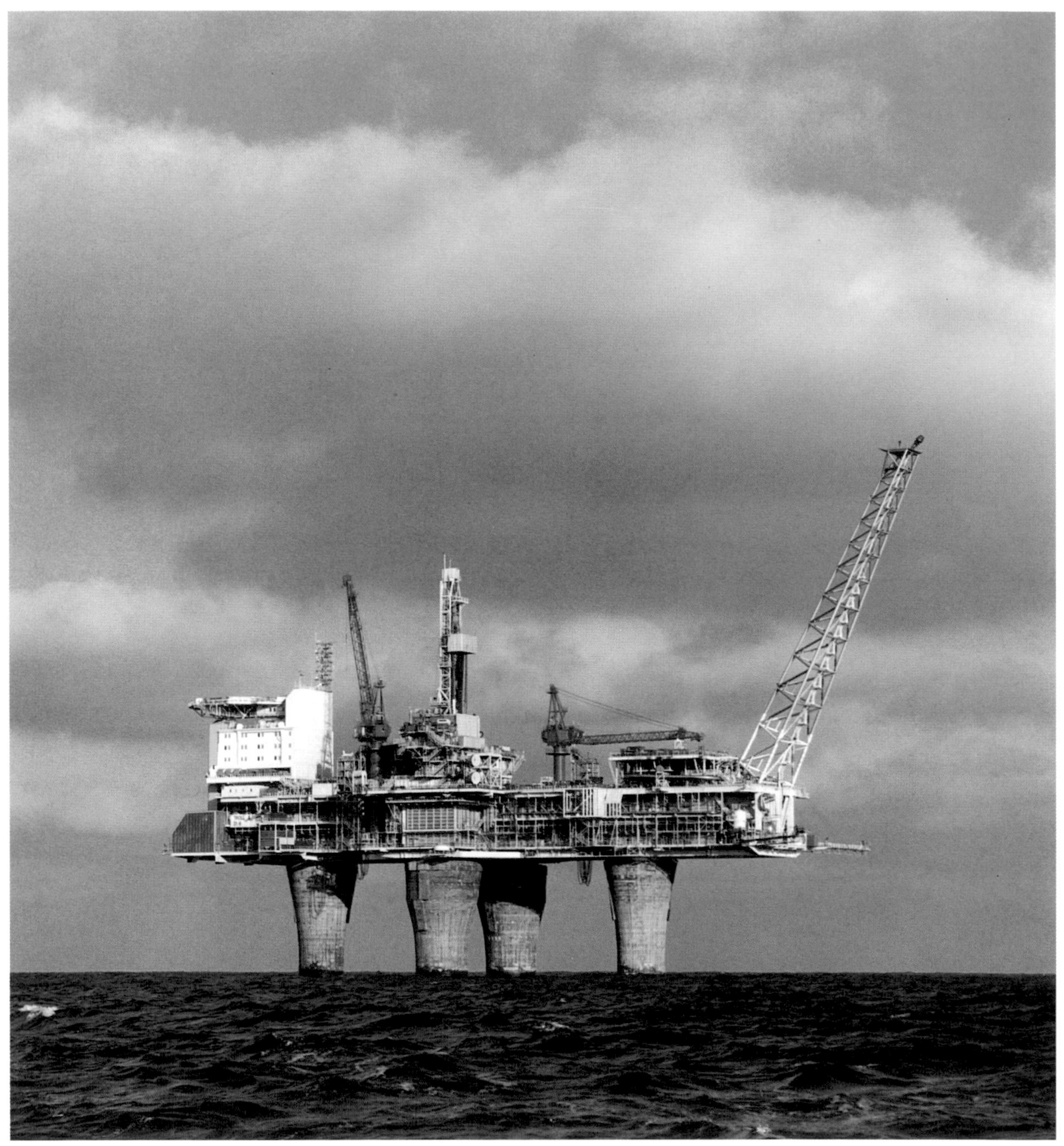

would involve a team of about 300 staff at its peak, with about 7,000 workers being involved in construction; it was calculated that in aggregate, the design, building and installation of the platform would take 17,000 man-years. As the Troll team worked away on one side of the world, another project on the other side of the world reached a momentous stage in its development.

The site was the Gulf of Mexico; the time was late 1987. The world's first commercially viable offshore oil well had commenced production there almost exactly 40 years earlier. Barely half a lifetime had elapsed, yet Shell was about to demonstrate in spectacular fashion the speed of technological advance. The new exploratory well was drilled in a staggering 7,520 feet of water – a depth of 2.3 kilometres, or not much less than a mile and a half. No one in the industry had ever drilled so deep before; and the previous world record (shallower by 688 feet, at 6,952 feet) had been set off the coast of New Jersey in 1984 by Shell.

'We won't survive unless we have and retain the technical edge...' Sir Peter Baxendell's warning seemed almost superfluous: Shell had superb technical mastery. But it was one thing to have mastery, and another to retain it, especially in a business where skills, knowledge and possibilities were still accelerating and where competition was not about to decline.

The Troll production platform

提高国民经济的整体素质！
十六铺
静安寺

CHAPTER NINETEEN

Deeper Still and Deeper: 1988–1992

For eight years, from 1 July 1985 to 30 June 1993, Shell Transport and Trading was led by a man who combined at professional standard the skills of soldier, mountaineer, photographer and author – and who was also, of course, highly experienced in the oil world.

With Lo van Wachem, president of Royal Dutch, as chairman of the CMD, Peter Holmes became Shell Transport's chief at the unusually early age of 52. Barely six months later, he and his colleagues had to manage Shell's response to the oil price collapse. When he took the chair for the first time at Shell Transport's subsequent AGM, he told shareholders that following the collapse, 'One immediate consequence was that all oil companies began urgent reviews of their capital investment programmes for oil and gas exploration and development...'

Shell's expenditure on exploration in 1985 had been £1,118 million; in 1986 it was reduced to £946 million. More (or perhaps one should say less) was to come: in 1987, 'exploration activity by the industry was sharply cut back' – in Shell's case, to a limit of £555 million. In just two years, its exploration budget had been halved. In the remaining ten years of Shell Transport's first century, the figure only once again exceeded £800 million – and then (at £809 million in 1991) only just, before retreating to £733 million in 1992.

But figures taken out of context can be misleading. In 1986 and '87, the budgetary cuts were indeed very severe; yet the short-term comparison is fairly deceptive. Two points place the cuts in some perspective.

Firstly, it was not as if exploration had been closed off. The £946 million of 1986 gave an average daily spend of £2.59 million, and even the 'half-rate' expenditure of 1987 of £555 million meant that more than £1.52 million was being spent every single day of the year. These were hardly negligible sums; and secondly, the comparison with 1985 was distorted, for the £1,118 million spent then was an abnormally high figure.

From 1979 (the time of the Second Price Shock) through 1985, Shell's average annual spend on exploration had been £772 million. Taking '86 and '87 together, a total of £1,501 million was spent, or an average of £750.5 million a year – that is to say, not very much less than the annual average for each of the preceding *seven* years. Moreover, the seven years after the collapse followed a very similar pattern, with the annual average exploration budget being £721 million. So, although Shell's exploration funding in 1985–7 was pretty spectacularly up and down, what is much more noticeable is its longer-term stability and steadiness.

In that light, the cuts of 1986–7 could almost be seen as nothing more than corrections; but one would have received small thanks for pointing this out at the time, for then the cuts were dramatic and real. With the abrupt curtailment of funds, many ambitious plans had suddenly to be abandoned, or (like the development of Shell's Gannet field in the North Sea, 112 miles east of Aberdeen in Scotland) at least temporarily shelved pending further study. Inevitably there was much confusion and disappointment; yet hard as it was, Holmes subsequently assessed the effects of the price fall as 'very healthy for the industry.'

> Quite clearly the days of getting new facilities on stream at any cost, say in the North Sea, were over, and quite clearly a new project had to be very carefully engineered. The real revolution which came out of this fall was for the upstream end of the business to re-engineer and to work out ways of doing things more cheaply. For example, in the North Sea today, the amount of steel on a typical platform is between one-quarter and one-fifth of what it was in 1980; so there has been a tremendous revolution.

The key to that revolution was research and development, 'working out ways of doing things more cheaply'. Every year the industry spends very large amounts on R&D. Shell in particular has traditionally regarded this as a valuable and worthwhile area, and has usually invested more in it than the other oil majors – as in 1990, when out of more than $3 billion spent by the majors collectively on R&D, BP spent in the order of $615 million and Exxon about $655 million, while Shell alone spent $845 million.

From its laboratories and those of others in the latter 1980s came many improvements in drilling techniques – for instance, slim-hole drilling, measurement-while-drilling, and directional drilling.

Slim-hole drilling was perhaps the most obvious of these ideas. In the course of drilling, a conventional borehole would be lined with a heavy steel pipe cemented in place, to prevent the sides from falling in and to keep out unwanted fluids from other strata. Since oil (if there was any) would emerge whatever the diameter of the borehole, it followed that a narrower diameter would be cheaper. Less steel, cement and drilling mud would be used; rigs could be smaller and lighter. Conventional drilling could constitute more than half the costs of exploration and field

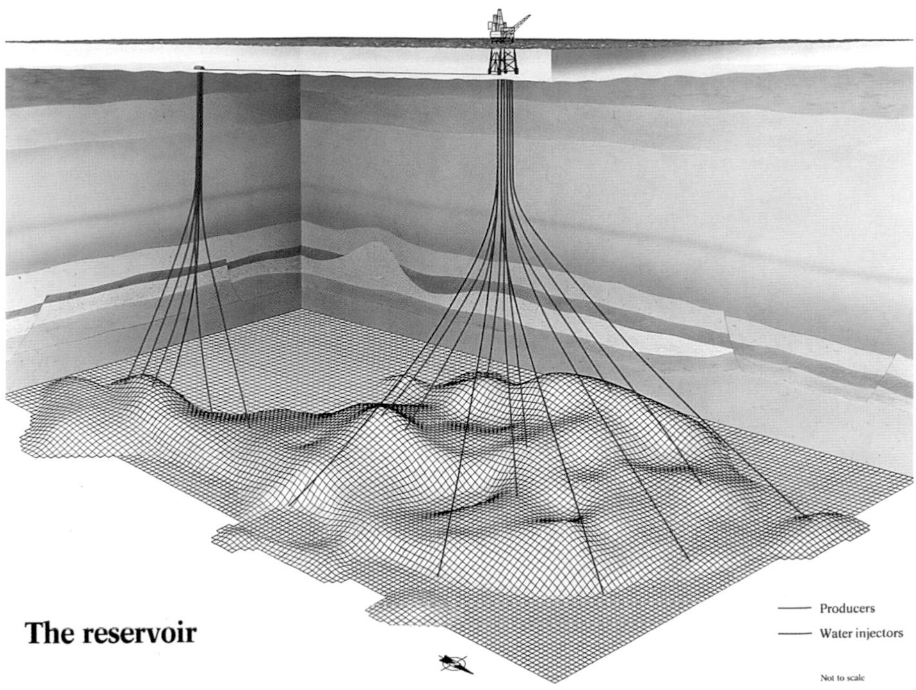

The reservoir

development, so slim-hole research was conducted eagerly – and with success. Prior to 1986, a conventional borehole could be 24 inches in diameter at the surface, tapering to seven inches at 4,500 metres depth. By 1990, those diameters had been halved, with corresponding cost reductions; and there was a welcome environmental benefit as well, because very much less waste mud and cuttings were produced.

Measurement-while-drilling (MWD) was a sophisticated advance: rather than stopping the drill periodically so that measuring devices could be lowered down the borehole, MWD could send back data from the drill bit while it was in operation, not only saving time but helping significantly to avoid damage to the drill. But directional drilling was still more impressive.

A traditional borehole was a relatively straightforward affair: a shaft sunk vertically into the earth and rock below the rig. But with directional drilling, the drill-bit could be guided as it advanced, and turned at will away from the vertical to any chosen angle – even to a right angle. In exploration, directional drilling enabled more accurate assessment of a well's geology and potential productivity; in production, the technique made it possible to link separate parts of an oil reservoir, or to penetrate very thin strata of low permeability, from which a conventional well could not economically extract oil. First used in 1984, the method was rapidly adopted by the industry from 1986: by 1990 about 700 horizontal wells, of which over 100 were Shell's, had been drilled world-wide.

It is strange to imagine how, if the rock were transparent, one would be able to watch the bit purposefully altering course like some sentient creature. And in a very practical sense, by using three-dimensional seismic surveys and computer graphics, the industry's scientists *did* make the rock transparent.

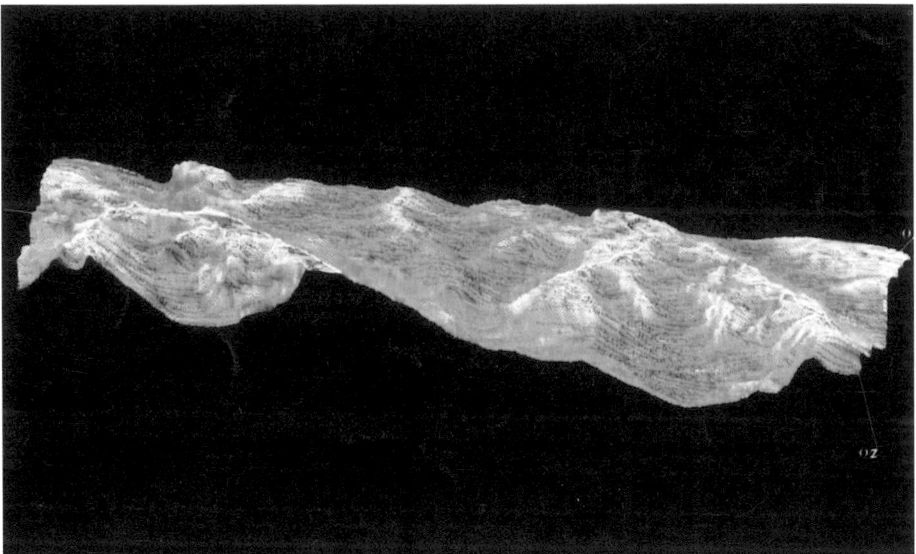

A traditional two-dimensional seismic survey provided a series of vertical slices of the subsurface. Three-dimensional seismic surveys, shot in a tight grid pattern with many sound sources and many recorders, were an expensive but enormously important technical advance. Shell's first use of the technique, in 1975, had covered a mere 16 square kilometres, with only small amounts being added over the following ten years; but after the price collapse of 1986, the use of 3D seismic was extended very swiftly. In the following five years, Shell alone surveyed over 60,000 square kilometres in this way, gaining enormous amounts of data – more precise than ever before – about the hidden reservoirs below the earth. Similarly, advances in computer graphics meant that from such data, highly accurate reservoir models could be created and rotated to any viewpoint on screen, with different rock types clearly marked. Operators on the surface could 'see' exactly where to start and where to turn, and know in advance what degree of resistance to expect.

There were numerous other concurrent interdisciplinary efforts towards the same end, the reduction of costs. In collaboration with other companies, Shell developed 'multi-phase' pipelines, capable of handling oil and gas simultaneously, along with multi-phase flowlines, boosters and meters. Improved maintenance planning and increased automation in offshore platforms reduced their previously high necessary manning levels, from perhaps 130 to only 50. More powerful offshore lifting capacity (bigger, stronger floating cranes that could pick up a 14,000-tonne weight) meant that larger parts of a new rig could be constructed onshore. Improved design and construction of the 'topsides', the above-water accommodation and working modules, reduced their weight by as much as 35%; and, with the concept of satellite production facilities, new subsea wells could be linked to existing platforms, rather than each requiring its own.

The collective effect of these and other innovations was of course very marked. To give just one striking example: as noted above, Shell's North Sea Gannet field, planned in the 'cost is no object' days of high-price oil, had gone on hold when the collapse made it simply uneconomic. Yet, by going literally back to the drawing-board, reappraising the entire project, and applying to it the new techniques and equipment as they became available, Shell was able to reduce the project costs by 45%. Not only did this make it economic once again, but, still more remarkably, the project was able to proceed without any deviation from its first schedule. Although it had been suspended for two years, Gannet came on stream in 1992, just as originally planned.

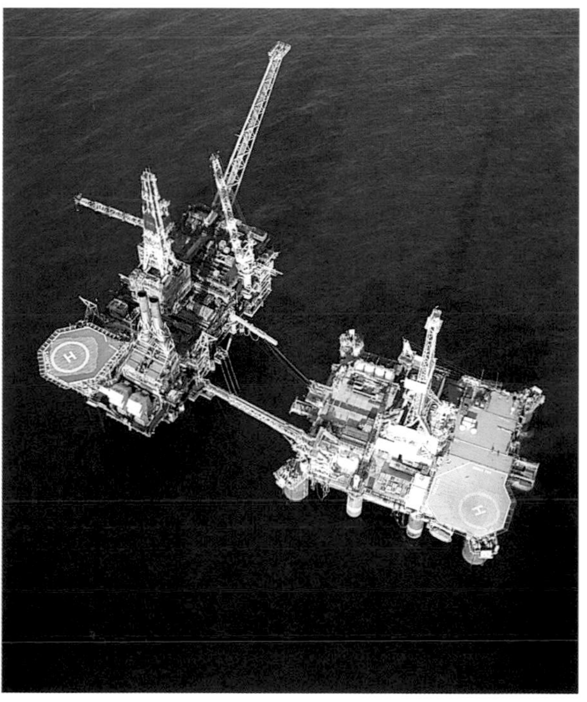

Such were a few of the main elements in the industry's technological revolution after the price collapse – a good example, overall, of how well people can cope with what at first appears to be a disaster. Certainly Shell and others did cope extremely well; and the year 1988 brought to Shell three more exceptional events.

First was an important alteration in the structure of the Royal Dutch/Shell Group. With effect from 1 January 1988, 'in recognition of the size and importance of present and future Group interest in the USA', the US-based company Shell Petroleum Incorporated became a Group holding company.

Strictly speaking this meant that the senior echelons of the Group now contained five holding companies. Two were the originals – Shell Transport and Trading on the one hand, Royal Dutch on the other. Now referred to as parent companies, they jointly owned all the shares of all the others and thereby, through dividends, derived their income. The next level, to which Shell Petroleum Inc. had been introduced, also included the British-based. The Shell Petroleum Company Ltd and the Dutch-based Shell Petroleum NV. Those three collectively owned all the shares of the

Group's subordinate but individual service companies and operating companies world-wide – and thereby, through dividends, derived *their* income, which was passed on in the correct 40:60 proportions to Shell Transport and Royal Dutch respectively. Unusual perhaps, but in the end fairly simple; enormously flexible; and with the full inclusion of Shell Petroleum Incorporated, thoroughly global.

Shell's second notable event of 1988 was the installation in the Gulf of Mexico of Bullwinkle, the world's tallest drilling and production platform. By any standard Bullwinkle was a pre-eminent technological achievement. Much photographed as it was towed out to its site, it was one of the biggest individual constructions mankind had ever produced, and at a depth of 1,350 feet would sit in the deepest water that any bottom-supported platform had ever known.

Given that just one year earlier, in 1987, Shell had established a world record of 7,520 feet water-depth for exploratory drilling, Bullwinkle's 1,350 feet of water might sound somewhat paltry; but to compare depths between exploratory drilling and bottom-supported production drilling is not to compare like with like. In sea-going exploratory drilling, a ship lowers string after string until the seabed is reached. To be able to reach 7,520 feet in itself calls for great skill and high technology: apart from anything else, the mothership must be able to maintain its exact position throughout. But when it comes to bottom-supported drilling, you do not have a ship – a moveable object – on the surface; you have, in effect, a skyscraper standing on the seabed.

The Group Managing Directors of The "Shell" Transport and Trading Company 1989. From left to right, H. de Ruiter, J.S. Jennings, Sir Peter Holmes, L.C. van Wachem, J.M.H. van Engelshoven, David Welham.

Such was the Bullwinkle platform, a creation as splendid in its own way as the Empire State Building or the Eiffel Tower (which some people said it resembled); and deservedly, it won for Shell the American Society of Civil Engineers' 'Outstanding Achievement' award. If that title sounds familiar, it is because Shell had won the same award exactly a decade earlier, when its Cognac platform was installed in the Gulf; yet before Bullwinkle, nobody had ever won the award twice.

Nor were the highlights of the year finished: there was also the thrilling matter of the Formula 1 championship. Shell had supported motor racing since the Peking-Paris rally of 1907, and Formula 1 racing since the beginning of the modern world championship in 1950. A brief absence from the circuits ended in 1984 when Shell became a major sponsor of McLaren International, and in 1988 began working closely with Honda, the team's engine supplier. It proved a tremendous partnership: Honda Marlboro McLaren cars won 15 out of the 16 Grand Prix races of 1988, taking both the Drivers' Championship and the Constructors' Championship. In the single race they lost, the team's two cars both fell out when one crashed and, very unusually, the other's engine failed. But out of the 32 first and second places available in the races, 24 went to Honda Marlboro McLaren, fuelled and lubricated by Shell.

The same team won the championship again in 1989; in 1990; and in 1991. And although Shell and Formula 1 sponsorship made a natural fit, for Shell there was a lot more to it than excitement, glamour, prestige and extremely costly but high-profile advertising. While all of that counted, the main attraction was that in any given year, Formula 1 cars contained the world's most advanced vehicle technology – technology which was not commercially available. Formula 1 thus provided an incomparable proving ground for developing fuels and lubricants. It was also a highly demanding and very public arena in which failure was just as clearly visible as success, so Shell's involvement was far from being limited to the 16 race days of the year. Rather, staff at the Thornton Research Centre in Britain and the Atsugi Research Centre in Japan worked on a year-round basis to formulate, test and improve the fuels and lubricants that McLaren would use for that crucial period. The outcome: in the eight years 1984–91, the McLaren team won seven out of eight Drivers' Championships and six out of eight Constructors' Championships; and since the start of Formula 1 in 1950, Shell has provided the fuel and lubricants for more winners than any other oil company.

David Welham

If 1988 ended with a glow of satisfaction for Shell Transport, 1989 was swiftly clouded.

With Holmes (who had been knighted in 1988) as its chairman, the company's board then included eight directors and two managing directors, David Welham and John Jennings. Welham, born on 11 September 1930, was almost exactly two years older than Holmes. He had joined Shell in 1958 after some years in government service, and had worked in South America, Africa, Asia and the UK. He and Holmes first met in 1977, and within weeks became firm friends. Holmes described him as a man of 'sound common sense coupled with a magnificent sense of humour ... I've never heard anyone in Shell create so much laughter.' He had been elected to the board in 1986, and was followed in 1987 by Jennings, Holmes' designated successor and the youngest managing director (he had just celebrated his 50th birthday). A small part of Shell's way of working is that an individual who is going to be proposed for board membership is only advised of this in the February, three months before the AGM at which their formal election is put to the shareholders' vote. Thus, when Welham was told in February 1986 about his selection for the board, he was a few months over 55 years old. At the time, he was Group Treasurer.

Given that he was two years older than Holmes, it was clear from the outset that Welham's election as a managing director would not be a step towards the chairmanship. Not that he nursed ambitions in that direction – indeed, quite the contrary: immediately before being told he had been chosen, he had been contemplating with pleasure the prospect of early retirement. He took a little time to consider the unexpected proposal, and Holmes became certain 'that although he recognized it was a great honour to be selected, he only agreed to take it on out of a sense of duty.' And so, through a sense of duty and the decree of fate, David Welham never did reach the retirement he desired: because on 27 February 1989, the car in which he and his wife were being driven was hit head-on by another. Their chauffeur survived, but both Mr and Mrs Welham were instantly killed.

To their friends in Shell, the shadow of the Welhams' deaths lengthened that of a tragedy the previous year. In May 1988, part of Shell's Norco refinery in Louisiana blew up, and, in the explosion and subsequent fire, seven employees had died. Oil wells, refineries and transport systems are inherently dangerous, and through line management responsibility and effective training, reporting and auditing, Shell had long sought to eliminate as completely as possible any unsafe practices or conditions which could combine to bring about serious accidents. The Norco explosion was a most unhappy setback for a Group of companies which individually and collectively prided themselves on setting and keeping very high standards of safety, both for their own employees and those of outside contractors. Yet without diminishing the misfortune of seven deaths, the fact remained that Shell consistently met its safety standards far more often than not, and many of its companies around the world racked up prodigious records of accident-free operation – such as the ten million working hours surpassed (also in 1988) by the Yokkaichi refinery in Japan. Nor were these high standards confined to the land: in 1989 the British Royal Society for the Prevention of Accidents awarded its gold medal to Shell in recognition of its fleets' safety performance, and in 1992 the fleets gained LRQA listing. LRQA – Lloyd's Register of Quality Assurance – assessed them against the two highest international standards: ISO 9002 (the manufacturing

industry gauge of the International Standards Organization) and IMO Resolution 647, the International Maritime Organization's top measurement of ship-board safety and pollution prevention. Though other merchant fleets could be as safe as Shell's (and some were), it was very hard to be safer.

Back in 1951, Shell Transport's then chairman Frederick Godber remarked to shareholders: 'It is a trite comment to say that we are passing through times of great uncertainty and even anxiety...' Harold Macmillan, Britain's prime minister 1957–63, used a similar but more witty phrase: 'We are passing through a period of social change – as Adam said to Eve.' But even in a world where change is the only constant, the year 1989 brought greater changes than most. In almost every communist country in the world, the ruling system came under powerful challenge from the people it ruled, until in November there occurred the single most significant event in a generation of world history: the political opening, and start of the physical destruction, of the Berlin Wall.

After being able only to market oil and chemicals in the communist areas of the world since the war, Shell could now return to direct investment. Following the 1984 agreement between Britain and the People's Republic of China about the handing back of Hong Kong in 1997, Shell in 1985 was able to open an industrial fuel depot and three filling stations in mainland China – its first re-entry there since 1949. In 1987 its renewed presence grew with the establishment of a lubricants blending plant in Shanghai; and simultaneously its links were strengthened by six-month

The end of the Berlin Wall

visits of staff members from China's national agrochemical laboratory to Shell's Sittingbourne laboratories, where experiments were conducted on the natural bases of traditional Chinese methods of crop pest control. In these, the properties of plants such as *Impatiens balsamina* ('Busy Lizzies') went through modern scientific analysis and produced promising results, keenly reported upon in in-house publications.

From 1989, Shell's moves into decaying communist states elsewhere accumulated steadily. In January of that year, building upon a previous franchising arrangement with a new joint venture in automotive retailing, Hungary became the site of its first major shareholding in any eastern European country for nearly half a century. By the end of 1990, this had grown to a total of fifty outlets. At the same time, with the dismantling of the Berlin Wall and all it represented, eastern Europe began to offer wider opportunities.

At Shell Transport's AGM on 16 May 1991, a nervous shareholder asked that caution should be exercised in entering any new markets beyond the former Wall. By the end of that year, wholly owned marketing companies had been established in Czechoslovakia (as it still was), Poland and Russia. Russia in addition held a new joint-venture oil production company, and a wholly owned oil and natural gas production company; and by the end of 1992 there were also wholly owned marketing companies in Bulgaria, Croatia and Romania, as well as a wholly owned exploration company in Romania. Whether or not this was the degree of caution envisaged by the nervous shareholder, it elicited no criticism; and it would have done Lord Godber's heart good. During the twelve months from May 1948 to May 1949, he as Shell Transport's chairman was forced to watch with powerless fury as, one after another, the then new communist states had nationalized Shell's assets. Now, although the assets had not been returned, deals had been struck producing some compensation, and all the nations involved were once again host to Shell.

Shell's instantly recognizable pecten now stood over petrol stations not only in several eastern European countries but in every country in western Europe too. The sole exception had been Italy, but after an absence of 14 years, its return to the Italian petrol retail market took place in 1987, in the form of a 50:50 joint venture called Monteshell. Created with an Italian company, Montedison, this provided Shell with an Italy-wide network of 3,000 service stations. Remembering the short-lived fuss in 1976 when it was wrongly accused of paying bribes there, the return was an event of much greater importance than economics alone.

Speaking from his retirement, 84-year-old John Loudon said in 1989, 'Ethics are terribly important. I think our reputation as a Group, that we don't go in for any sort of side payments, is excellent.' It was, he added, 'the only right policy' for Shell Transport, Royal Dutch and their daughter companies to follow. A few years later, after he too had retired, Holmes enlarged on this in private conversation. Shell, in his view, possessed three great strengths. Firstly, its decentralization enabled the chief executive in a given country to respond immediately in any negotiations with the host government. 'You can actually respond to governments on the spot. That's important; governments like it.' Second was Shell's degree of internationalization, unique in the commercial world: 'We're the only company in the world which has, currently, people of over 65 nationalities working outside their country of origin. Not training, *working*.' And the third, 'on which everything else rests', was integrity.

President Nelson Mandela

The return to the oil products marketplace in Italy was undertaken not only because of the assessed likelihood of renewed profitability there, but also because of the permeating and positive influence of the *Statement of General Business Principles*. The return was not just to a country, but to a Group-wide state of moral self-confidence. However, there remained the constant test of public scrutiny and perception: to be believed as real, integrity had to be seen to be consistently in action.

By 1990 this was beginning to be so once again. Shell's good reputation had been winged by the Italian bribery allegations of 1976; but that had been as nothing in comparison to the long, complex and often confusing investigation of Rhodesian sanctions. That spotlight had in turn drawn considerable attention to Shell's presence, as one of numerous other large western businesses, in apartheid South Africa. Amid the world-wide opponents of that inhumane regime were many who believed that to work in South Africa was of itself a tainted act. If that was a simplistic belief, it was nonetheless deeply and sincerely held, prompting vociferous demands for Shell to quit South Africa. Its refusal to do so was correspondingly dismaying. Yet within South Africa at least, Shell was known to be not even sympathetic towards apartheid – so much so that when the African National Congress, apartheid's most bitter opponent, needed a new headquarters, it was happy to take over the lease of the Johannesburg office building which Shell itself had recently vacated; and what was more, though it had the right to change the building's name, it did not act to do so but kept the existing name: Shell House.

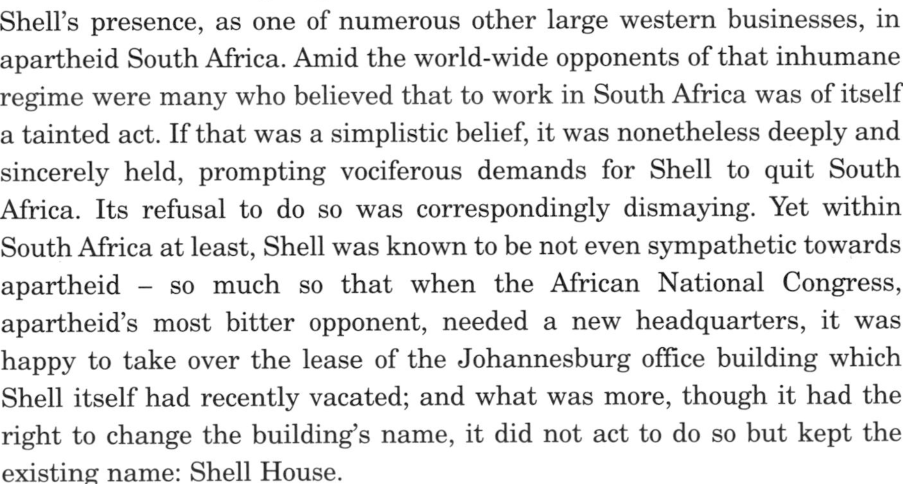

Nevertheless, Holmes later regarded this period as the hardest of his chairmanship:

> Outside South Africa we were hated by the anti–apartheid groups, because we were determined not to leave under duress, indeed not

> to leave at all – we felt we could do more good by being there than by making an empty gesture, and leaving the country would have been an empty gesture. My first AGM was quite noisy; my second was noisier; my third and fourth were *really* noisy, with protestors blowing whistles and letting off balloons, almost like football crowds. At the same time, *inside* South Africa, we were hated by the right-wingers, because we'd spent a lot of money on educating blacks and teachers of blacks. And we were very unpopular with the South African government of the time, because we were so overtly supporting black aspirations. For a long time in the *Mail* in South Africa, a major left-wing newspaper, we took out full-page advertisements on a regular basis, supporting free speech, unions, education, everything. But it was a very difficult road to tread.

What was at issue, between Shell on the one hand and anti-apartheid activists on the other, was not the end but the chosen means. Inevitably there were some who would never accept this, choosing instead to see Shell as the ogre; they would do so whatever the circumstances, simply because it was a multinational oil company. However, after release from his long incarceration when apartheid fell in 1990, Nelson Mandela himself said to

PET bottles

Shell, 'We're glad you stayed.' His was a moral authority which could influence almost anyone, and at least some of those who had long despised Shell's stance in South Africa were big enough to recognize and accept in public that they had, after all, been seeking the same end.

> After the political situation changed, my last two or three AGMs were very quiet. And in my penultimate AGM [in May 1992], one of the demonstrators from years before actually got up and said, 'If I'd

> known three years ago what I'm going to say now, I'd have had a heart attack; but I have to congratulate Shell on what they've done in South Africa.'

It was a welcome and in its way a gracious acknowledgement. Yet life for Shell could never be simple. It never had been, even when Shell Transport and Trading was young and independent. The centenary of its sister, Royal Dutch, had been marked in 1990; Shell Transport was approaching its own 100th birthday, 18 October 1997. Separately and together, they had faced, surmounted and learned from many technical and economic challenges. Those never stopped, and no one wanted them to stop. Though often very hard, they were part of the fascination. Without them one could become complacent; and if complacency set in then incompetence was sure to follow, and entry to the retirement home for aged major oil companies would beckon. Gulf Oil was a resident there already, mumbling between feeble gums the rusks of its memories. No: each generation needed technical and economic challenges. But since the 1960s, Shell had been presented with additional challenges, the active input of a wary world. 'In the 1960s,' said Holmes,

> we came under fire simply for being a multinational, but there was nothing specific about that. Then we came under fire for our alleged role in supplying oil to Rhodesia during UDI; then we came under fire for a long period – seven, eight, nine years – because of South Africa. I think there'll always be something; the very fact you're big makes you a target.

The last eighteen months of Holmes' chairmanship saw numerous continuing proofs of Shell's technical mastery and scientific leadership. A prime example from the chemicals sector occurred in 1992 in Italy, when Shell began producing polyethylene terephthalate (PET) – a strong, lightweight, transparent, recyclable polymer that was increasingly replacing glass in bottle manufacture. Inside a year its first plant had been augmented by others in the UK and USA, and Shell had become the world's second largest PET supplier.

Simultaneously further progress was being made in meeting the world's energy demands, both through conventional and not-yet-conventional means. The conventional included, most notably, world leadership in natural gas; the not-yet-conventional included biomass (renewable organic fuels, of which the most common and important was wood) and the Shell Middle Distillate Synthesis system, or SMDS.

The use of biomass was being steadily developed in Brazil. Plant life traps enormous quantities of energy from the sun – each year, somewhere between five and eight times more energy than mankind uses from all sources. But releasing that energy by simple burning is extremely inefficient, and sends a great deal of the 'greenhouse gas' carbon dioxide into the atmosphere. Working with various Brazilian authorities, Shell was seeking to devise a cost-effective method of generating electricity from turbines fuelled by the gases from burning wood. Researching the 'biomass integrated gasification/gas turbine' process, or BIG/GT, was a long-term five-phase project. If successful, it would provide an energy source with two important and attractive environmental benefits: the fuel would be renewable, and would not contribute to atmospheric carbon dioxide accumula-

tion, because as they grew, the trees planted for future fuel would absorb the carbon dioxide released by combustion. (Incidentally, in the separate but related field of forestry, Shell by now was routinely planting 25 million trees every year – a real environmental benefit in itself.) By March 1992 the project's first phase, preliminary investigation, was complete and the second phase, the development of processes and equipment, had begun.

While the Brazilian BIG/GT project was breaking new ground both technically and commercially, turning biomass into electricity, the town of Bintulu in Sarawak was the scene for SMDS, a process designed to turn natural gas into high-grade, environmentally clean, liquid fuels.

Based on the Fischer-Tropsch conversion method (discovered in 1923), SMDS was first researched by Shell in the late 1940s. The process fell out of favour when crude oil from the Middle East became abundantly available, but interest revived in the early 1970s, after the First Price

Shock made it clear that to be over-reliant on the Middle East would be unwise. By the late 1980s the process had been fully tested both in the laboratory and pilot-plant stages, and the Bintulu SMDS plant came on stream in 1993.

Bintulu was chosen as the site for this pioneering plant because the region was rich in natural gas, which could be economically extracted and processed with the technology of the day. But the same SMDS process could be applied to coal – indeed, during World War II Germany had used the original Fischer-Tropsch chemistry to produce synthetic fuel. The Shell Middle Distillate Synthesis system therefore offered great promise for future generations, because coal is far and away the world's most abundant and widespread fossil fuel. With the technology and prices of the early 1990s, only one-tenth of the globe's estimated coal reserves are economically recoverable, yet at the contemporary rate of consumption, even that one-tenth would be sufficient for over 230 years. With SMDS as a method of liquid fuel production, and Shell's award-winning parallel process of coal gasification as a means of producing electricity, coal could be viewed as a reasonably secure longer-term source of energy for the world, probably for several centuries to come. In the 1990s, however, much of the energy industry's thought and action was focused on Bintulu's feedstock, natural gas.

There were several good reasons for this. Natural gas was a clean fuel, easy to apply to domestic or industrial heating purposes; it was a highly efficient source of electricity generation, and its large proven reserves were growing year on year. Customers liked it very much – almost too much. In 1990 Roland Williams (Shell's Co-ordinator, Natural Gas and Coal, 1987-95) co-wrote a paper entitled *Energy – the end of crises?*, extolling the virtues of and prospects for natural gas. Hurrah, said clients, and queued up to buy such quantities that in 1993 Williams felt obliged to write another paper: *Natural gas – the world supply challenge.*

In 1990 Shell sold 6 billion cubic feet of natural gas a day. In 1993 the figure exceeded 7 billion, and continued to rise steadily. Before the break-up of the Soviet Union, only Soyuzgaz-Export recorded higher figures – but this was hardly a competitor working under similar conditions, for Soyuzgaz was a state-owned monopoly, and the USSR contained over one-third of the world's proven gas reserves. In contrast, Shell drew its own natural gas from every other part of the world, with fields from Norway to New Zealand and from Brunei to Canada; and worldwide, its sales, production and proved reserves of gas were greater than those of any of the other private oil majors. In short, putting it simply if immodestly, Shell was tops in anything to do with natural gas, the cleanest and most efficient of all the fossil fuels. But as Holmes had observed, 'the very fact you're big makes you a target.' In the next chapter we shall see the truth of those words.

No one who wanted their life to be an easy breeze would take on the chairmanship of a world-wide business. For those few who are ever in a position to accept such a role, the hurdles are part of the attraction; but sometimes an unanticipated hurdle can appear just as one is sprinting confidently along the home straight. Lo van Wachem, president of Royal Dutch, retired on 30 June 1992; the following day, as next most senior member of the CMD, Sir Peter Holmes became its chairman and commenced the eighth and last year of his chairmanship of Shell Transport and Trading. It was during that period that the various major projects noted above – PET, BIG/GT and SMDS – came to fruition; but in the same period a very serious contrast occurred, an unexpected hurdle of quite astounding dimensions.

At the end of 1992 Showa Shell, a Japanese-based joint venture subsidiary, disclosed that a small group on its staff had run up a loss of over one billion US dollars, or about £700,000,000 sterling. Given the staggering size of the loss, Holmes's only comment to shareholders, made at the AGM shortly before he retired in 1993, was a model of restraint: he was 'shocked and disappointed'. Who would not have been?

The loss had come about from 'foreign exchange dealings outside normal business practice' – or in a word, gambling, by a handful of individuals acting without authorization. Those concerned had pledged company money in the foreign exchange futures market and lost. They had then sought to cover the deed by the same method, and compounded the misfortune until at last it could not longer be concealed. The phenomenon would be repeated in the rapidly developing electronic money markets of the 1990s, threatening and sometimes bringing down large and long-established businesses of different sorts, such as Baring's, Britain's oldest private bank.

Shell Transport and Royal Dutch together had to shoulder half the loss which, even when reduced to the minimum, still amounted to £196 million. Holmes's terse comment contained an echo of the age of Marcus Samuel: in 1906, when Shell Transport had been about to lose its independence, Samuel had declared himself 'a disappointed man'. Yet Samuel had gone on to see his company grow, in its alliance with Royal Dutch, to a health and vigour he had not imagined possible, and in 1992–3 it could even survive the Showa Shell deception. Discussing the matter, Holmes and his colleagues in the Committee of Managing Directors knew the loss could not merely be shrugged off; but it could be managed, and was. Though the experience was shocking, surviving it – clearing the hurdle – was a proof of both Shell's underlying financial strength and the skill of its collective management.

CHAPTER TWENTY

The Changing Shape of Shell: 1993–1997

Sir Peter Holmes retired from the offices both of chairman of the Committee of Managing Directors and chairman of Shell Transport and Trading on 30 June 1993. The following day Dr John Jennings became tenth chairman of Shell Transport as well as vice-chairman of the CMD, while the incumbent president of Royal Dutch, Cor Herkströter, became the CMD's new chairman.

The collegiate nature of the CMD meant that the Group was unlikely to endure any sudden lurch in policy or strategy. A managing director might retire and a new one be appointed, but at any given time the majority of the CMD had worked together for several years; so, after a change of chairmanship, naturally enough they could be relied upon to continue with the decisions they had jointly reached in the past. But this did not necessarily imply that everything would be the same for ever, and in the 16 months between August 1994 and January 1996 the new CMD brought to the Group wider-ranging changes than any in the previous 35 years: structural alterations of such importance that before describing them, it is appropriate to revisit the concept of the CMD itself.

As Jennings pointed out in conversation, a traditional board of management contains the obvious risk of dominance by one powerful character. 'Individuals', he remarked, 'can carry a company forward spectacularly in a short period; but it becomes very hard, and they can lead the company in very much the wrong direction too.' With management by committee, an equally obvious risk is that healthy change might be unhealthily slow in coming. Yet though the CMD was called a committee, its collective responsibility and collegiate nature meant it did not suffer from the delays that committees can entail – in John Loudon's words, 'If any managing director was travelling around and something happened in his particular field and you could not get hold of him, the other managing directors would decide.'

For many people outside the Royal Dutch/Shell Group, the CMD's collegiality remained a somewhat awkward idea to grasp. Ever since its establishment in 1959/60, the general public and many shareholders (whether of Shell Transport or Royal Dutch) had continued to regard their company chairman as the ultimate authority of the given company, when in fact he should have been seen more accurately as its emblem and public spokesman – though by no means a mere figurehead. Shell Transport's previous leaders in that period (Stephens, Barran, McFadzean, Pocock, Baxendell and Holmes) had all in their day been extremely active managing directors, and no less active as chairmen.

Sir John Jennings, Chairman of Shell Transport and Trading 1993–7

Of course collectivity has two sides. In the event of outside criticism, each company chairman in his turn has had to stand up in public and take the flak on behalf of the whole CMD; but when the time comes for praise, they do not seek personal commendation. It is simply inappropriate: although good ideas originate with someone, they are evolved by the CMD. So it was in 1994–6: 'New Shell' was evolved collectively. However, just as John Loudon is credited with the original idea of the CMD, it is fitting to register that in this instance the perception originated from Cor Herkströter and John Jennings. Neither man sought the credit, but it was a concept of such potentially enormous influence that it seems right to identify them for the record. The way it came about and was developed provides an intriguing insight into the creative thinking of a multinational oil major, and will form a properly large proportion of this chapter; yet without wishing to tantalize, it is probably best to defer that account until closer to its chronological position in this narrative – that is, towards the end of this book, and the end of Shell Transport's first century.

Meanwhile, the period of John Jennings' chairmanship of Shell Transport saw significant developments in Shell's staffing profile; numerous technical advances, particularly in deep-sea exploration and production and in natural gas; two short but dramatic periods of public criticism; and the world-wide introduction (outside the United States) of a strikingly re-designed pecten.

Last undertaken in 1971, the 1990s redesign of the pecten actually involved far more than the pecten alone. Within Shell the process was called Retail Visual Identity, or RVI, and it entailed a complete overhaul of the total design of Shell's service stations. The professional press described it as 'the world's largest re-imaging programme' – with justice, for outside the US, which would retain its own pattern, the new look was to be applied to all Shell's 38,000 service stations world-wide in a ten-year programme, at a cost of £500 million. Obviously it could not be undertaken lightly, and when applied, it had to be right.

Beginning in 1989, the new RVI was four years in the planning, under a team led by Paddy Briggs, a marketing specialist of 25 years' standing with Shell. Market research in seventeen countries established that motorists perceived Shell petrol stations as places staffed by friendly, caring, trustworthy professionals, and that Shell red and yellow were popular colours; but it also established that Shell as a whole was beginning to be seen as old-fashioned and rather undynamic. Moreover, the 1970s design had lacked detailed guidelines: individual operating companies had only artists' impressions of architecture and signage to go on, and stations in neighbouring countries could look confusingly different, with no common elements apart from red, yellow and the pecten, which was often scattered randomly around the site. The new appearance was intended to change all those negative elements while maintaining and enhancing the positive, in an evolutionary rather than revolutionary manner; and like all really good design, the result – formally launched in the UK in May 1994 – looked as though it must have been quite simple to determine and achieve. But the fact that its achievement took four busy years indicates how much hard work and careful thought was required.

1995

There were ten main elements in the new RVI, all working together to produce a harmonious, attractive and welcoming setting in which customers felt relaxed and safe – 'not the greasy garage,' as someone said, 'but the fully fitted kitchen.' Of the ten elements, the single most noticeable was that the pecten, hitherto always two-dimensional, became three-dimensional. Otherwise the numerous alterations were so subtle that their total effect was almost subliminal, and people only really recognized the differences when photographs of sites old and new were placed side by side; but if that was done, it became obvious that in contrast, the older design just did seem old-fashioned. How effectively RVI would achieve its main aim (the preservation of Shell's position as the world's leading petrol retailer) in the longer term remained to be seen, but it looked like a winner at once. An encouraging level of success was apparent as soon as it was launched: the new look not only earned a European Sign Design Society award, which was gratifying, but also something much more valuable – the warm approval of customers. And if a 3D pecten did for Shell's sales what 3D seismic had done for its exploration, its £500 million cost would prove a wise investment.

Although the RVI task was targeted at the public's perception of Shell, it was typical that the considerable effort and enormous cost were undertaken without great fanfare. Shell has been the author of so many record-breaking technological feats and achievements, and is in every respect so large, that modesty is not a word one associates with it; yet in many ways its people are very modest, and perhaps unduly so. For example, each year Shell gives routinely and unconditionally – absolutely no strings attached – tens of millions of pounds in support of education, intermediate technology, conservation, medical care, exploration, cultural events, sporting events and youth projects. Routinely its ships save the lives of castaways

in distress; the average figure is well over a hundred souls saved each year. And routinely its doctors administer health care free of charge to hundreds of thousands of people world-wide. Such things may be small acts of goodness individually, but they are many, they are customary and cumulatively they are important; yet Shell does not seek great publicity for them. Perhaps it should, but, backed up by the confidence that its decisions and actions are both commercially and socially principled, the preference is simply to get on with the job. So when the North Sea storage buoy Brent Spar came to the end of its useful life and was decommissioned, Shell simply got on with the job of figuring out the best means of its disposal, assuming that if all proper methods were followed as usual, then public trust would be assured. This turned out to be a mistake.

After 15 years of safe service, Brent Spar was decommissioned in 1991. As described in chapter 16, environmental safety was a key factor in the original decision to use the Spar rather than the cheaper alternative of a permanently moored tanker. The same concern guided the choice of how best to dispose of the buoy. Thirteen different options were considered in a scientific and economic review spanning 30 months, and the conclusion was that by far the safest option would be to tow the Spar out into the deep Atlantic and sink it in 6,000 feet of water. There is not space here to run through the science; it is suffice to say that the analysis was undertaken conscientiously, that Shell was confident the most responsible decision had been reached, and that both then and later, most independent scientists agreed. Permission to proceed with the disposal was applied for and received from the British government, which informed the other governments in the European Union. The other EU governments offered no objection, and on 17 February 1995 the plan was announced. Then on 30 April a team from Greenpeace, the environmental campaigners, boarded the redundant buoy in protest against its planned disposal.

So began an intense and hectic 51 days. Greenpeace gained an unexpected benefit when TV news programmes accepted their film footage and broadcast it unedited and without critical commentary. It was good television, but some editors later recognized they had only aided an increasingly simplistic understanding of the issue. Much of the information given out was misleading or even wrong. In particular, there was soon a wide-spread misconception that Shell planned to sink the Spar in the shallow North Sea rather than the deep Atlantic, and that the Spar

HAABULL

contained large quantities of highly toxic waste. Neither was true, but Shell service stations, especially in Germany, were boycotted and subjected to threatened or actual violence, including fire-bombing and shooting.

In Britain, some newspapers hinted that Greenpeace was mistaken and that Shell's published statistics and environmental analysis were correct. Elsewhere there was no such doubt. Though provided once again with the original information from Shell, most of the EU governments which hitherto had raised no objection now chose to respond instead to the inflamed mood of their electorates and began to exert great pressure. On 21 June, Shell decided to abandon the deep Atlantic sinking, at least for the time being, and with Norway's permission the Brent Spar was towed across the North Sea and anchored in a sheltered fjord, pending an informed decision reflecting public dialogue on its ultimate fate.

A few weeks later, Greenpeace issued a public acknowledgement that crucial parts of its data had been wrong and that the Spar's contents were, after all, not significantly toxic. Independent confirmation came from Det Norske Veritas, the globally respected Norwegian maritime insurance agency; but these facts were widely ignored, for the second – and much more complex – of Shell's two major shocks had arrived. The Nigerian region of Ogoniland was now in the headlines.

The crux of the matter was that Nigeria, a nation beset by long-standing violent tribal rivalries, was ruled by an unpredictable military government which denied to tribal minorities various fundamental human rights, and would not provide the peoples of Nigeria's oil-producing areas with their legal proportion of the proceeds from the national oil industry.

Under the terms of a 1979 agreement with one of Nigeria's rare democratic governments, Shell conducted onshore oil operations nationwide on behalf of a consortium consisting of Agip (5%), Elf (10%) and Shell (30%) with NNPC, the Nigerian National Petroleum Corporation, as the dominant partner (55%). (It was actually a relatively unprofitable operation for the minor partners. In 1995 Shell's share of Nigerian crude oil production represented 14% of its world-wide total, but only 4.5% of its profits.)

Niger Delta

Throughout Nigeria, Shell employed over 5,000 staff directly, of whom only 300 were non-Nigerian. Located in south-eastern Nigeria, Ogoniland covered about 700 square miles, of which Shell had bought, and operated in, an area of less than five square miles. Numbering approximately 500,000 people, the Ogoni constituted about 0.5 per cent of Nigeria's population. Over the years Shell had provided numerous tangible assets to Ogoniland, as it would anywhere else in the developing world - schools, a hospital, roads, water schemes, support to farmers, scholarships and employment. But the Ogoni were well aware that the government was denying them a proper share of federal revenues. They also wished for much greater autonomy, and in 1990 addressed an Ogoni Bill of Rights to the actively antagonistic government. Shell also communicated its view that historically, not enough federal revenue had been passed to the oil-producing areas, and in 1992 the government agreed to double the figure to 3%. However, the actual payment of this remained largely theoretical.

Though Ogoniland and its population were tiny in relation to Nigeria as a whole, they found in the writer Ken Saro-Wiwa a gifted spokesman who sought to place their grievances on the world stage by tapping the widespread, half-subconscious public wariness of oil companies. There was no point in engaging the state-owned company NNPC; such a campaign would attract little international attention and could expect no success at home. Agip and Elf were too small to be effective targets. Shell was the obvious choice.

People in Shell felt considerable sympathy for Ogoni aspirations. This was particularly true of the head of Shell Nigeria, Brian Anderson, who had been born and brought up in Nigeria. But it was difficult for anyone in Shell to sympathize with the tactics adopted. Physical attacks against personnel and plant reached such a level that in 1993 Shell ceased its operations in Ogoniland. Concurrent accusations of wholesale environmental despoliation and collaboration with the military government were untrue and offensive, yet difficult to disprove. Anderson willingly met critics face to face, and responded to even the most hostile accusations with impressive patience and unfailing courtesy. Shell agreed that the Nigerian oil industry as a whole had been responsible for some environmental degradation, and the industry's profits were indeed the government's lifeblood. But on the environmental question, contrary to widespread assumptions, much of the damage in Nigeria was unrelated to the oil industry, and most Ogoniland oil pollution (over 60%) was created by saboteurs seeking compensation. As to the charge of collaboration – an emotive term – Shell had no choice but to work with the government; the fulfilment of legally binding obligations was not collaboration. The heart of Shell's dilemma was that a lawful contract made with a democracy had to be fulfilled with a dictatorship.

In July 1994 the protest movement prompted an oilworkers' strike. If the military government had had its way, the outcome would have been direct armed confrontation. Shell refused to work 'behind the guns', and after two months managed to defuse the strike. Violence between the government and the Ogoni continued to rise, commanding greater international attention; but not all the Ogoni people believed Saro-Wiwa's confrontational approach was correct. In challenging it, some were killed, whereupon Saro-Wiwa was arrested on a charge of inciting murder. On 2 November 1995, the London *Times* said in its leading article:

> Since neither wealth nor power is distributed equitably in Nigeria, the Ogoni people have seen nothing of the money which oil exports earn for the 'Giant of Africa'. Instead, they constitute perhaps the poorest ethnic group in the country, their environment despoiled in the extraction of petroleum deposits – for which blame must lie, squarely, with the Nigerian Government...The military junta has responded with iron aggression to calls for Ogoni independence. This has included Mr Saro-Wiwa's imprisonment and serious ill-treatment in jail; it now includes [his] death sentence...

By then there had been huge international condemnation of, and many appeals to, the Nigerian government. Among the appellants favouring 'quiet diplomacy' were Shell, and South Africa's president Nelson Mandela – an alliance of opinion which Shell's critics twenty years earlier would have found inconceivable. But on 10 November 1995 Saro-Wiwa and eight of his colleagues were hanged.

Some critics declared that if Shell had intervened directly, the hangings could have been avoided. Yet even the almost universal castigation of the Nigerian government had been useless. If Shell had broken one of its fundamental tenets and interfered with national politics, other critics (or even the same ones) would certainly and rightly have seen that as scandalous.

The protesting governments then considered imposing an oil export embargo on Nigeria, but rejected it as being counter-productive. This meant that none of the foreign oil companies could claim *force majeure* and legally cease production. However, when the governments retreated, outfaced by Nigeria, Shell's critics placed the moral onus upon it and interpreted its continued production as mere cynicism. And by ill-starred coincidence, Shell's Committee of Managing Directors had to consider within weeks whether or not to conclude and confirm long-standing negotiations with Nigeria on the development of the country's largely wasted production of natural gas.

This posed an absolute dilemma. Nigerian crude oil contained large proportions of associated natural gas for which there had never been a viable domestic or international market. For many years the only option had been to flare off the excess gas. To continue doing so would be denounced as environmentally irresponsible; to begin to turn the waste into sellable LNG would be denounced as supporting the regime, even though there would be no benefit for the government of the day, whatever its political complexion, until several years into the next century.

Certain that it would be criticized whatever it did, the CMD weighed four factors in its deliberations. Firstly, an LNG industry would bring thousands of jobs to the people of Nigeria. Secondly, the national economy urgently needed to be diversified away from its almost total dependence on oil. Thirdly, the progressive reduction of flaring-off would be environmentally beneficial; and fourthly, if the current opportunity were missed, those benefits would be delayed by many years. In support of the project, all parties involved had placed large amounts of money (in the case of the Nigerian government, a billion dollars) in an escrow account. The account had an apt nickname – the 'glass box'. Inside it, the money was temptingly visible, but if any party were to withdraw its money, the project would collapse and all the long-term benefits would be indefinitely postponed. For all these reasons, Shell's CMD concluded that there should be no hint

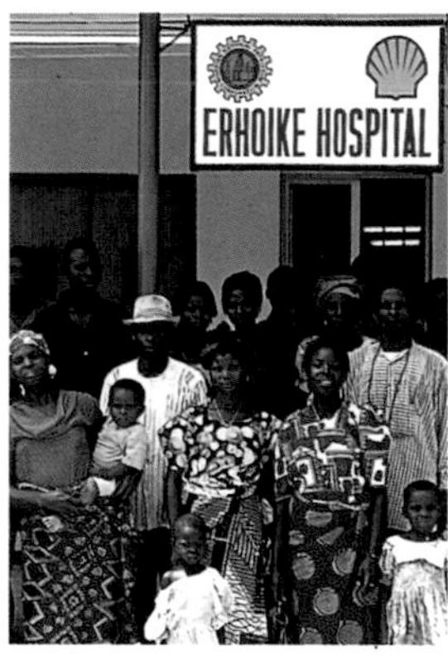

of hesitation. In December 1995, Shell resolved to commit to the establishment of a Nigerian LNG industry.

The Brent Spar and Nigerian issues made 1995 the hardest year of John Jennings' chairmanship of Shell Transport. Some of those who knew him well described him as one of the Group's deepest-thinking people, but not a man to show his feelings in public. However, in a most unusual way, he demonstrated his unhappiness about the Ogoni situation. Responding to a question at Shell Transport's AGM on 15 May 1996, Jennings stated his conviction that reconciliation within Nigeria was essential; and then he led the meeting in a brief silence as a mark of respect for all who had died in Ogoniland.

From a dispassionate point of view, one of the most conspicuous elements of the Ogoni episode was the way it underlined characteristics of the Brent Spar affair. Both were sharply focused single-issue campaigns, well-conducted and highly dramatic, and both demonstrated how easily a highly visible multinational oil company could be made into a useful target. Another very striking element emerged if one compared Nigeria and the Sultanate of Oman, where Shell was also just beginning a national LNG industry.

The two states had several similar aspects, and one extremely vivid difference. The similar aspects were firstly that, in a manner very like the Nigerian consortium dominated by NNPC, Petroleum Development Oman (PDO) was now owned by a consortium consisting of the Omani government (60%), Shell (34%), Total (4%) and Partex (2%). Secondly, in the same way as the vast majority of Shell's Nigerian employees were Nigerian, PDO's staff in 1995 was already 66% Omani, with that percentage planned to increase. Thirdly, Oman was just as dependent on oil as Nigeria and urgently needed to diversify its national product base into gas. Fourthly, Oman was no more a democracy than Nigeria: the Sultan, HRH Qaboos bin Said, was an absolute monarch. There, however, the similarities with the Nigerian situation ended, for the far-sighted Sultan was a wise and generous man, intelligent, well educated and sensitive. He had brought the sultanate from a feudal condition to great prosperity; he was determined to ensure its continued prosperity on behalf of his people; and naturally, his people respected and admired him. Despite the parallels between the countries, it was not the presence of Shell that brought misery or gladness; the difference was how the ruler chose to rule, and that made all the difference in the world.

In the effect upon their nations, there could scarcely be a sharper contrast than between the Sultan's enlightened policies and the policies of the Nigerian military government. In the middle 1930s,

Sir Henri Deterding remarked of Shell:

> Everywhere we come we bring our experience, our work and our capital, and we are happy when we are received as sincere and faithful allies, who succeed in finding a satisfactory profit for ourselves, as well as assuring prosperity and progress for our neighbours, thanks to the natural riches of the country, the work of the population side by side with us, and a community of interests and reciprocal good feeling.

Oman was an excellent example of the vital triangle – the mutually beneficial relationship between company, government and population – that Shell sought everywhere. It was far from being a solitary example. The 'community of interests and reciprocal good feeling' between Shell, the host government and the populace was so often a matter of fact that (as with routine safe practices at work) no one else noticed. Indeed, as a parent of the Group, in 1995 Shell Transport was present in no fewer than 128 countries, and with the sole exception of Nigeria, the vital triangle was intact in each.

Thus Shell Transport was present on 1 April 1995 when the world's largest manufacture of polyolefins (plastics such as polypropylene) was

created in the shape of Montell, a 50:50 partnership between Montedison and the Royal Dutch/Shell Group. It was present in the Norwegian city of Stavanger in mid-May when the Troll platform ('the largest man-made object ever to have moved across the face of the earth') was towed out to its North Sea destination. And, again in 1995, it was present in Gambia when Shell Marketing Gambia earned a governmental Clean/Safe award for its environmental and safety work.

It was present too in another region where the natural gas industry was making great progress. After years of preparation, Australia's giant North-West Shelf Gas Project (Shell 22%) had commenced exports to Japan in August 1989, two months ahead of schedule. Within a year the project had exceeded design performance. Inside two years a second gas production platform, named Goodwyn, was under construction and a fifth dedicated tanker had joined the exporting fleet, with a willing buyer for every cubic foot of gas produced. By 1995, Goodwyn was in production; the fleet held eight gas tankers; Spain and South Korea had become customers; and in a long-term additional contract Japan had significantly increased its annual order.

And of course Shell Transport, as a parent of the Group, was present in many parts of the world in the continuing search for and development of oil fields. In 1993, permission was received for a redevelopment of the North Sea's most famous oil and gas field, Brent. Shell's planned new expenditure of £1.3 billion would extend Brent's productive life by more than ten years. In the same year and the same region, the Shell-operated Draugen field came on stream to provide Norway's first oil production north of the 62nd parallel; and in the Gulf of Mexico, 214 miles south of New Orleans, it was also in 1993 that Shell installed the Auger production platform, at a water-depth of 2,860 feet.

Shell's deep-sea achievements were particularly impressive in this period. The Bullwinkle production platform had set a world record (1988) when it was installed in 1,350 feet of water. Auger more than doubled that depth, through using a different kind of technology. Instead of being a bottom-supported platform, it was a 'tension leg platform' or TLP – that is, it was a platform anchored to the seabed with immensely thick vertical steel cables. The principle had been researched in the 1970s; Conoco had been the first oil company to put it into practice, in 1984, and engineers calculated that TLPs could be successfully used in waters as deep as 10,000 feet. Having installed Auger, Shell took a further step towards the theoretical maximum in 1995, setting yet another new world record with the develop-

ment (again in the Gulf of Mexico) of the Ursa TLP in 3,950 feet of water. If that rate of progress were maintained, it would not be many years before the 10,000-foot TLP came into being. But there was no likelihood that the oceans would soon be sprinkled with dozens of TLPs. Ten thousand feet is about three kilometres, or nearly two miles – very deep water, yet when placed in relation to the oceans as a whole, not as deep as it sounds. Apart from the technical and economic difficulties in creating even one TLP of such a size, and leaving aside the elementary fact that not every part of the ocean bed contains workable hydrocarbon reservoirs, the world's oceans are much deeper than most people suppose: their *average* depth is over 12,000 feet – about 3.75 kilometres or 2.3 miles. Even a 10,000-foot TLP would be somewhat lost.

On the other hand, the technology was already in use for deeper offshore production than anything an existing TLP could match. Using remote handling and control systems, wellheads could be installed directly onto the seabed and linked to existing offshore facilities. Inspection, maintenance and repair could all be done remotely, at depths far beyond maximum human diving limits. Shell had led the industry in subsea technology since 1961, with an installation off California; and in the Gulf of Mexico in 1995, Shell commenced subsea development of the Mensa natural gas field in 5,400 feet of water – a depth of one mile. If at some distant point in the future the market required wellheads at two miles' depth, or even three, Shell's technical designers and engineers would probably be the people to meet the need first.

'The people' – not 'the men'. The social and cultural changes that have taken place since World War II have naturally had their effect on everyone associated with Shell, and in 1993 Shell commissioned an independent confidential survey of over 11,000 past, current and potential future expatriate staff and spouses in 35 countries. Reckoned to be the most comprehensive survey of expatriation ever done, it was virtually a social profile of the times. It disclosed that more than ever before, many employees had travelled widely and cheaply when they were young, and had less desire to live abroad; far fewer than before were willing to leave their children in boarding schools in their home country; and, most importantly, many now had partners with their own careers. As a consequence, some of the most talented employees had turned down senior foreign opportunities because their partner was unwilling to change or abandon an established personal career structure, and because the employee placed family above business ambition.

With those and numerous other conclusions, the survey promised to be a valuable tool in reviewing Shell's expatriation policy. Yet the changing patterns, possibilities and expectations of post-war society were already bringing about a new staff profile, as women in Shell gained more varied and increasingly senior posts. Compared to other less technically and scientifically based organizations it was a slow process, because there was no wish for tokenism or positive discrimination. With recruitment and promotion based solely on ability, the only need was for top-class candidates, so a time-lag was probably inevitable: in engineering, for example, potential female recruits were rare until recent decades. But there have been women lawyers in Shell since 1969, women engineers since the early 1970s, and women sailors since 1975. By the late 1970s women employees were starting careers in research, manufacturing, international marketing, and supply planning; in 1986 the first woman Secretary to the CMD was appointed; and in 1993 the same person became Shell Transport's first woman Company Secretary.

Jyoti Munsiff, Company Secretary

Still more recently, in 1996, a woman was appointed vice president of Group External Affairs. Her work in Shell's international marketing had already taken her, her husband (a businessman, but not a Shell employee) and their family to Brazil, Uruguay and Texas. Certainly this showed that the two-career family, Shell and non-Shell, was practical, and did not inhibit career development. And it was one small part of the changing shape of Shell.

Shell's massive world-wide programme of restructuring began as a moment of reflective unease. In November 1993, less than five months after he had become chairman of the CMD, Cor Herkströter gave a presentation in the United States to an audience of leading financial analysts. For the Group it was an annual event; for himself it was a first time, and afterwards he did not feel entirely happy with the prepared material he had presented. But until he saw his CMD colleagues again, he kept to himself the doubts which his text had raised in his own mind.

Herkströter's background was somewhat unusual for a chairman of the CMD. By training he was both an economist and an accountant, and by employment he had been, from 1967 until 1980, a member of the staff of the metals-mining company Billiton: the source, within the Royal Dutch/Shell Group, of long and resonant historical echoes. The facts of his background – that he was neither a scientist nor a geologist nor a Shell man 'from the cradle' – may have contributed to his perception; one cannot say for certain. What is certain is that the prepared material he presented to the American analysts left him uneasy, and that after careful thought he expressed his unease to his colleagues on the CMD. Despite the record profits he had announced, he felt there was less room than might be supposed for self-congratulation. His concern was that Shell should be proactive rather than reactive. It was neither the past nor the present that worried him, but the future. Shell was doing well; its position in the commercial world was strong. But was Shell doing well enough to maintain its strong position? What might be the case in ten or fifteen years' time? With new competitors and new methods of competition emerging almost daily, could you really be sure of Shell?

In August 1994 it was announced that McKinsey & Co. had been engaged as advisors in a review of the Group's service companies, the review being part of a long-term strategy to reduce costs. Since the 1986 oil price collapse, every part of the industry had suffered from over-capacity in refineries, storage and transport capability. In the days when OPEC dictated the price of oil and dictated it high, heavy capital investment had seemed justified and sensible. By now, though, at any rate within Shell's scenarios of the future, it seemed sure that there would be no early return to high-cost oil and that for the foreseeable future supplies would outstrip demand.

The analysis was reflected in Shell's diminished returns on capital: as recently as 1991, Sir Peter Holmes had described the 12% return of 1990 as 'modest'; since then, 12% had become a target to be achieved. Various refineries and other facilities could be and had been sold, and others would be – including Shell Centre's Downstream building. In the three decades since it had been fully opened for business, so much routine work had been automated and computerized that its 265,000 square feet of office space had become superfluous. (The 13-storey building was put on the market in March 1996, and was soon sold for conversion to residential apartments.) Yet the sale of unneeded facilities was only a comparatively small part of the reshaping process that was under way. On 6 December 1994 the CMD convened for a two-day meeting with members of the small team that had been formed for the service companies' review. This meeting proved to be a brain-storming session of crucial importance.

The location, Cannizaro House, was 'off-site' – not a part of Shell's premises but an hotel on Wimbledon Common in west London. As usual with CMD matters, the discussion was collegiate, the outcome consensual and the responsibility collective. As well as the eight other people present, all members of the CMD contributed – Cor Herkströter, John Jennings, Mark Moody-Stuart (elected as a managing director of Shell Transport in 1991 and, like Jennings, the holder of a doctorate in geology) and Maarten van den Bergh (elected as a managing director of Royal Dutch in 1992 and, like Herkströter, an economist). But though the outcome was the fruit of the thoughts of all concerned, in this instance – as noted at the beginning of this chapter, and without diminishing the contributions of the others – it is right to register that after Herkströter had provided the spark, it was Jennings who fanned it into flame with, of all things, a doodle on a piece of paper. He threw away the original, and it might have been forgotten altogether as an individual input. But the original was never forgotten by at least one of the others who read it, because what it sketched was nothing less than Shell's new shape.

Looking back it is clear that if any one person was going to think this thought, then Jennings was the most likely person to do so. In the contemporary CMD, he had the longest experience of Shell, and he was the

Committee's longest-standing member; and of course one of the prime purposes of the CMD was to remove its members from the hurly-burly, to take them from the mind-filling daily demands of running a given company, and to provide them instead with time to think – to register, ponder, analyze, consider, connect and envisage in a Group-wide manner.

Group Managing Directors 1995 (left to right): Dr J.S. Jennings, M.A. van den Bergh, C.A.J. Herkströter, M. Moody-Stuart.

That was precisely what John Jennings had done. It would be an exaggeration to suggest that the whole plan was there in his sketch, like Pallas Athena springing fully formed and armoured from the forehead of Zeus. Much work followed before the sketch became reality, and it is also important to point out that the thought had a background: it was not so much a flash of insight as a mature reflection on and development of various *ad hoc* arrangements that had arisen in parts of the Group. Nevertheless, no one else had thought of it before, and for all present at Cannizaro House the idea was so exciting that on the second day of this marathon meeting, almost every sentence seemed to begin with 'What kept me awake last night was this...'

Once it had been thought of, the central concept seemed blindingly simple: the collegiate nature of the CMD should be extended to the service companies by means of business committees.

Why had this not been done before? Only because, when the CMD was established in 1959/60, collegiate leadership and collective responsibility were so novel and unusual in the business world. They could have been applied Group-wide from the beginning of the system. They had not been, simply because they were then too different, too revolutionary. But 35 years of existence at the highest level had proved their value and viability. Collegiality and collective responsibility worked. Why not apply the same principles to a new arrangement of new businesses with global reach?

That was, in short, the result. There is not space here to detail all the other meetings, briefings, discussions and staff consultations which ensued, but one meeting is worth mentioning: the occasion on 8 February 1995 when the CMD presented their collective thinking to The Conference, the Group's supervisory body. Although it had no existence in law, The Conference held considerable power of moral suasion over the CMD's decisions; and when it heard the proposals, several of its members were initially very sceptical. Knowing and accepting the limits of their authority and responsibility, most were eventually persuaded – but, while letting it pass, some remained unconvinced. Much the same had happened when the original collegiate CMD strategies had been put in place 35 years earlier. There had been no foregone conclusion then; nor was there in 1995. But after the crucial meeting with The Conference on 8 February, one of the members of the CMD was heard to say with relief, 'I think we're all right.'

Planning continued throughout 1995, with the new world-wide structure coming into official existence on 1 January 1996. Although this had never been intended, there was something very tidy about the date. The alliance of Shell Transport and Trading and Royal Dutch, forged so painfully in the negotiations between Marcus Samuel and Henri Deterding, had been created exactly 89 years earlier. The Group was now in its 90th year, and Shell Transport and Trading was on the verge of its own centenary in 1997. At nearly 100 years old, Shell Transport was being reborn; its alliance with Royal Dutch was being reinvigorated; their Group was being refocused. These actions were the apotheosis of the CMD, the transference of all its best qualities to the whole of Shell. And Shell Transport and Trading approached its second century with confidence and eagerness, because there was no doubt about it: you *could* be sure of Shell.

Page numbers in *italics* refer to illustrations